An Annotated Checklist of
WOODY ORNAMENTAL PLANTS
of California, Oregon, & Washington

The authors are:
Elizabeth McClintock,
Research Associate, Department of Botany,
University of California, Berkeley,
and
Andrew T. Leiser,
Professor, Department of Environmental Horticulture,
University of California, Davis.

This publication replaces Manual 32, *A Checklist of Woody Ornamental Plants of California.*

TO ORDER ADDITIONAL COPIES:

An Annotated Checklist of Woody Ornamental Plants of California, Oregon, and Washington (Number 4091). $4.00.

Please include number as well as title with order. Make check or money order payable to: The Regents of the University of California. All sales are final; no returns for cash or credit.

California residents: Please include sales tax.

Residents outside the United States: Please request Pro Forma Invoice and postal book rate desired—air mail or surface mail.

Address: Agricultural Sciences Publications
Division of Agricultural Sciences
University of California
Berkeley, California 94720

No part of this publication may be reproduced, stored in a retrieval system, or transmitted, in any form or by any means, electronic, mechanical, photocopying, recording, or otherwise, without the written permission of the publisher and the author.

Copyright © by The Regents of the University of California.

Printed in the United States of America.

Library of Congress Catalog Card Number: 78-73983
International Standard Book Number: 0-931876-28-1

6m—2/79—BT/FB

AN ANNOTATED CHECKLIST OF WOODY ORNAMENTAL PLANTS OF CALIFORNIA, OREGON, AND WASHINGTON

The preface to *A Checklist of Woody Ornamental Plants of California* by Mildred E. Mathias and Elizabeth McClintock (Manual 32, California Agricultural Experiment Station, University of California, Berkeley, April 1963) included, in part, the following:

"The diverse and abundant ornamental flora of California ranges from many of the widely cultivated, temperate-zone plants to those of subtropical and tropical origin. Consequently no single reference is available which lists all the names of California ornamental plants. The need for a checklist, long recognized by botanists and horticulturists, became even more urgent in 1957 with the addition to the *Agricultural Code of California* of Chapter 9, section 1148.2, which states: 'Ornamentals, except roses and annual or herbaceous perennial ornamental plants, shall be labeled with the botanical name.'[1] . . .

"The list is designed to serve as a guide to nurserymen and others in the labeling of nursery stock in compliance with the *Agricultural Code of California*. Uniform labeling throughout the nursery trade will also aid the customer. In addition, the list provides a ready reference to correct botanical names, synonyms, common names, and botanical authorities for all who work with plants—botanists, horticulturists, entomologists, landscape architects, and the like."

An Annotated Checklist of Woody Ornamental Plants of California, Oregon, and Washington contains about 12,000 entries, nearly double the number in the 1963 checklist. The alphabetical listing includes the correct botanical names, botanical and commercial synonyms, cultivar names, and many of the common names used for woody ornamental plants of the Pacific states.

Although prepared for the Pacific states, this checklist will be useful to persons outside the Pacific region who are concerned with the correct names of woody ornamentals. We have included a number of species and cultivars that do not appear in *Hortus Third* or other recent publications. Many of these plants also are grown in areas east of the Pacific states. However, our listing of cultivars of some species will be incomplete for the eastern part of the country, where many more cultivars of such species as **Taxus × media** and **Ilex crenata** are grown.

The present checklist is partially based on the 1963 checklist but has been expanded to include Oregon and Washington. A number of suffrutescent (half-woody) perennials that are used as permanent landscape plantings have been added, such as **Chrysanthemum frutescens** and **Impatiens sodenii** (*I. oli-*

---

[1] *Agricultural Code of California*, p. 459. Sacramento: California State Department of Agriculture, 1957.

veri), as well as agaves and certain other succulents, bamboos, and tree ferns. We have also added annotations to correct many misidentifications and misapplied names.

We have not listed woody plants grown primarily for commercial orchard use and not normally used as ornamentals, nor have we included bulbous plants or house plants, except a few marginally hardy ones. For example, some plants grown out of doors in southern California are used as indoor foliage plants in other areas. No attempt has been made to list all cultivars in such genera as **Camellia, Fuchsia, Rhododendron,** and **Rosa,** because checklists and registration lists are available for these genera.

This checklist is primarily a compilation from nursery catalogs and lists of California, Oregon, and Washington, with the addition of plants known to have been in the trade intermittently in the past and some that may be expected to become popular in the future. However, not enough field work has been done to verify all identifications in some difficult genera, such as **Juniperus** and **Cotoneaster** and others where there are problems. It is hoped that our annotations will call attention to some of these problems.

In addition to the list itself, this publication includes: a discussion of the use of the international botanical and horticultural codes of nomenclature for naming plants; notes on how to use the list; suggestions for labeling and for preparing nursery lists and catalogs; an appendix listing the authors of botanical names; and references giving the major publications used in compiling the checklist.

BOTANICAL PLANT NAMES

How plants are named

The names used for ornamental plants have been and continue to be a problem for those who buy, sell, and use them. Plant nomenclature is concerned with determining the correct name for a known plant according to a system. The naming of all plants, whether they grow in the wild or in cultivation, follows rules given in two international codes of nomenclature. However, strict application of the two codes means that plant names must sometimes be changed to achieve the goal of stability in nomenclature.

The codes are the *International Code of Botanical Nomenclature* (1972, Utrecht, Netherlands, for the International Association for Plant Taxonomy) and the *International Code of Nomenclature of Cultivated Plants* (1969, Utrecht, Netherlands, formulated and adopted by the International Commission for the Nomenclature of Cultivated Plants). In the following discussion the first code will be referred to simply as the botanical code and the second as the cultivated code. Neither code deals with the common names of plants, but, because they play an important role in communication about plants, we shall comment on them briefly after discussing the two codes.

Plant names are a means of communication. Ideally we should be able to talk about the plants around us without causing confusion as to just what plant is being referred to. Each plant should have only one name, which should be accepted by everyone around the world. This goal should be accomplished by using the two codes of nomenclature.

People began to use plants long before there were written records, but even then they must have had names for their plants. In more recent times, the Greeks and the Romans had names for plants known to them; we still use some of these names for the same plants.

Modern plant nomenclature goes back only about 200 years. The first attempt to arrive at international agreement on nomenclature was in 1867, the year of the first botanical congress. However, it was not until the fifth international congress in 1930 that the botanical code became international in agreement and function, as well as in name.

According to the botanical code, every individual plant belongs to a series of categories (or ranks) in descending order in the hierarchy of botanical classification—family, genus, and species. We are concerned here with only two of these ranks, the species and the genus.

Genus. Composed of one or more species, it contains plants recognizable as being more or less related. (Plural: genera.) Example: **Ceanothus.**

Species. The basic unit of classification, a species consists of a group of individuals that can be referred to as a kind of plant. (The word is either singular or plural; when singular it is abbreviated as sp., when plural as spp. Another word, specie, which refers to coined money, is sometimes erroneously used for the singular of species.) Each kind (species) of plant is presumed to be distinct from other kinds in marked or essential features, which provide characters for its identification and for distinguishing it from other species in the same genus. Related species are combined into a genus. Example: **Ceanothus griseus, C. thyrsiflorus.**

The individuals making up a species are not always uniform, and recognizable variants may be distinguished nomenclaturally by giving them names in the following categories: (1) **subspecies** (abbreviated subsp. or ssp.), example: **Hydrangea anomala** subsp. **petiolaris;** (2) **varietas,** more commonly known by the English word, variety (abbreviated var.), example: **Ceanothus griseus** var. **horizontalis;** (3) **forma,** or form (may be abbreviated f.), example: **Cotoneaster buxifolius** forma **vellaeus.**

Each plant name, which is the name of a species, consists of two words: (1) the generic name, a proper noun, followed by (2) the specific epithet, a descriptive adjective that modifies the generic name. Many such epithets are used and reused to describe plants with similar characters in different genera, such as **Hebe buxifolia** and **Cotoneaster buxifolius.** The two epithets refer to the boxlike leaves of the two shrubs.

Specific epithets cannot stand alone but must always be used with the generic name; generic names, however, can stand alone. The generic name, when used with the specific epithet, should be spelled out the first time it appears in the text, but later may be abbreviated to its capitalized initial letter if the meaning is clear. Example: **Ceanothus griseus; C. thyrsiflorus.**

The two-word combination is called a binomial and the system is credited to Linnaeus (Carl von Linné or Carolus Linnaeus), who used binomial plant names, consistently for the first time, in 1753 in his *Species Plantarum.* In this monumental work on the species of plants Linnaeus described and named about 8,000 plants—all of the plants known to the European world at that time. Linnaeus assigned names to these plants according to certain rules (his own). Botanists after Linnaeus tried to follow his rules, but, as time went on and many more plants became known through discoveries and exploration, the state of botanical nomenclature became so confused that in 1867 European botanists met for the first botanical congress.

The aim of all international botanical congresses has been to standardize nomenclatural practices, which will eventually stabilize plant names and prevent name changes other than those brought about by differences of opinion among botanists and by new knowledge of plant relationships.

The cornerstone of the botanical code is the principle of priority, which provides that the oldest available name or epithet published according to the rules is the one to be used. Other principles are that: each plant can bear only one correct name; names of plants must be written in latinized form regardless of their derivation; and names of all plants must be based on type specimens, which are dried, pressed, and deposited in a herbarium (a place where such specimens are filed according to a systematic arrangement). The code also provides rules on the valid publication of names, such as the requirements that a Latin description accompany all names published after 1935 and that publication be by printed matter available to the public in botanical libraries.

Plants grown in cultivation are mostly of two kinds from the standpoint of nomenclature. First are those which are brought into cultivation from the wild, and which in their cultivated habitat do not differ significantly from their wild counterparts. Such plants must carry the same names as the plants growing in the wild. The nomenclature of these plants at the species level is based on the rules of the botanical code.

The second group includes: (1) individual variants

found in wild populations; (2) variants found in cultivated populations of species, subspecies, varieties, and forms; and (3) hybrids between related species (and even genera) not known in the wild. (The interspecific or intergeneric hybrids may be given Latin descriptions and names and treated according to the rules of the botanical code.) Some of these variants and hybrids are selected for their superior qualities, vegetatively propagated, and eventually maintained in cultivation.

Man-selected variants, until recent years, were named in the same way as variants that originated in the wild. However, because man-selected variants differ from botanical varieties by reason of their origin, maintenance in cultivation, and particular characters that make them useful to man, the term *cultivar*, a contraction of *culti*vated *var*iety, was proposed for them. The cultivated code, the first edition of which was published in 1952, regulates the naming of cultivars. Formulated under the auspices of both the Seventh International Botanical Congress held in 1950 in Stockholm and the Thirteenth International Horticultural Congress held in London in 1952, the cultivated code was revised in 1958, 1961, and 1969.

The two codes are both botanical, and both apply to plants in cultivation, but they perform different functions. The botanical code governs the use of botanical names in Latin form for wild and cultivated plants. Its aim is to provide a stable method of naming plants, to avoid and reject "the use of names which may cause error or ambiguity or throw science into confusion." The cultivated code aims to promote uniformity, accuracy, and fixity in the naming of cultivars, that is, cultivated plants below the rank of species. It supplements the botanical code by dealing with special situations applying only to cultivated plants and carries on for horticultural, agricultural, and sylvicultural plants where the botanical code stops.

One provision of the cultivated code is that a cultivar name published on or after January 1, 1959, be a "fancy name" in a modern language, such as English, and not a name in Latin form. However, a name in Latin form published before that date (in accordance with the code) for a plant considered to be a cultivar is to be retained.

A second provision of this code is that a "cultivar name, when immediately following a botanical or a common name, must be distinguished clearly from the latter, either by placing the abbreviation cv. before the cultivar name, or by some typographical device, such as enclosing it within single quotation marks." It should be noted that neither double quotation marks nor the abbreviation var. should be used to distinguish cultivar names. Also, the single quotation marks are considered to be part of the cultivar name and therefore, in this special use, are never separated from the name by other punctuation, such as periods or commas.

Why names change

A name change may be brought about by one of four situations: misidentification; erroneous uses of names; differences in taxonomic opinions; and application of the rules of the codes.

The case of mistaken identity may be illustrated by the well-known black-eyed heather, **Erica canaliculata,** which has been called **E. melanthera,** a species not known in cultivation. The two names represent two different South African species of **Erica** that have been confused. The confusion apparently goes back to 1824, when an illustration was published in England of **Erica canaliculata** to which the name **E. melanthera** was mistakenly applied.

Another case of misidentification involves two privets, the Japanese privet, **Ligustrum japonicum,** and the glossy or tree privet, **Ligustrum lucidum.** Although readily distinguishable, these two have often been confused in commercial usage. Two other examples are the use of the name **Juniperus excelsa 'Stricta'** for **J. chinensis 'Stricta'** and the name **Cotoneaster glaucophyllus** for **C. buxifolius.**

In numerous instances names, sometimes of long standing, must be corrected because the original use of the name was erroneous, inappropriate, or even intentionally modified. *Juniperus scopulorum* 'Blue Haven' was a deliberate corruption of **J. scopulorum 'Blue Heaven'.** *Cotoneaster pannosus* 'Nanus' was apparently applied to a plant because it bore a superficial resemblance to **C. pannosus,** when actually its true identity was **C. buxifolius.** The use of the cultivar names, 'Blue Rug', 'Blue Carpet', 'Wilton Carpet', for **Juniperus horizontalis 'Wiltonii'** are cases of careless nomenclatural practice.

In considering how differences in taxonomic opinions may result in name changes, we must point out that in this case we are dealing with two variables. First, plants vary, and, second, the opinions of those working on them vary, often resulting in differences in the names used. Examples are three pairs of closely related genera, **Mahonia** and **Berberis, Hebe** and **Veronica, Diplacus** and **Mimulus.** In each case the two genera of each pair are similar enough, some botanists believe, to combine them into a single genus, but, in the opinion of other botanists, the two members of each pair represent separate genera. The rules do not cover such differences of opinion. In this checklist the three pairs of genera are considered to represent six separate genera, regardless of differing opinions.

Another example of a pair of related genera is that of **Rhododendron** and *Azalea*. After many years of differing opinions, the two are now considered to represent one genus, which is called **Rhododendron.**

In some instances botanists and horticulturists have apparently more readily accepted new taxonomic opinions. On the basis of research in several disciplines, the separation of **Sequoiadendron** from **Se-**

quoia (*Sequoia gigantea* becoming **Sequoiadendron giganteum**) and **Calocedrus** from **Libocedrus** (*Libocedrus decurrens* becoming **Calocedrus decurrens**) have been accepted without a delay of many years.

Name changes due to application of the rules have often resulted when the same species has been named and described more than once. According to the rule of priority, the oldest name is the correct one and must be used. Later names are referred to as synonyms. For many years, we have used the name *Plumbago capensis* for the blue-flowered Cape plumbago. Carl Thunberg collected a specimen of this plant at the Cape of Good Hope and in 1794 published the name *Plumbago capensis*. But unknown to Thunberg, a specimen had been sent to Lamarck in Paris, who in 1786 published the name **Plumbago auriculata**. This is the older name and the correct one according to the rules.

The cultivated code states that "the naming of cultivars is based on priority of publication." Although cultivar names are not documented with the precision of botanical names, it is possible to trace the priority of some. The twisted Chinese juniper, known for so long as *Juniperus chinensis* 'Torulosa', had been given an earlier cultivar name **J. chinensis 'Kaizuka';** therefore, it must be known by the older name.

COMMON NAMES

Common names also allow us to communicate about plants. But their use does not follow any rules or regulations, and they are not always precise. Nor are they international; rather, they are almost entirely local in their use. However, they cannot be disregarded or ignored, because they are a part of the language to which they belong. When given recognition as local names, they may often be useful and may even be more precise and stable than some botanical names.

Douglas fir is one of the most frequently used common names on the Pacific Coast for the forest tree with the botanical name of **Pseudotsuga menziesii**. Following the rules of the botanical code, this tree has had several name changes, but through all of these the use of the name Douglas fir has never changed. Unfortunately not all common names are so precise. The user of this checklist needs only to look up such common names as myrtle, laurel, broom, and bottlebrush to see that each is used for different and sometimes even unrelated plants.

Although some plants in the checklist have more than one common name, we are not recommending the use of one in preference to another. Neither have we attempted to list every common name in the world literature for all plants in the checklist. Many plants have no common names, and for these the only names are the botanical ones. For some of these the generic name through usage has also become the common name, such as rhododendron, eucalyptus, banksia, and protea.

USE OF THE CHECKLIST

The approximately 12,000 entries represent different kinds of plant names in a single alphabetical listing. Each kind of name in this checklist may be recognized by the type face in which it is written.

1. Correct botanical name, a Latin name: **Abies bracteata** (D. Don) D. Don ex Poit. (bold italics).
2. Synonym, a second name, but not the correct one, for the same plant, also a Latin name: *Abies venusta* (Dougl.) K. Koch (light italics).
3. Correct cultivar name, either a Latin name or an English name: **Abies balsamea 'Nana'; Acer platanoides 'Crimson King'** (bold roman type, enclosed in single quote marks, never double quotes).
4. Synonym cultivar name: *Acer platanoides* 'Nigrum' (light roman type, enclosed in single quote marks).
5. Common name, an English name: NORWAY MAPLE (light roman type, all capital letters).

All specific epithets in this checklist begin with a lowercase letter, a procedure that is always correct. It has been the practice sometimes to capitalize the initial letter of those epithets derived from personal names and formerly used generic names. Anyone wishing to follow this practice may use *Hortus Third* as a guide. The rule to remember is that, although the first letter of a generic name must always be capitalized, the first letter of a specific epithet does not need to be capitalized. All major words of cultivar names are always capitalized (first letter only).

The naming of hybrids follows the botanical code, which recognizes two kinds: interspecific and intergeneric. An interspecific hybrid is one between two species of the same genus. The botanical name of such a hybrid consists of the generic name followed by the descriptive epithet with a multiplication sign ($\times$) separating the two: **Abelia $\times$ grandiflora**. In this checklist the hybrid parentage, when known, is designated by enclosing the parental species in square brackets: [**A. chinensis $\times$ A. uniflora**].

An intergeneric hybrid is one between species of two (or more) genera. The botanical name of such a hybrid consists of a generic name, usually formed by combining parts of the names of the two parent genera. The hybrid generic name thus formed is preceded by a multiplication sign ($\times$) and followed by the descriptive specific epithet: $\times$ **Cupressocyparis leylandii**. Here again, the parental species are designated as for the interspecific hybrid: [**Chamaecyparis nootkatensis $\times$ Cupressus macrocarpa**].

Botanical authorities have been included with the names in the checklist for scientific reference only. It is not necessary to use these personal names when labeling plants in nurseries or listing plant names in catalogs.

Often only one author is given for each name—for example, **Abies alba** Mill. Mill. is the abbreviation for Philip Miller (see Appendix), who first named and described this fir tree. In other cases, two authors are given, one in parentheses: **Abies balsamea** (L.) Mill. The author whose name is in parentheses (L. for Linnaeus) first described this species but in a different genus. Miller, the second author, transferred the species to the genus **Abies.** In a few cases an author's name is followed by ex and a second author's name: **Eucalyptus perriniana** F. J. Muell. ex Rodw. This indicates that the name **Eucalyptus perriniana** was proposed but not validly published by F. J. Mueller, then later was validly published by Rodway.

Annotations, the brief notes added to some of the entries, call attention to problems in the nomenclature or identification of plants of some species. For example, the commonly cultivated parlor palm, **Chamaedorea elegans,** is often erroneously listed as *Neanthe bella*, a name not validly published according to the botanical code and therefore one that should not be used. Our entries under these two names include:

Chamaedorea elegans Mart. Plants called *Neanthe bella* belong here.
Neanthe bella. Listed name for **Chamaedorea elegans.**

There is confusion regarding the identification of plants presumed to be **Cotoneaster buxifolius,** and two names are erroneously used for them. Our entries under **C. buxifolius** include:

Cotoneaster buxifolius Wallich ex Lindl. Plants called *C. pannosus* 'Nanus', and in Calif. **C. glaucophyllus,** belong here.
Cotoneaster glaucophyllus Franch. Probably not cult. in Calif. Plants so called are **C. buxifolius.**
Cotoneaster pannosus 'Nanus'. Listed name for **C. buxifolius.**

We have tried to correct the confusion regarding the correct name for plants of **Hebe odora** by the following entries:

Hebe buxifolia (Benth.) Cockayne & Allan. Probably not cult. Plants so called are **H. odora.**
Hebe odora (Hook. f.) Cockayne. Plants called **H. buxifolia** and *Veronica buxifolia* belong here.
Veronica buxifolia Benth. See **Hebe buxifolia.** Plants called *Veronica buxifolia* are probably **Hebe odora.**

DEFINITIONS, ABBREVIATIONS, AND SYMBOLS

Author citations are listed in the Appendix and are alphabetized as they are abbreviated.
Cult. Cultivated.
Cv. Cultivar. An international term defined by the cultivated code as "an assemblage of cultivated plants which is clearly distinguished by any characters (morphological, physiological, cytological, chemical, or others), and which, when reproduced (sexually or asexually), retains its distinguishing characters." This term is derived from *culti*vated *var*iety.
Fl. Flourished. Used in listings of authors for whom we have no dates.
Hort. Hortorum, literally *of the gardens*. Refers to a plant name used by some horticulturists and often for a cultivated plant of unknown origin. Such plants frequently have not been adequately described, and the name has no botanical standing.
Listed name. A name taken from nursery catalogs or less formal listings. These names have no botanical standing and, because they are a source of confusion, should not be used. When such a name can be identified as referring to a particular species, it is placed under the correct name for that species.
Subsp. Subspecies.
Var. *Varietas*, variety (botanical variety of a species).
? (with a name following a main entry). A questionable name that cannot be referred to a particular species. The name has no botanical standing and should not be used. The question mark and name may represent an unidentified plant that should be identified or a new cultivar for which adequate descriptions are not available. When used with one of the parents of a hybrid (with a name within brackets—[**Berberis ?sargentiana** × **Mahonia aquifolium**]—it indicates doubt regarding one of the parents of × **Mahoberberis aquisargentii.** [**?Salix babylonica** × **S. fragilis**] indicates doubt regarding both parents for **Salix** × **blanda.**
×. Sign for a sexual hybrid. When preceding a generic name, it indicates a hybrid between species of two genera. Example: × **Cupressocyparis leylandii** [**Chamaecyparis nootkatensis** × **Cupressus macrocarpa**]. Preceding a specific epithet, it indicates a hybrid between two species within the same genus. Example: **Malus** × **arnoldiana** [**Malus baccata** × **M. floribunda**].
+. Sign for an asexual hybrid (graft chimera). This sign is used in the same manner as the × is used for sexual hybrids: +**Laburnocytisus adamii** [**Cytisus purpureus** + **Laburnum anagyroides**].

SUGGESTIONS FOR PREPARATION OF NURSERY LABELS, LISTS, AND CATALOGS

The authorities for the botanical names (L. for Linnaeus, Franch. for Franchet, and others) are given in the checklist for precise identification of the names. They do not need to be used in nursery lists, catalogs, or labels.

When cultivars have been selected from a subdivision within a species (subspecies, variety, or forma) the checklist includes the full botanical designation for information and clarity. A nursery list or label need only give the name of the species followed by the cultivar name: **Gleditsia triacanthos 'Shademaster'** and not **Gleditsia triacanthos** var. **inermis 'Shademaster'**, unless ambiguity might result. The listing for **Juniperus chinensis** var. **sargentii 'Glauca'** should be used in full, because the name 'Glauca' is used for other cultivars in **J. chinensis**.

A nursery list or catalog should use the correct name as the main entry (as in this checklist); the synonym, if needed, should follow but preferably should be in a different type face. Cross references may be made (as in this checklist):

Carissa macrocarpa. Syn.: *C. grandiflora*

or as an alternative:

Carissa macrocarpa (*C. grandiflora*).

A separate entry for the synonym may also be used:

Carissa grandiflora. See **C. macrocarpa.**

Bold and light-face type will further distinguish between the correct name and synonym.

It is recommended that the name of a species, which is in Latin form, be set in a type face (such as italics or bold face) different from that of the cultivar name or common name. If the Latin names are italicized, cultivar names and common names should not be. In addition, the cultivar name is distinguished by capitalizing the initial letter or letters and using single quote marks or the abbreviation cv. preceding it. Common names may be set in the same type face as the cultivar names, but they need not be capitalized unless they are names of persons or places.

ACKNOWLEDGMENTS

We gratefully acknowledge the assistance and encouragement of the following: David M. Bates, Fredrick C. Boutin, Dennis E. Breedlove, Philip E. Chandler, William J. Dress, Theodore R. Dudley, Thomas C. Fuller, Alwyn Gentry (Bignoniaceae), Paul C. Hutchison, Myron Kimnach (palms), Frederick G. Meyer, Brian O. Mulligan, Hadley Osborn (*Rhododendron*), Thomas R. Soderstrom (bamboos).

We thank the nurserymen of California, Oregon, and Washington who responded to our request for lists and catalogs and who assisted in the verification of plant materials, the California Association of Nurserymen, and the American Association of Nurserymen for their interest and encouragement.

The Elvina J. Slosson Endowment Fund of the University of California provided financial assistance in preparation of this checklist.

A

AARON'S-BEARD: **Hypericum calycinum.**
ABELE: **Populus alba.**
ABELIA, GLOSSY: **Abelia × grandiflora.**
  GOUCHER: **Abelia 'Edward Goucher'.**
  MEXICAN: **Abelia floribunda.**
  PINK: **Abelia 'Edward Goucher'.**
  PROSTRATE: **Abelia × grandiflora 'Prostrata'.**
  PROSTRATE WHITE: **Abelia × grandiflora 'Prostrata'.**
  RED: **Abelia floribunda.**
  SCHUMANN: **Abelia schumannii.**
  SHERWOOD DWARF: **Abelia × grandiflora 'Sherwoodii'.**
  WHITE: **Abelia × grandiflora.**
Abelia chinensis R. Br. Syn.: *A. rupestris.*
Abelia 'Edward Goucher'. [**A. schumannii** × **A. × grandiflora**]. PINK ABELIA. GOUCHER ABELIA.
Abelia floribunda (Martens & Galeotti) Decne. MEXICAN ABELIA. RED ABELIA.
Abelia × grandiflora (André) Rehd. [**A. chinensis** × **A. uniflora**]. WHITE ABELIA. GLOSSY ABELIA.
Abelia × grandiflora 'Prostrata'. PROSTRATE WHITE ABELIA. PROSTRATE ABELIA.
Abelia × grandiflora 'Sherwoodii'. SHERWOOD DWARF ABELIA.
Abelia rupestris Lindl. See **A. chinensis.**
Abelia schumannii (Graebn.) Rehd. SCHUMANN ABELIA.
Abelia 'Sherwoodii'. See **A. × grandiflora 'Sherwoodii'.**
Abelia uniflora R. Br. ex Wallich.
ABELIA-LEAF, KOREAN: **Abeliophyllum distichum.**
Abeliophyllum distichum Nakai. KOREAN ABELIA-LEAF. WHITE FORSYTHIA.
Aberia caffra Hook. f. & Harv. See **Dovyalis caffra.**
Abies alba Mill. EUROPEAN SILVER FIR.
Abies amabilis Forbes. SILVER FIR. CASCADE FIR. PACIFIC SILVER FIR.
Abies arizonica Merriam. See **A. lasiocarpa** var. **arizonica.**
Abies arizonica 'Compacta'. See **A. lasiocarpa 'Compacta'.**
Abies balsamea (L.) Mill. BALSAM FIR.
Abies balsamea 'Compacta Nana'?
Abies balsamea 'Nana'.
Abies bracteata (D. Don) D. Don ex Poit. Syn.: *A. venusta.* SANTA LUCIA FIR. BRISTLE-CONE FIR.
Abies candicans Fisch. ex Henkel & W. Hochst. See **A. nordmanniana.**
Abies cephalonica Loud. GREEK FIR.
Abies concolor (Gord.) Lindl. ex Hildebr. WHITE FIR.
Abies concolor 'Argentea'.
Abies concolor 'Candicans'. See **A. concolor 'Argentea'.**
Abies concolor 'Sherwood Blue'.
Abies firma Siebold & Zucc. MOMI FIR. JAPANESE FIR.
Abies fraseri (Pursh) Poir. FRASER FIR.
Abies grandis (D. Don ex Lamb.) Lindl. GIANT FIR. LOWLAND FIR.
Abies koreana Wils. KOREAN FIR.
Abies lasiocarpa (Hook.) Nutt. ALPINE FIR. ROCKY MOUNTAIN FIR.
Abies lasiocarpa var. *arizonica* (Merriam) Lemmon. Syn.: *A. arizonica.* CORKBARK FIR.
Abies lasiocarpa 'Compacta'.
Abies magnifica A. Murr. RED FIR.
Abies magnifica var. *shastensis* Lemmon. Syn.: *A. shastensis.* SHASTA FIR.
Abies nobilis (Dougl. ex D. Don) Lindl. See **A. procera.**
Abies nordmanniana (Steven) Spach. Syn.: *A. candicans.* NORDMANN FIR.
Abies numidica De Lannoy ex Carrière. ALGERIAN FIR.
Abies pinsapo Boiss. SPANISH FIR.
Abies pinsapo 'Glauca'.
Abies procera Rehd. Syn.: *A. nobilis.* NOBLE FIR.
Abies shastensis (Lemmon) Lemmon. See **A. magnifica** var. **shastensis.**
Abies veitchii Lindl. VEITCH'S SILVER FIR.
Abies venusta (Dougl.) K. Koch. See **A. bracteata.**
ABSINTHIUM: **Artemisia absinthium.**
Abutilon 'Ashford Red'.
Abutilon hybridum Hort. CHINESE-LANTERN. FLOWERING MAPLE. Some plants called **A. speciosum** belong here.
Abutilon 'Mariana'.
Abutilon megapotamicum (K. Spreng.) St.-Hil. & Naud. FLOWERING MAPLE.
Abutilon × milleri Hort. [Presumably **A. megapotamicum × A. pictum**].
Abutilon pictum (Gillies ex Hook. & Arn.) Walp. Syn.: *A. striatum.* Some plants called **A. speciosum** belong here.
Abutilon pictum 'Pleniflorum'.
Abutilon pictum 'Thompsonii'. Syn.: *A. thompsonii.* VARIEGATED FLOWERING MAPLE.
Abutilon 'Red Monarch'. FLOWERING MAPLE.
Abutilon speciosum G. Don. May not be cult. Plants so called are usually **A. hybridum** or **A. pictum.**
Abutilon striatum G. Dickson ex Lindl. See **A. pictum.**
Abutilon thompsonii Veitch. See **A. pictum 'Thompsonii'.**
Abutilon vitifolium (Cav.) K. Presl. See **Corynabutilon vitifolium.**
ACACIA, BLACK: **Acacia melanoxylon.**
  BLACKWOOD: **Acacia melanoxylon.**
  BROAD-LEAF: **Acacia longifolia.**
  BUSH: **Acacia longifolia.**
  CATCLAW: **Acacia greggii.**
  EVERBLOOMING: **Acacia retinodes.**
  FALSE: **Robinia pseudoacacia.**
  FERN-LEAF: **Acacia baileyana.**
  KNIFE: **Acacia cultriformis.**
  KNIFE-BLADE: **Acacia cultriformis.**
  PEARL: **Acacia podalyriifolia.**
  PINK: **Albizia julibrissin.**
  PURPLE-LEAF: **Acacia baileyana 'Purpurea'.**
  RED-STEM: **Acacia rubida.**
  ROSE: **Robinia hispida.**
  SMOOTH ROSE: **Robinia hispida** var. **macrophylla.**
  STAR: **Acacia verticillata.**
  SWEET: **Acacia farnesiana.**
  SYDNEY: **Acacia floribunda.**
  WILLOW: **Acacia saligna.**
Acacia alata R. Br.
Acacia aneura F. J. Muell. MULGA.
Acacia armata R. Br. KANGAROO THORN.
Acacia baileyana F. J. Muell. SILVER WATTLE. FERN-LEAF ACACIA. COOTAMUNDRA WATTLE.
Acacia baileyana 'Purpurea'. PURPLE-LEAF ACACIA.
Acacia calamifolia Sweet ex Lindl. Syn.: *A. pulverulenta.* BROOM WATTLE.
Acacia cardiophylla A. Cunn. ex Benth.
Acacia caven (Molina) Molina.
Acacia conferta. Benth.
Acacia cultriformis A. Cunn. KNIFE-BLADE ACACIA. KNIFE ACACIA.

Acacia cyanophylla Lindl. BLUE-LEAF WATTLE. BLUE WATTLE. Some plants called **A. saligna** belong here.
Acacia cyclops A. Cunn. COASTAL WATTLE.
Acacia dealbata Link. Syn.: *A. decurrens* var. *dealbata*. SILVER WATTLE. FLORIST'S MIMOSA.
Acacia decora Rchb. GRACEFUL WATTLE.
Acacia decurrens (Wendl.) Willd. Syn.: *A. decurrens* var. *normalis*. GREEN WATTLE.
Acacia decurrens var. *dealbata* (Link) F. J. Muell. See **A. dealbata**.
Acacia decurrens var. *mollis* Lindl. See **A. mearnsii**.
Acacia decurrens var. *normalis* Benth. See **A. decurrens**.
Acacia dietrichiana F. J. Muell. Plants so called may be **A. suaveolens**.
Acacia dodoniifolia (Pers.) Willd.
Acacia doratoxylon A. Cunn.
Acacia drummondii Lindl.
Acacia elata A. Cunn. ex Benth. See **A. terminalis**.
Acacia farnesiana (L.) Willd. SWEET ACACIA. Some plants so called are **A. smallii**.
Acacia fimbriata A. Cunn. ex G. Don.
Acacia floribunda (Vent.) Willd. Syn.: *A. longifolia* var. *floribunda*. SYDNEY WATTLE. SYDNEY ACACIA. GOSSAMER WATTLE. Some plants so called are **A. retinodes**.
Acacia giraffae Burchell. CAMEL THORN.
Acacia gladiiformis A. Cunn. ex Benth.
Acacia glandulicarpa Reader.
Acacia glaucoptera Benth.
Acacia graffiana F. J. Muell.
Acacia greggii Gray. CAT'S-CLAW. CATCLAW ACACIA.
Acacia howittii F. J. Muell.
Acacia iteaphylla F. J. Muell.
Acacia jonesii F. J. Muell.
Acacia laricina Meissn.
Acacia latifolia Benth.
Acacia latifolia 'Columns'?
Acacia latifolia floribunda. Listed name, probably for **A. floribunda**.
Acacia linearis (H. L. Wendl.) Macbr.
Acacia lineata A. Cunn. ex G. Don.
Acacia longifolia (Andr.) Willd. SYDNEY GOLDEN WATTLE. BUSH ACACIA. BROAD-LEAF ACACIA.
Acacia longifolia var. *floribunda* (Vent.) F. J. Muell. See **A. floribunda**.
Acacia longifolia var. *sophorae* (Labill.) F. J. Muell. Syn.: *A. sophorae*.
Acacia mearnsii De Wild. Syn.: *A. decurrens* var. *mollis*. *A. mollissima*. BLACK WATTLE.
Acacia meissneri Lehm.
Acacia melanoxylon R. Br. BLACKWOOD ACACIA. BLACK ACACIA.
Acacia microbotrya Benth.
Acacia mollissima of authors, not Willd. See **A. mearnsii**.
Acacia montana Benth.
Acacia myrtifolia (Sm.) Willd.
Acacia notabilis F. J. Muell.
Acacia nutans?
Acacia obliqua A. Cunn. ex Benth.
Acacia "Ongerup," listed name for **A. redolens**.
Acacia oxycedrus Sieber ex DC.
Acacia pendula A. Cunn. ex G. Don. WEEPING MYALL.
Acacia pendula glauca. Listed name for **A. pendula**.
Acacia penninervis Sieber ex DC.
Acacia podalyriifolia A. Cunn. PEARL ACACIA. QUEENSLAND GOLDEN WATTLE.
Acacia prominens A. Cunn. GOLDEN-RAIN WATTLE.
Acacia pruinosa A. Cunn. ex Benth. FROSTY WATTLE. Some plants so called are **A. schinoides**.
Acacia pulverulenta A. Cunn. See **A. calamifolia**.
Acacia pycnantha Benth. GOLDEN WATTLE.
Acacia redolens Maslin. Plants called *Acacia* "ongerup" belong here.
Acacia retinodes Schlechtend. EVERBLOOMING ACACIA. Some plants called **A. floribunda** belong here.
Acacia riceana Henslow.
Acacia 'Rosemary'?
Acacia rostellifera Benth.
Acacia rotundifolia Hook.
Acacia rubida A. Cunn. RED-STEM ACACIA.
Acacia saligna (Labill.) H. Wendl. WILLOW ACACIA. Some plants so called are **A. cyanophylla**.
Acacia schinoides Benth. Some plants called **A. pruinosa** belong here.
Acacia smallii Isely. Some plants called **A. farnesiana** belong here.
Acacia sophorae (Labill.) R. Br. ex Ait. See **A. longifolia** var. **sophorae**.
Acacia suaveolens (Sm.) Willd. Some plants called **A. dietrichiana** belong here.
Acacia suberosa A. Cunn.
Acacia subporosa F. J. Muell.
Acacia subulata Bonpl.
Acacia terminalis (Salisb.) Macb. Syn.: *A. elata*. CEDAR WATTLE.
Acacia trineura F. J. Muell.
Acacia undulifolia C. Fraser ex Lodd.
Acacia verniciflua A. Cunn.
Acacia verticillata (L'Hér.) Willd. STAR ACACIA.
Acacia visco Lorentz ex Griseb.
Acalypha californica Benth. CALIFORNIA COPPERLEAF.
Acalypha godseffiana M. T. Mast. See **A. wilkesiana** 'Godseffiana'.
Acalypha hispida Burm. f. CHENILLE PLANT.
Acalypha 'Macafeeana'. See **A. wilkesiana** 'Macafeeana'.
Acalypha tricolor Seem. See **A. wilkesiana**.
Acalypha wilkesiana Muell.-Arg. Syn.: *A. tricolor*. PAINTED COPPERLEAF.
Acalypha wilkesiana 'Butterfly'.
Acalypha wilkesiana 'Godseffiana'. Syn.: *A. godseffiana*.
Acalypha wilkesiana 'Macafeeana'.
Acalypha wilkesiana 'Tricolor'. See **A. wilkesiana**.
Acanthopanax ricinifolius (Siebold & Zucc.) Seem. See **Kalopanax septemlobus**.
Acanthopanax sieboldianus Makino.
Acanthus mollis L. BEAR'S-BREECH. GRECIAN PATTERN PLANT.
Acer 'Aconitifolium'. See **A. japonicum** 'Aconitifolium'.
Acer aidzuense (Franch.) Nakai. See **A. ginnala** var. **aidzuense**.
Acer 'Atropurpureum'. See **A. palmatum** 'Atropurpureum' or **A. pseudoplatanus** 'Atropurpureum'.
Acer 'Atropurpureum Burgundy Lace'. See **A. palmatum** 'Burgundy Lace'.
Acer buergeranum Miq. TRIDENT MAPLE.
Acer buergeranum 'Goshiki Kaede'?
Acer buergeranum 'Postalense'.
Acer buergeranum 'Pulverulentum'.
Acer 'Butterfly'. See **A. palmatum** 'Butterfly'.
Acer campestre L. HEDGE MAPLE. ENGLISH MAPLE.
Acer campestre 'Compactum'.
Acer capillipes Maxim.
Acer cappadocicum Gleditsch. COLISEUM MAPLE.
Acer cappadocicum 'Rubrum'. RED COLISEUM MAPLE.
Acer caudatum Wallich. Syn.: *A. papilio*.

Acer caudatum subsp. **ukurunduense** (Trautv. & C. A. Mey.) E. Murr. Syn.: *A. ukurunduense*.
Acer 'Chikushigata'. See **A. palmatum 'Chikushigata'**.
Acer 'Chishio'. See **A. palmatum 'Sanguineum'**.
Acer circinatum Pursh. VINE MAPLE.
Acer circinatum 'Monroe'?
Acer 'Cleveland'. See **A. platanoides 'Cleveland'**.
Acer crataegifolium Siebold & Zucc. HAWTHORN MAPLE.
Acer crataegifolium 'Beni-uri'?
Acer 'Crimson King'. See **A. platanoides 'Crimson King'**.
Acer crispum Lauth. See **A. platanoides 'Crispum'**.
Acer dasycarpum Ehrh. See **A. saccharinum**.
Acer dasycarpum forma *pyramidale* Spaeth. See **A. saccharinum 'Pyramidale'**.
Acer dasycarpum forma *wieri* Schwerin. See **A. saccharinum 'Wieri'**.
Acer davidii Franch. DAVID MAPLE.
Acer dissectum Thunb. See **A. palmatum 'Dissectum'**.
Acer 'Dissectum Atropurpureum'. See **A. palmatum 'Dissectum Atropurpureum'**.
Acer 'Dissectum Nigrum'. See **A. palmatum 'Dissectum Nigrum'**.
Acer 'Dissectum Purpureum'. See **A. palmatum 'Dissectum Purpureum'**.
Acer 'Dissectum Viride'. See **A. palmatum 'Dissectum Viride'**.
Acer 'Emerald Queen'. See **A. platanoides 'Emerald Queen'**.
Acer 'Faassen'. See **A. platanoides 'Faassen's Black'**.
Acer 'Faassen Red Leaf'. See **A. platanoides 'Faassen's Black'**.
Acer 'Faassen's Black'. See **A. platanoides 'Faassen's Black'**.
Acer franchetii Pax.
Acer ginnala Maxim. GINNALA MAPLE. AMUR MAPLE.
Acer ginnala var. **aidzuense** (Franch.) Pax. Syn.: *A. aidzuense*.
Acer ginnala 'Durand Dwarf'.
Acer glabrum Torr. ROCKY MOUNTAIN MAPLE.
Acer glabrum subsp. **douglasii** (Hook.) Wesm.
Acer griseum (Franch.) Pax. PAPERBARK MAPLE.
Acer grosseri Pax.
Acer grosseri var. **hersii** (Rehd.) Rehd.
Acer japonicum Thunb. JAPANESE MAPLE. FULL-MOON MAPLE.
Acer japonicum 'Aconitifolium'. FERN-LEAF FULL-MOON MAPLE.
Acer japonicum 'Aureum'. GOLDEN FULL-MOON MAPLE.
Acer japonicum 'Filicifolium'. See **A. japonicum 'Aconitifolium'**.
Acer japonicum 'Green Cascade'?
Acer japonicum 'Junihitoe'.
Acer japonicum 'Microphyllum'.
Acer macrophyllum Pursh. BIG-LEAF MAPLE. OREGON MAPLE.
Acer maximowiczianum Miq. Syn.: *A. nikoense*. NIKKO MAPLE.
Acer mono Maxim. See **A. truncatum** subsp. **mono**.
Acer monspessulanum L. MONTPELIER MAPLE.
Acer morrisonense Hayata.
Acer negundo L. BOX ELDER. ASH-LEAVED MAPLE.
Acer negundo 'Argenteo-variegatum'. See **A. negundo 'Variegatum'**.
Acer negundo subsp. **californicum** (Torr. & Gray) Wesm.
Acer negundo subsp. **interius** (Britt.) A. & D. Löve.
Acer negundo 'Variegatum'. SILVER VARIEGATED BOX ELDER.
Acer nigrum Michx. f. See **A. saccharum** subsp. **nigrum**.
Acer nikoense of authors, not (Miq.) Maxim. See **A. maximowiczianum**.
Acer oblongum Wallich ex DC. OBLONG MAPLE. EVERGREEN MAPLE.
Acer oblongum var. *biauritum* W. W. Sm. See **A. paxii**.
Acer palmatum Thunb. DWARF JAPANESE MAPLE. JAPANESE MAPLE.
Acer palmatum 'Aoyagi'.
Acer palmatum 'Asahi-zuru'.

Acer palmatum 'Atro-lineare'. See **A. palmatum 'Linearilobum Atropurpureum'**.
Acer palmatum 'Atropurpureum'. BLOODLEAF MAPLE. RIBBONLEAF PURPLE MAPLE. RED JAPANESE MAPLE.
Acer palmatum 'Aureum'.
Acer palmatum 'Azuma-murasaki'.
Acer palmatum 'Beni-shichi-henge'.
Acer palmatum 'Bloodgood'.
Acer palmatum 'Bonfire'.
Acer palmatum 'Burgundy Lace'.
Acer palmatum 'Butterfly'.
Acer palmatum 'Chiba'.
Acer palmatum 'Chikushigata'.
Acer palmatum 'Chishio'. See **A. palmatum 'Sanguineum'**.
Acer palmatum 'Chitoseyama'.
Acer palmatum var. **coreanum** Nakai.
Acer palmatum 'Crispum'.
Acer palmatum 'Cristatum'. See **A. palmatum 'Shishi-gashira'**.
Acer palmatum 'Deshojo'?
Acer palmatum 'Dissectum'. Syn.: *A. dissectum*. LACE-LEAF JAPANESE MAPLE. CUT-LEAF GREEN MAPLE. THREAD-LEAF MAPLE.
Acer palmatum 'Dissectum Atropurpureum'.
Acer palmatum 'Dissectum Crimson Queen'.
Acer palmatum 'Dissectum Ever Red'.
Acer palmatum 'Dissectum Filigree'?
Acer palmatum 'Dissectum Garnet'.
Acer palmatum 'Dissectum Inaba Shidare'?
Acer palmatum 'Dissectum Inazuma'.
Acer palmatum 'Dissectum Julian Pendulum'.
Acer palmatum 'Dissectum Nigrum'.
Acer palmatum 'Dissectum Omuroyama'.
Acer palmatum 'Dissectum Ornatum'.
Acer palmatum 'Dissectum Palmatifidum'. See **A. palmatum 'Dissectum'**.
Acer palmatum 'Dissectum Pendulum Angustilobum Atropurpureum'. See **A. palmatum 'Dissectum Atropurpureum'**.
Acer palmatum 'Dissectum Purpureum'.
Acer palmatum 'Dissectum Seiryu'?
Acer palmatum 'Dissectum Variegatum'.
Acer palmatum 'Dissectum Viride'.
Acer palmatum 'Dissectum Waterfall'.
Acer palmatum 'Elegans'.
Acer palmatum 'Filiferum Purpureum'.
Acer palmatum 'Hagoromo'. See **A. palmatum 'Sessilifolium'**.
Acer palmatum var. **heptalobum** Rehd.
Acer palmatum 'Hessei'.
Acer palmatum 'Higasayama'.
Acer palmatum 'Hillieri'.
Acer palmatum 'Hogyuko'.
Acer palmatum 'Ichigyoin'.
Acer palmatum 'Iijima-sunago'.
Acer palmatum 'Inazuma'. See **A. palmatum 'Dissectum Inazuma'**.
Acer palmatum 'Kamagata'?
Acer palmatum 'Kasen-nishiki'.
Acer palmatum 'Kashima'?
Acer palmatum 'Katsura-yatsubusa'?
Acer palmatum 'Ki-hachojo'?
Acer palmatum 'Kinran'?
Acer palmatum 'Kiyohime-yatsubusa'?
Acer palmatum var. *koreanum*. See **A. palmatum** var. **coreanum**.
Acer palmatum 'Koshibori-nishiki'.
Acer palmatum 'Koshimino'. See **A. palmatum 'Sessilifolium'**.

Acer palmatum 'Kotohime'?
Acer palmatum 'Kurui-jishi'.
Acer palmatum 'Laceleaf'.
Acer palmatum 'Laciniatum'.
Acer palmatum 'Linearilobum'.
Acer palmatum 'Linearilobum Atropurpureum'.
Acer palmatum 'Lutescens'.
Acer palmatum 'Mama'?
Acer palmatum 'Mino-yatsubusa'?
Acer palmatum 'Murogawa'.
Acer palmatum 'Nomura'.
Acer palmatum 'Novum'.
Acer palmatum 'Nuresagi'.
Acer palmatum 'Okushimo'. See **A. palmatum 'Crispum'**.
Acer palmatum 'Omuroyama'. See **A. palmatum 'Dissectum Omuroyama'**.
Acer palmatum 'Oridomo-nishiki'. See **A. palmatum 'Versicolor'**.
Acer palmatum 'Ornatum'. RED LACE-LEAF JAPANESE MAPLE.
Acer palmatum 'Osakazuki'.
Acer palmatum 'Oshio-beni'.
Acer palmatum 'Purpureum Superbum'.
Acer palmatum 'Red Leaf'.
Acer palmatum 'Red Pygmy'.
Acer palmatum 'Reticulatum'.
Acer palmatum 'Ribesifolium'. See **A. palmatum 'Shishi-gashira'**.
Acer palmatum 'Roseo-marginatum'.
Acer palmatum 'Rubrum'. RED JAPANESE MAPLE.
Acer palmatum 'Sagara-nishiki'.
Acer palmatum 'Samidare'.
Acer palmatum 'Sangokaku'.
Acer palmatum 'Sanguineum'.
Acer palmatum 'Sanguineum Chishio'. See **A. palmatum 'Sanguineum'**.
Acer palmatum 'Scolopendriifolium'.
Acer palmatum 'Sekimori'.
Acer palmatum 'Sessilifolium'.
Acer palmatum 'Sherwood Flame'?
Acer palmatum 'Shindeshojo'?
Acer palmatum 'Shishi-gashira'.
Acer palmatum 'Shishio Improved'?
Acer palmatum 'Sinuatum'.
Acer palmatum 'Suminagashi'.
Acer palmatum 'Superbum'? May be **A. palmatum 'Purpureum Superbum'**.
Acer palmatum 'Taihai'.
Acer palmatum 'Tana'?
Acer palmatum 'Threadleaf'. See **A. palmatum 'Dissectum'**.
Acer palmatum 'Trompenburg'?
Acer palmatum 'Tsuchigumo'.
Acer palmatum 'Tsumegaki'.
Acer palmatum 'Versicolor'.
Acer palmatum 'Villa Taranto'.
Acer palmatum 'Volubile'.
Acer palmatum 'Waka-momiji'.
Acer palmatum 'Wou-nishiki'.
Acer palmatum 'Yezo-nishiki'. See **A. palmatum 'Sinuatum'**.
Acer papilio King. See **A. caudatum**.
Acer paxii Franch. Syn.: *A. oblongum* var. *biauritum*. EVERGREEN MAPLE.
Acer pensylvanicum L. STRIPED MAPLE. PENNSYLVANIA MAPLE. MOOSEWOOD.
Acer platanoides L. NORWAY MAPLE.
Acer platanoides 'Cleveland'.
Acer platanoides 'Columnare'. PYRAMIDAL NORWAY MAPLE.
Acer platanoides 'Crimson King'. CRIMSON KING MAPLE.
Acer platanoides 'Crispum'. Syn.: *A. crispum*.
Acer platanoides 'Drummondii'. SILVER VARIEGATED NORWAY MAPLE.
Acer platanoides 'Emerald Queen'.
Acer platanoides 'Faassen's Black'. FAASSEN REDLEAF MAPLE.
Acer platanoides 'Globosum'. GLOBE MAPLE.
Acer platanoides 'Royal Red'.
Acer platanoides 'Schwedleri'. SCHWEDLER PURPLE MAPLE. SCHWEDLER NORWAY MAPLE.
Acer platanoides 'Schwedleri Nigrum'. See **A. platanoides 'Crimson King'**.
Acer platanoides 'Summershade'. SUMMERSHADE NORWAY MAPLE.
Acer platanoides 'Variegatum'. VARIEGATED NORWAY MAPLE.
Acer pseudoplatanus L. SYCAMORE MAPLE.
Acer pseudoplatanus 'Atropurpureum'. SPAETH MAPLE. Many plants so called are **A. pseudoplatanus 'Purpureum'**.
Acer pseudoplatanus 'Purpureum'. Some plants called **A. pseudoplatanus 'Atropurpureum'** belong here.
Acer pseudoplatanus 'Spaethii'. See **A. pseudoplatanus 'Atropurpureum'**.
Acer pseudo-sieboldianum (Pax) Kom.
Acer rubrum L. RED MAPLE. SCARLET MAPLE.
Acer rubrum 'Schlesinger'.
Acer rufinerve Siebold & Zucc. RED-VEIN MAPLE.
Acer rufinerve 'Hatsuyuki'?
Acer saccharinum L. Syn.: *A. dasycarpum*. SILVER MAPLE. WHITE MAPLE. SOFT MAPLE.
Acer saccharinum 'Laciniatum'. See **A. saccharinum 'Wieri'**.
Acer saccharinum 'Pyramidale'. Syn.: *A. dasycarpum* forma *pyramidale*.
Acer saccharinum 'Wieri'. Syn.: *A. dasycarpum* forma *wieri*. WIER MAPLE. WIER CUTLEAF MAPLE.
Acer saccharum Marsh. SUGAR MAPLE. HARD MAPLE. ROCK MAPLE.
Acer saccharum 'Globosum'. GLOBE SUGAR MAPLE.
Acer saccharum forma *monumentale* (Temple) Rehd. See **A. saccharum 'Temple's Upright'**.
Acer saccharum subsp. **nigrum** (Mich. f.) Demarais. Syn.: *A. nigrum*. BLACK MAPLE.
Acer saccharum 'Temple's Upright'. Syn.: *A. saccharum* forma *monumentale*.
Acer 'Sangokaku'. See **A. palmatum 'Sangokaku'**.
Acer 'Sanguineum Chishio'. See **A. palmatum 'Sanguineum'**.
Acer 'Schwedleri'. See **A. platanoides 'Schwedleri'**.
Acer seminowii Regel & Herder.
Acer shirasawanum G. Koidz.
Acer shirasawanum 'Palmatifolium'?
Acer sieboldianum Miq.
Acer sieboldianum var. **microphyllum** Maxim.
Acer sieboldianum 'Ogurayama'?
Acer sieboldianum 'Sode-no-uchi'.
Acer 'Spaethii'. See **A. pseudoplatanus 'Atropurpureum'**.
Acer spicatum Lam. MOUNTAIN MAPLE.
Acer syriacum Boiss. & Gaillardot. SYRIAN MAPLE.
Acer tataricum L. TATARIAN MAPLE.
Acer tegmentosum Maxim.
Acer truncatum Bunge. PURPLEBLOW MAPLE. SHANTUNG MAPLE.
Acer truncatum subsp. **mono** (Maxim.) E. Murr. Syn.: *A. mono*.
Acer ukurunduense Trautv. & C. A. Mey. See **A. caudatum** subsp. **ukurunduense**.
Acer 'Wieri'. See **A. saccharinum 'Wieri'**.
Acmena smithii (Poir.) Merrill & L. M. Perry. Syn.: *Eugenia smithii*. LILLY-PILLY TREE.

Acoelorraphe wrightii (Griseb. & H. Wendl.) H. Wendl. ex Becc. Syn.: *Paurotis wrightii*. SAW CABBAGE PALM. SILVER SAW PALMETTO.
Acokanthera oblongifolia (Hochst.) Codd. Syn.: *A. spectabilis*. WINTERSWEET.
Acokanthera oppositifolia (Lam.) Codd. Syn.: *A. venenata*. BUSHMAN'S POISON.
Acokanthera spectabilis (Sond.) Hook. f. See **A. oblongifolia**.
Acokanthera venenata of authors, not G. Don. See **A. oppositifolia**.
Acrocarpus fraxinifolius Arn. PINK CEDAR.
Acrocomia mexicana Karw. ex Mart. COYOLI PALM.
Acrocomia totai Mart. GRUGRU PALM.
ACTINIDIA, BOWER: **Actinidia arguta**.
Actinidia arguta (Siebold & Zucc.) Planch. ex Miq. BOWER ACTINIDIA. TARA VINE.
Actinidia chinensis Planch. KIWI. KIWI VINE. CHINESE GOOSEBERRY. YANGTAO.
Actinidia kolomikta (Maxim. & Rupr.) Maxim.
Actinophloeus macarthurii (H. Wendl.) Becc. See **Ptychosperma macarthurii**.
Actinotus helianthi Labill. FLANNEL FLOWER.
ADAM'S NEEDLE: **Yucca filamentosa**.
Adenia glauca Schinz.
Adenium obesum (Forssk.) Roemer & Schult.
Adenocarpus foliolosus (Ait.) DC. CANARY ISLAND FLATPOD.
Adenocarpus viscosus (Willd.) Webb & Berth.
Adenostoma fasciculatum Hook. & Arn. CHAMISE. GREASEWOOD.
Adenostoma sparsifolium Torr. RIBBONWOOD.
Adhatoda vasica (L.) Nees. See **Justicia adhatoda**.
Aeonium arboreum (L.) Webb & Berth.
Aeonium arboreum 'Atropurpureum'.
Aeonium arboreum 'Schwarzkopf'. Sometimes cv. name incorrectly spelled as 'Swartkop' or 'Swartzkop'.
Aeonium decorum Webb.
Aeonium haworthii Webb & Berth.
Aeonium holochrysum Webb & Berth.
Aeschynanthus lobbianus Veitch ex Hook. See **A. radicans**.
Aeschynanthus marmoratus T. Moore.
Aeschynanthus radicans Jack. Syn.: *A. lobbianus*. *Trichosporum lobbianum*. LIPSTICK PLANT.
Aeschynanthus speciosus Hook.
Aesculus 'Briotii'. See **A. × carnea 'Briotii'**.
Aesculus californica (Spach) Nutt. CALIFORNIA BUCKEYE.
Aesculus × carnea Hayne. [*A. hippocastanum* × *A. pavia*]. RED HORSE CHESTNUT.
Aesculus × carnea 'Briotii'. RUBY HORSE CHESTNUT.
Aesculus × carnea 'O'Neill Red'.
Aesculus glabra Willd. OHIO BUCKEYE.
Aesculus hippocastanum L. HORSE CHESTNUT. EUROPEAN HORSE CHESTNUT.
Aesculus octandra Marsh. YELLOW BUCKEYE.
Aesculus parviflora Walter. DWARF HORSE CHESTNUT.
Aesculus pavia L. Syn.: *A. pavia* var. *humilis*. RED BUCKEYE.
Aesculus pavia var. *humilis* (Lodd. ex Lindl.) Mouillefert. See **A. pavia**.
Agapetes incurvata (W. Griffith) Sleumer. Syn.: *Pentapterygium rugosum*.
Agapetes 'Ludgvan Cross'. [**A. incurvata** × **A. serpens**].
Agapetes serpens (Wight) Sleumer. Syn.: *Pentapterygium serpens*.
Agathis australis Lindl. KAURI PINE. KAURI.
Agathis robusta (F. J. Muell.) F. M. Bailey. QUEENSLAND KAURI.
Agathosma corymbosa (Montin) G. Don. Syn.: *A. villosa*.
Agathosma ovata (Thunb.) Pillans. Syn.: *Barosma ovata*.
Agathosma villosa (Willd.) Willd. See **A. corymbosa**.

Agati grandiflora (L.) Desv. See **Sesbania grandiflora**.
Agave americana L. CENTURY PLANT. MAGUEY.
Agave americana 'Variegata'.
Agave attenuata Salm-Dyck.
Agave victoriae-reginae T. Moore.
Aglaonema commutatum Schott.
Aglaonema crispum (Pitcher & Manda) Nichols. Syn.: *A. roebelinii*.
Aglaonema modestum Schott ex Engler. CHINESE EVERGREEN.
Aglaonema roebelinii Pitcher & Manda. See **A. crispum**.
Agonis flexuosa (K. Spreng.) Schauer. PEPPERMINT TREE. AUSTRALIAN WILLOW MYRTLE.
Agonis juniperina Schauer. JUNIPER WILLOW MYRTLE.
Agonis marginata (Labill.) Schauer.
Ailanthus altissima (Mill.) Swingle. Syn.: *A. glandulosa*. TREE-OF-HEAVEN.
Ailanthus glandulosa Desf. See **A. altissima**.
AKEBIA, FIVELEAF: **Akebia quinata**.
Akebia quinata (Houtt.) Decne. FIVELEAF AKEBIA.
Alberta magna E. H. Mey.
ALBIZIA, PLUME: **Albizia lophantha**.
Albizia distachya (Vent.) Macb. See **A. lophantha**.
Albizia julibrissin Durazz. MIMOSA. PINK ACACIA. SILK TREE.
Albizia julibrissin 'Rosea'.
Albizia julibrissin 'Rubra'. See **A. julibrissin 'Rosea'**.
Albizia lophantha (Willd.) Benth. Syn.: *A. distachya*. PLUME ALBIZIA.
Albizia polyphylla Fournier.
Albizzia. See **Albizia**.
ALDER, AFRICAN RED: **Cunonia capensis**.
  BLACK: **Alnus glutinosa**.
  CALIFORNIA: **Alnus rhombifolia**.
  EUROPEAN: **Alnus cordata**.
  EUROPEAN GREEN: **Alnus viridis**.
  EVERGREEN: **Alnus nepalensis**.
  ITALIAN: **Alnus cordata**.
  JAPANESE: **Alnus japonica**.
  OREGON: **Alnus oregona**.
  RED: **Alnus oregona**.
  SIERRA: **Alnus rhombifolia**.
  SITKA: **Alnus sinuata**.
  WHITE: **Alnus rhombifolia**.
Alectryon excelsus Gaertner.
Alectryon subcinereus (Gray) Radlk. SMOOTH RAMBUTAN.
Aleurites fordii Hemsl. TUNG-OIL TREE.
ALFALFA, TREE: **Medicago arborea**.
ALGARROBA: **Ceratonia siliqua**.
Allamanda cathartica L.
Allamanda cathartica 'Hendersonii'.
ALLSPICE, CAROLINA: **Calycanthus floridus**.
ALMOND: **Prunus dulcis**.
  BITTER: **Prunus dulcis** var. **amara**.
  DESERT: **Prunus fasciculata**.
  DWARF FLOWERING: **Prunus glandulosa**.
  DWARF RUSSIAN: **Prunus tenella**.
  DWARF SWEET: **Prunus dulcis 'Nana'**.
  FLOWERING: **Prunus jacquemontii**.
  FLOWERING: **Prunus triloba**.
  INDIAN: **Sterculia foetida**.
  SWEET: **Prunus dulcis**.
Alnus californica Winkler. See **A. rhombifolia**.
Alnus cordata (Loisel.) Duby. ITALIAN ALDER. EUROPEAN ALDER.
Alnus glutinosa (L.) Gaertner. BLACK ALDER.
Alnus japonica (Thunb.) Steud. JAPANESE ALDER.
Alnus nepalensis D. Don. EVERGREEN ALDER.
Alnus oregona Nutt. Syn.: *A. rubra*. RED ALDER. OREGON ALDER.

Alnus rhombifolia Nutt. Syn.: *A. californica*. CALIFORNIA ALDER. SIERRA ALDER. WHITE ALDER.
Alnus rubra Bong. See **A. oregona**.
Alnus sinuata (Regel) Rydb. Syn.: *A. viridis* var. *sinuata*. SITKA ALDER.
Alnus viridis (Chaix) DC. EUROPEAN GREEN ALDER.
Alnus viridis var. *sinuata* Regel. See **A. sinuata**.
Alocasia macrorrhiza (L.) Schott. ELEPHANT'S-EAR. TARO. APÉ. GIANT APÉ.
ALOE, BARBADOS: **Aloe barbadensis**.
   CAPE: **Aloe ferox**.
   CORAL: **Aloe striata**.
   MEDICINAL: **Aloe barbadensis**.
   TIGER: **Aloe variegata**.
   TREE: **Aloe arborescens**.
Aloe africana Mill.
Aloe arborescens Mill. TREE ALOE.
Aloe aristata Haw.
Aloe bainsii Dyer.
Aloe barbadensis Mill. Syn.: *A. vera*. BARBADOS ALOE. MEDICINAL ALOE.
Aloe brevifolia Mill.
Aloe ciliaris Haw.
Aloe excelsa Berger.
Aloe ferox Mill. CAPE ALOE.
Aloe marlothii Berger.
Aloe nobilis Haw.
Aloe reitzii Reynolds.
Aloe saponaria (Ait.) Haw.
Aloe striata Haw. CORAL ALOE.
Aloe thraskii Baker.
Aloe vaombe Decorse & Poisson.
Aloe variegata L. TIGER ALOE. PARTRIDGE-BREAST.
Aloe vera (L.) Webb & Berth. See **A. barbadensis**.
Aloe wickensii Pole Evans.
Aloysia triphylla (L'Her.) Britt. Syn.: *Lippia citriodora*. LEMON VERBENA.
Alpinia nutans (Andr.) Roscoe. Not cult. Plants so called are **A. zerumbet**.
Alpinia speciosa (Wendl.) Schum. See **A. zerumbet**.
Alpinia zerumbet (Pers.) B. L. Burtt & R. M. Sm. Syn.: *A. speciosa*. SHELLFLOWER. SHELL GINGER. PINK PORCELAIN LILY. Some plants called **A. nutans** belong here.
Alsophila australis R. Br. May not be cult. Plants so called are usually **Sphaeropteris cooperi**.
Alsophila cooperi F. J. Muell. See **Sphaeropteris cooperi**.
Alstonia scholaris (L.) R. Br. Probably not cult. in Calif. Plants so called are **Rauvolfia samarensis**.
ALTHAEA, SHRUB: **Hibiscus syriacus**.
Althaea 'Alba'. See **Hibiscus syriacus 'Albus'**.
Althaea 'Ardens'. See **Hibiscus syriacus 'Ardens'**.
Althaea 'Pulcherrima'. See **Hibiscus syriacus 'Pulcherrimus'**.
Althaea syriaca Hort., and listed cultivars. See **Hibiscus syriacus**.
Alyogyne hakeifolia (Giordano) Alef. Syn.: *Cienfuegosia hakeifolia*.
Alyogyne huegelii (Endl.) Fryx. Syn.: *Hibiscus huegelii*. *H. wrayae*. BLUE HIBISCUS.
Amaracus dictamnus (L.) Benth. See **Origanum dictamnus**.
AMAZON VINE: **Stigmaphyllon ciliatum**.
Amelanchier alnifolia Nutt. WESTERN SERVICEBERRY. SASKATOON SERVICEBERRY.
Amelanchier canadensis (L.) Medic. SHADBLOW SERVICEBERRY.
Amelanchier × grandiflora Rehd. [**A. arborea** (Michx. f.) Fern. × **A. laevis**].
Amelanchier laevis Wiegand. SHADBUSH.
Amelanchier pumila Nutt.
Amelanchier stolonifera Wiegand.
Amomyrtus luma (Molina) D. Legrand & Kausel. Syn.: *Myrtus lechlerana*.
Amorpha fruticosa L. FALSE INDIGO. INDIGO BUSH.
Ampelopsis brevipedunculata (Maxim.) Trautv. BLUEBERRY CLIMBER.
Ampelopsis brevipedunculata var. **maximowiczii** (Regel) Rehd. Syn.: *Vitis heterophylla*.
Ampelopsis henryana (Hemsl.) Rehd. See **Parthenocissus henryana**.
Ampelopsis 'Lowii'. See **Parthenocissus tricuspidata 'Lowii'**.
Ampelopsis quinquefolia (L.) Michx. See **Parthenocissus quinquefolia**.
Ampelopsis quinquefolia var. *engelmannii* Rehd. See **Parthenocissus quinquefolia 'Engelmannii'**.
Ampelopsis sempervirens Hort. See **Cissus striata**.
Ampelopsis tricuspidata Siebold & Zucc. See **Parthenocissus tricuspidata**.
Ampelopsis tricuspidata 'Veitchii'. See **Parthenocissus tricuspidata 'Veitchii'**.
Ampelopsis veitchii. Listed name for **Parthenocissus tricuspidata 'Veitchii'**.
Ampelopsis veitchii 'Beverly Brooks'. See **Parthenocissus tricuspidata 'Beverly Brooks'**.
Ampelopsis virginiana Dipp. See **Parthenocissus quinquefolia**.
Amygdalus persica L. See **Prunus persica**.
ANAQUA: **Ehretia anacua**.
ANCHOR PLANT: **Colletia paradoxa**.
ANDROMEDA: **Pieris floribunda**.
ANDROMEDA: **Pieris japonica**.
   CHINESE: **Pieris formosa** var. **forrestii**.
Andromeda floribunda Pursh. See **Pieris floribunda**.
Andromeda glaucophylla Link.
Andromeda polifolia L. BOG ROSEMARY.
Andromeda polifolia 'Compacta'.
Andromeda polifolia 'Glauca'. Listed name, probably for **A. polifolia** or **A. glaucophylla**.
Andromeda polifolia 'Nana'.
ANEMONE, BUSH: **Carpenteria californica**.
Anemopaegma chamberlaynii (Sims) Bur. & K. Schum. Syn.: *Bignonia chamberlaynii*. YELLOW TRUMPET VINE.
ANGEL'S HAIR: **Artemisia schmidtiana**.
ANGELICA TREE: **Aralia spinosa**.
   CHINESE: **Aralia chinensis**.
   JAPANESE: **Aralia elata**.
ANGEL'S-TRUMPET: **Brugmansia suaveolens**.
   RED: **Brugmansia sanguinea**.
   WHITE: **Brugmansia suaveolens**.
Angophora cordifolia Cav.
Angophora costata (Gaertner) Britten. Syn.: *A. lanceolata*. GUM MYRTLE.
Angophora floribunda (Sm.) Sweet. Syn.: *A. intermedia*. GUM MYRTLE.
Angophora intermedia DC. See **A. floribunda**.
Angophora lanceolata Cav. See **A. costata**.
Angophora subvelutina F. J. Muell.
Anigosanthus flavidus DC. ALBANY KANGAROO-PAWS.
Anigosanthus manglesii D. Don. GREEN KANGAROO-PAWS.
Anisacanthus thurberi (Torr.) Gray. May not be cult. in Calif. Plants so called are **Justicia leonardii**.
ANISE, PURPLE: **Illicium floridanum**.
Anisostichus capreolatus (L.) Bur. See **Bignonia capreolata**.
Annona cherimola Mill. CUSTARD APPLE. CHERIMOYA.
Anredera baselloides (HBK) Baill. Syn.: *Boussingaultia baselloides* HBK, not Hook. May not be cult. Plants so called are probably **Anredera cordifolia**.

Anredera cordifolia (Ten.) Steenis. Syn.: *Boussingaultia baselloides* Hook., not HBK. *B. gracilis*. MADEIRA VINE. MIGNONETTE VINE. Some plants called **A. baselloides** belong here.
ANTELOPE BRUSH: **Purshia tridentata.**
Antigonon leptopus Hook. & Arn. CORAL VINE. CORALLITA. QUEEN'S WREATH. ROSA DE MONTAÑA.
APACHE-PLUME: **Fallugia paradoxa.**
APÉ: **Alocasia macrorrhiza.**
  GIANT: **Alocasia macrorrhiza.**
Aphelandra 'Louisae'. See **A. squarrosa** 'Louisae'.
Aphelandra squarrosa Nees. ZEBRA PLANT.
Aphelandra squarrosa 'Louisae'.
APOOS: **Gigantochloa apus.**
APPLE: **Malus sylvestris.**
  ALDENHAM CRAB: **Malus 'Aldenhamensis'.**
  AMERICAN CRAB: **Malus coronaria.**
  ARNOLD CRAB: **Malus × arnoldiana.**
  BECHTEL CRAB: **Malus ioensis 'Plena'.**
  CARMEN CRAB: **Malus × atrosanguinea.**
  CHINESE: **Malus prunifolia.**
  CHINESE FLOWERING: **Malus spectabilis.**
  COMMON: **Malus domestica.**
  CRAB: **Malus sylvestris.**
  CUSTARD: **Annona cherimola.**
  ECHTERMEYER CRAB: **Malus 'Oekonomierat Echtermeyer'.**
  ELEY CRAB: **Malus × eleyi.**
  JAPANESE FLOWERING CRAB: **Malus floribunda.**
  KAIDO CRAB: **Malus micromalus.**
  KANGAROO: **Solanum laciniatum.**
  KEI: **Dovyalis caffra.**
  LOVE: **Solanum aculeatissimum.**
  MANCHURIAN CRAB: **Malus baccata** var. **mandschurica.**
  MEXICAN: **Casimiroa edulis.**
  MIDGET CRAB: **Malus micromalus.**
  PARADISE: **Malus pumila.**
  PARKMAN CRAB: **Malus halliana 'Parkmanii'.**
  PERUVIAN: **Cereus peruvianus.**
  PLUM-LEAVED: **Malus prunifolia.**
  POSSUM: **Diospyros virginiana.**
  PRAIRIE CRAB: **Malus ioensis.**
  RED-VEIN CRAB: **Malus niedzwetzkyana.**
  ROSE: **Syzygium jambos.**
  SARGENT CRAB: **Malus sargentii.**
  SCHEIDECKER CRAB: **Malus × scheideckeri.**
  SIBERIAN CRAB: **Malus baccata.**
  SQUAW: **Peraphyllum ramosissimum.**
  TEA CRAB: **Malus hupehensis.**
  TORINGO CRAB: **Malus sieboldii.**
  WEEPING CRAB: **Malus 'Oekonomierat Echtermeyer'.**
  WILD SWEET CRAB: **Malus coronaria.**
APRICOT: **Prunus armeniaca.**
  CHARLES ABRAHAM FLOWERING: **Prunus armeniaca 'Charles Abraham'.**
  JAPANESE: **Prunus mume.**
  JAPANESE FLOWERING: **Prunus mume.**
  ROSEMARY CLARKE FLOWERING: **Prunus mume 'Rosemary Clarke'.**
APRICOT VINE: **Passiflora incarnata.**
ARALIA, CHINESE: **Aralia chinensis.**
  FALSE THREAD-LEAF: **Dizygotheca elegantissima.**
  FERN-LEAF: **Aralia chinensis.**
  FERN-LEAF: **Polyscias filicifolia.**
  FINGER: **Dizygotheca elegantissima.**
  GERANIUM-LEAF: **Polyscias guilfoylei.**
  JAPANESE: **Fatsia japonica.**
  SNOWFLAKE: **Trevesia palmata 'Micholitzii'.**
Aralia californica S. Wats. ELK CLOVER.
Aralia chinensis L. FERN-LEAF ARALIA. CHINESE ARALIA. CHINESE ANGELICA TREE.
Aralia elata (Miq.) Seem. JAPANESE ANGELICA TREE.
Aralia elegantissima Veitch. See **Dizygotheca elegantissima.**
Aralia guilfoylei Bull. See **Polyscias guilfoylei.**
Aralia japonica Thunb. See **Fatsia japonica.**
Aralia papyrifera Hook. See **Tetrapanax papyriferus.**
Aralia sieboldii K. Koch. See **Fatsia japonica.**
Aralia spinosa L. DEVIL'S-WALKING-STICK. HERCULES'-CLUB. ANGELICA TREE.
Araucaria araucana (Molina) K. Koch. Syn.: *A. imbricata*. MONKEY-PUZZLE TREE.
Araucaria bidwillii Hook. BUNYA-BUNYA.
Araucaria columnaris (G. Forst.) Hook. Syn.: *A. cookii*. *A. excelsa*. NEW CALEDONIAN PINE.
Araucaria cookii R. Br. ex Endl. See **A. columnaris.**
Araucaria cunninghamii D. Don. HOOP PINE. MORETON BAY PINE.
Araucaria excelsa (Lamb.) R. Br. See **A. columnaris.** Most plants called *A. excelsa* are **A. heterophylla.**
Araucaria heterophylla (Salisb.) Franco. STAR PINE. NORFOLK ISLAND PINE. Most plants called *A. excelsa* belong here.
Araucaria imbricata Pav. See **Araucaria araucana.**
Araujia sericofera Brot. Syn.: *Physianthus albens*. WHITE BLADDER FLOWER.
ARBORVITAE, BAKER'S: **Platycladus orientalis 'Bakeri'.**
  BERCKMAN'S: **Platycladus orientalis 'Aureus Nanus'.**
  BEVERLY HILLS: **Platycladus orientalis 'Beverleyensis'.**
  DWARF GOLDEN: **Platycladus orientalis 'Aureus Nanus'.**
  EASTERN: **Thuja occidentalis.**
  ELLWANGER: **Thuja occidentalis 'Ellwangerana'.**
  ELLWANGER GOLDEN: **Thuja occidentalis 'Ellwangerana Aurea'.**
  FALSE: **Thujopsis dolabrata.**
  FRUITLAND: **Platycladus orientalis 'Fruitlandii'.**
  GIANT: **Thuja plicata.**
  GLOBE: **Thuja occidentalis 'Globosa'.**
  GOLDEN PYRAMID: **Platycladus orientalis 'Beverleyensis'.**
  GOLDSPOT: **Thuja plicata 'Aurea'.**
  HETZ'S MIDGET: **Thuja occidentalis 'Hetz's Midget'.**
  HIBA: **Thujopsis dolabrata.**
  HOHMAN: **Platycladus orientalis 'Hohman'.**
  LITTLE GEM: **Thuja occidentalis 'Little Gem'.**
  LITTLE GLOBE: **Thuja occidentalis 'Compacta'.**
  ORIENTAL: **Platycladus orientalis.**
  PYRAMIDAL: Sometimes applied to **Thuja occidentalis 'Columna'.**
  PYRAMIDAL: **Thuja occidentalis 'Fastigiata'.**
  PYRAMIDAL: **Thuja occidentalis 'Pyramidalis'.**
  PYRAMIDAL: Sometimes applied to **Thuja occidentalis 'Pyramidalis Compacta'.**
  RAFFLES: **Platycladus orientalis 'Raffles'.**
  ROSEDALE: **Platycladus orientalis 'Rosedale'.**
  UMBRELLA: **Thuja occidentalis 'Umbraculifera'.**
  WOODWARD: **Thuja occidentalis 'Woodwardii'.**
ARBUTUS, TRAILING: **Epigaea repens.**
Arbutus andrachne L. Some plants called **A. canariensis** belong here.
Arbutus canariensis Duhamel. Some plants so called are **A. andrachne.**
Arbutus menziesii Pursh. MADRONE. PACIFIC MADRONE.
Arbutus unedo L. STRAWBERRY MADRONE. STRAWBERRY TREE.
Arbutus unedo 'Compacta'. COMPACT STRAWBERRY TREE.
Arbutus unedo 'Elfin King'.

Archontophoenix alexandrae (F. J. Muell.) H. Wendl. & Drude. Syn.: *Ptychosperma alexandrae*. KING PALM. ALEXANDRA PALM.
Archontophoenix cunninghamiana (H. Wendl.) H. Wendl. & Drude. Syn.: *Ptychosperma cunninghamianum*. KING PALM. Some plants called *Seaforthia elegans* belong here.
Arctostaphylos andersonii Gray. Syn.: *A. regismontana*. HEART-LEAF MANZANITA.
Arctostaphylos auriculata Eastw. MT. DIABLO MANZANITA.
Arctostaphylos bakeri Eastw. See **A. stanfordiana** subsp. **bakeri**.
Arctostaphylos bakeri 'Louis Edmunds'. See **A. stanfordiana** subsp. **bakeri** 'Louis Edmunds'.
Arctostaphylos canescens Eastw. HOARY MANZANITA.
Arctostaphylos columbiana Piper. HAIRY MANZANITA.
Arctostaphylos cushingiana Eastw. See **A. glandulosa** var. **cushingiana**.
Arctostaphylos densiflora M. S. Baker. VINE HILL MANZANITA.
Arctostaphylos densiflora 'Harmony'.
Arctostaphylos densiflora 'Howard McMinn'.
Arctostaphylos densiflora 'James West'. JAMES WEST MANZANITA.
Arctostaphylos densiflora 'Sentinel'.
Arctostaphylos diversifolia Parry. See **Comarostaphylis diversifolia**.
Arctostaphylos edmundsii J. T. Howell. LITTLE SUR MANZANITA.
Arctostaphylos edmundsii 'Danville'.
Arctostaphylos 'Emerald Carpet'. [**A. uva-ursi** × **A. nummularia**].
Arctostaphylos franciscana Eastw. See **A. hookeri** subsp. **franciscana**.
Arctostaphylos glandulosa Eastw. EASTWOOD MANZANITA.
Arctostaphylos glandulosa var. **cushingiana** (Eastw.) Adams ex McMinn. Syn.: *A. cushingiana*.
Arctostaphylos glauca Lindl. BIG-BERRY MANZANITA.
Arctostaphylos 'Greensphere'. [**A. edmundsii** × **A. sp.**]
Arctostaphylos hispidula T. J. Howell. See **A. stanfordiana** subsp. **hispidula**.
Arctostaphylos hookeri G. Don. HOOKER MANZANITA. MONTEREY MANZANITA.
Arctostaphylos hookeri subsp. **franciscana** (Eastw.) Munz. Syn.: *A. franciscana*. LAUREL HILL MANZANITA.
Arctostaphylos hookeri 'Monterey Carpet'.
Arctostaphylos hookeri 'Wayside'.
Arctostaphylos 'Howard McMinn'. See **A. densiflora** 'Howard McMinn'.
Arctostaphylos insularis Greene. ISLAND MANZANITA.
Arctostaphylos manzanita Parry. PARRY MANZANITA.
Arctostaphylos manzanita 'Doctor Hurd'.
Arctostaphylos × *media* Greene. [**A. columbiana** × **A. uva-ursi**]. STATE HYBRID MANZANITA.
Arctostaphylos morroensis Wiesl. & Schreib. MORRO BAY MANZANITA.
Arctostaphylos myrtifolia Parry. IONE MANZANITA.
Arctostaphylos nevadensis Gray. PINE-MAT MANZANITA.
Arctostaphylos nummularia Gray. FORT BRAGG MANZANITA.
Arctostaphylos obispoensis Eastw. SERPENTINE MANZANITA.
Arctostaphylos otayensis Wiesl. & Schreib. OTAY MANZANITA.
Arctostaphylos pajaroensis Adams. PAJARO MANZANITA.
Arctostaphylos pechoensis Dudl. ex Abrams var. **viridissima** Eastw. Syn.: *A. viridissima*. LOMPOC MANZANITA.
Arctostaphylos 'Point Reyes'. See **A. uva-ursi** 'Point Reyes'.
Arctostaphylos pumila Nutt. SAND-MAT MANZANITA. DUNE MANZANITA.
Arctostaphylos pungens HBK. MEXICAN MANZANITA. POINT-LEAF MANZANITA.
Arctostaphylos 'Radiant'. See **A. uva-ursi** 'Radiant'.
Arctostaphylos regismontana Eastw. See **A. andersonii**.
Arctostaphylos rudis Jeps. & Wiesl. SHAGBARK MANZANITA.
Arctostaphylos stanfordiana Parry. STANFORD MANZANITA.
Arctostaphylos stanfordiana subsp. **bakeri** (Eastw.) Adams. Syn.: *A. bakeri*. BAKER MANZANITA.
Arctostaphylos stanfordiana subsp. **bakeri** 'Louis Edmunds'.
Arctostaphylos stanfordiana 'Fred Oehler'.
Arctostaphylos stanfordiana subsp. **hispidula** (T. J. Howell) Adams. Syn.: *A. hispidula*. HOWELL MANZANITA.
Arctostaphylos stanfordiana 'Trinity'.
Arctostaphylos 'Sunset'. [**A. pajaroensis** × **A. hookeri**].
Arctostaphylos tomentosa (Pursh) Lindl. SHAGGY-BARK MANZANITA. WOOLLY MANZANITA.
Arctostaphylos uva-ursi (L.) K. Spreng. BEARBERRY. KINNIKINNICK. CREEPING MANZANITA. PROSTRATE MANZANITA.
Arctostaphylos uva-ursi 'Compacta'. COMPACT KINNIKINNICK.
Arctostaphylos uva-ursi 'Point Reyes'.
Arctostaphylos uva-ursi 'Radiant'.
Arctostaphylos uva-ursi 'Wood's Red'.
Arctostaphylos viridissima (Eastw.) McMinn. See **A. pechoensis** var. **viridissima**.
Arctostaphylos viscida Parry. WHITE-LEAF MANZANITA.
Arctotis stoechadifolia Bergius. AFRICAN DAISY.
ARDISIA, JAPANESE: **Ardisia japonica**.
Ardisia crenata Sims. Syn.: *A. crenulata*.
Ardisia crenulata Lodd. See **A. crenata**.
Ardisia crispa (Thunb.) A. DC.
Ardisia japonica (Hornstedt) Blume. JAPANESE ARDISIA.
Areca lutescens Bory. See **Chrysalidocarpus lutescens**.
Areca madagascariensis. Listed name for **Chrysalidocarpus madagascariensis**.
Arecastrum romanzoffianum (Cham.) Becc. Syn.: *Cocos plumosa*. QUEEN PALM. PLUME PALM. COCOS PALM.
Arecastrum romanzoffianum var. **australe** (Mart.) Becc. Syn.: *Cocos australis*.
Arenga engleri Becc.
Arenga microcarpa Becc.
Argania spinosa (L.) Skeels. ARGAN-OIL TREE.
ARGAN-OIL TREE: **Argania spinosa**.
Aristea ecklonii Baker.
Aristolochia californica Torr. CALIFORNIA DUTCHMAN'S-PIPE. CALIFORNIA PIPE VINE.
Aristolochia durior. J. Hill. Syn.: *A. macrophylla*. DUTCHMAN'S-PIPE.
Aristolochia macrophylla Lam. See **A. durior**.
Aristolochia tomentosa Sims. DUTCHMAN'S-PIPE.
Aristotelia chilensis (Molina) Stuntz.
Aristotelia racemosa Hook f. See **A. serrata**.
Aristotelia serrata (J. R. Forst. & G. Forst.) W. Oliver. Syn.: *A. racemosa*. NEW ZEALAND WINEBERRY.
Aronia arbutifolia (L.) Pers. RED CHOKE BERRY.
Aronia arbutifolia 'Brilliantissima'.
Aronia arbutifolia 'Erecta'.
ARROWROOT, FLORIDA: **Zamia pumila**.
ARROWWOOD: **Viburnum dentatum**.
  SOUTHERN: **Viburnum dentatum**.
ARTEMISIA, SILVER KING: **Artemisia ludoviciana** var. **albula**.
Artemisia abrotanum L. SOUTHERN-WOOD. OLD-MAN.
Artemisia absinthium L. COMMON WORMWOOD. ABSINTHIUM.
Artemisia albula Woot. See **A. ludoviciana** var. **albula**.
Artemisia arborescens L.
Artemisia californica Less. CALIFORNIA SAGEBRUSH.
Artemisia caucasica Willd. CAUCASIAN ARTEMISIA.
Artemsia dracunculus L. FRENCH TARRAGON. TRUE TARRAGON.
Artemisia frigida Willd. FRINGED WORMWOOD.
Artemisia lactiflora Wallich ex DC. WHITE MUGWORT.

Artemisia ludoviciana Nutt. WESTERN MUGWORT.
Artemisia ludoviciana var. *albula* (Woot.) Shinn. Syn.:
A. albula. SILVER KING ARTEMISIA.
Artemisia pontica L. ROMAN WORMWOOD.
Artemisia pycnocephala (Less.) DC. SANDHILL SAGEBRUSH.
  BEACH SAGEBRUSH.
Artemisia schmidtiana Maxim. ANGEL'S HAIR.
Artemisia stellerana Besser. BEACH WORMWOOD. OLD-WOMAN.
  DUSTY-MILLER.
Artemisia tridentata Nutt. SAGEBRUSH. GREAT BASIN SAGEBRUSH.
Arthropodium cirrhatum (G. Forst.) R. Br.
Arthrostylidium longifolium of authors, not (Fournier) E. G.
  Camus. Plants so called are **Yushania aztecorum.**
Arundinaria argenteo-striata (Regel) Ohwi. SOFT SILVER-STRIPE
  BAMBOO.
Arundinaria argenteo-striata 'Akebono'. SUNRISE BAMBOO.
Arundinaria disticha (Mitf.) Pfitzer. Syn.: *Bambusa disticha.*
  Pleioblastus distichus. Sasa disticha. DWARF FERN-LEAF
  BAMBOO.
Arundinaria falcata Nees. See **Chimonobambusa falcata.**
Arundinaria graminea (Bean) Makino.
Arundinaria humilis Mitf. Syn.: *Sasa humilis.* PIN-STRIPE
  BAMBOO.
Arundinaria japonica Siebold & Zucc. ex Steud. See **Sasa
  japonica.**
Arundinaria japonica 'Tsutsumiana'. See **Sasa japonica
  'Tsutsumiana'.**
Arundinaria marmorea (Mitf.) Makino. See **Chimonobambusa
  marmorea.**
Arundinaria murielae Gamble. See **Sinarundinaria murielae.**
Arundinaria nitida Mitf. See **Sinarundinaria nitida.**
Arundinaria pygmaea (Miq.) Aschers. & Graebn. Syn.:
  Bambusa pygmaea. Sasa pygmaea. DWARF BAMBOO. PYGMY
  BAMBOO. Plants called *Pleioblastus viridi-striatus* var. *vagans*
  may belong here.
Arundinaria simonii (Carrière) A. & C. Rivière. SIMON BAMBOO.
  MEDAKE.
Arundinaria simonii 'Variegata'.
Arundinaria tessellata (Nees) Munro.
Arundinaria variegata (Siebold ex Miq.) Makino. Syn.:
  Bambusa variegata. Pleioblastus variegatus. Sasa variegata.
  DWARF WHITE-STEM BAMBOO. Plants called *Sasa fortunei* may
  belong here.
Arundinaria viridi-striata (Siebold ex André) Makino ex Nakai.
Arundo donax L. GIANT REED GRASS. GIANT REED.
Arundo donax 'Variegata'.
ASH, ALPINE: **Eucalyptus delegatensis.**
  ARIZONA: **Fraxinus velutina.**
  ARIZONA: **Fraxinus velutina** var. **glabra.**
  BLUE: **Fraxinus quadrangulata.**
  CLARET: **Fraxinus oxycarpa 'Raywood'.**
  EUROPEAN: **Fraxinus excelsior.**
  EUROPEAN MOUNTAIN: **Sorbus aucuparia.**
  EVERGREEN: **Fraxinus uhdei.**
  FLOWERING: **Fraxinus dipetala.**
  FLOWERING: **Fraxinus ornus.**
  FOOTHILL: **Fraxinus dipetala.**
  GREEN: **Fraxinus pennsylvanica** var. **lanceolata.**
  MANNA: **Fraxinus ornus.**
  MARSHALL SEEDLESS GREEN: **Fraxinus pennsylvanica** var.
    lanceolata 'Marshall'.
  MODESTO: **Fraxinus velutina** var. **glabra 'Modesto'.**
  MONTEBELLO: **Fraxinus velutina** var. **coriacea.**
  MORAINE: **Fraxinus 'Moraine'.**
  OREGON: **Fraxinus latifolia.**
  PRICKLY: **Zanthoxylum americanum.**
  RAYWOOD: **Fraxinus oxycarpa 'Raywood'.**
  RED: **Fraxinus pennsylvanica.**
  SHAMEL: **Fraxinus uhdei.**
  SINGLE-LEAF: **Fraxinus anomala.**
  SMOOTH ARIZONA: **Fraxinus velutina** var. **glabra.**
  STINKING: **Ptelea trifoliata.**
  TOMLINSON'S: **Fraxinus uhdei 'Tomlinson'.**
  TWO-PETAL: **Fraxinus dipetala.**
  VELVET: **Fraxinus velutina.**
  WATER: **Ptelea trifoliata.**
  WEEPING EUROPEAN: **Fraxinus excelsior 'Pendula'.**
  WHITE: **Fraxinus americana.**
Asimina triloba (L.) Dunal. PAWPAW.
ASPARAGUS, BASKET: **Asparagus crispus.**
  BASKET: **Asparagus scandens.**
  SICKLE-THORN: **Asparagus falcatus.**
  SPRENGER: **Asparagus densiflorus 'Sprengeri'.**
Asparagus crispus Lam. BASKET ASPARAGUS.
Asparagus densiflorus (Kunth) Jessop. Syn.: *A. myriocladus,*
  A. sarmentosus.
Asparagus densiflorus 'Myers'.
Asparagus densiflorus 'Sprengeri'. Syn.: *A. sprengeri.*
  SPRENGER ASPARAGUS.
Asparagus falcatus L. SICKLE-THORN ASPARAGUS.
Asparagus laracinus Burchell.
Asparagus macowanii Baker. Some plants called
  A. myriocladus and *A. retrofractus* belong here.
Asparagus meyeri. Listed name for **A. densiflorus 'Myers'.**
Asparagus myersii. Listed name for **A. densiflorus 'Myers'.**
Asparagus myriocladus Baker. See **A. densiflorus.** Some plants
  called *A. myriocladus* may be **A. macowanii.**
Asparagus plumosus Baker. See **A. setaceus.**
Asparagus retrofractus L. Some plants so called may be
  A. macowanii.
Asparagus sarmentosus Baker. See **A. densiflorus.**
Asparagus scandens Thunb. BASKET ASPARAGUS.
Asparagus scandens var. *deflexus* Baker.
Asparagus setaceus (Kunth) Jessop. Syn.: *A. plumosus.*
  ASPARAGUS FERN.
Asparagus sprengeri Regel. See **A. densiflorus 'Sprengeri'.**
Asparagus virgatus Baker.
ASPEN, EUROPEAN: **Populus tremula.**
  QUAKING: **Populus tremuloides.**
  TREMBLING: **Populus tremuloides.**
Aspidistra elatior Blume. CAST-IRON PLANT. BARROOM PLANT.
  IRON PLANT.
Aspidistra elatior 'Variegata'.
ASTER, NEW ENGLAND: **Aster novae-angliae.**
Aster × *frikartii* Frikart. [A. amellus L. × A. thomsonii
  C. B. Clarke].
Aster × *frikartii* 'Jungfrau'.
Aster fruticosus L. See **Felicia fruticosa.**
Aster novae-angliae L. NEW ENGLAND ASTER.
Asteriscus maritimus (L.) Less. Syn.: *Odontospermum*
  maritimum.
Asteriscus sericeus (L. f.) DC. Syn.: *Odontospermum sericeum.*
Asystasia bella Hort. See **Mackaya bella.**
Athanasia parviflora L.
ATHEL: **Tamarix aphylla.**
Atriplex breweri S. Wats. See **A. lentiformis** subsp. **breweri.**
Atriplex californica Moq. COAST SALTBUSH. CALIFORNIA SALTBUSH.
Atriplex canescens (Pursh) Nutt. FOUR-WING SALTBUSH.
Atriplex cuneata A. Nels. See **A. nuttallii** subsp. **cuneata.**
Atriplex gardneri (Moq.) Standl. See **A. nuttallii** var. **gardneri.**
Atriplex halimus L. MEDITERRANEAN SALTBUSH.
Atriplex hymenelytra (Torr.) S. Wats. DESERT HOLLY.

Atriplex lentiformis (Torr.) S. Wats. SALTBUSH. QUAIL-BUSH.
Atriplex lentiformis subsp. *breweri* (S. Wats.) Hall & Clements. Syn.: *A. breweri*. BREWER SALTBUSH.
Atriplex muelleri Benth.
Atriplex nummularia Lindl.
Atriplex nuttallii S. Wats. subsp. *cuneata* (A. Nels.) Hall & Clements. Syn.: *A. cuneata*.
Atriplex nuttallii S. Wats. var. *gardneri* (Moq.) Hall & Clements. GARDNER VALLEY SALTBUSH.
Atriplex polycarpa (Torr.) S. Wats. CATTLE SPINACH.
Atriplex rhagodioides F. J. Muell.
AUCUBA, CROTON-LEAF: *Aucuba japonica* 'Crotonifolia'.
   DWARF: *Aucuba japonica* 'Nana'.
   HIMALAYAN: *Aucuba himalaica*.
   JAPANESE: *Aucuba japonica*.
   LANCELEAF: *Aucuba japonica* 'Longifolia'.
   SAN JOSE DWARF: *Aucuba japonica* 'San Jose'.
   WHITE-FRUITED: *Aucuba japonica* 'Fructu-albo'.
Aucuba 'Croton'. See *A. japonica* 'Crotonifolia'.
Aucuba 'Crotonifolia'. See *A. japonica* 'Crotonifolia'.
Aucuba 'Goldieana'. See *A. japonica* 'Goldieana'.
Aucuba himalaica Hook. f. & Thoms. HIMALAYAN LAUREL. HIMALAYAN AUCUBA.
Aucuba japonica Thunb. JAPANESE LAUREL. JAPANESE AUCUBA.
Aucuba japonica 'Croton'. See *A. japonica* 'Crotonifolia'.
Aucuba japonica 'Crotonifolia'. CROTON-LEAF AUCUBA.
Aucuba japonica 'Fructu-albo'. WHITE-FRUITED AUCUBA.
Aucuba japonica 'Goldieana'.
Aucuba japonica 'Longifolia'. LANCELEAF AUCUBA.
Aucuba japonica 'Maculata'. See *A. japonica* 'Variegata'.
Aucuba japonica 'Nana'. DWARF AUCUBA.
Aucuba japonica 'Nana Femina'. See *A. japonica* 'Nana'.
Aucuba japonica 'Picturata'.
Aucuba japonica 'Salicifolia'.
Aucuba japonica 'San Jose'. SAN JOSE DWARF AUCUBA.
Aucuba japonica 'Serratifolia'.
Aucuba japonica 'Sulphur'?
Aucuba japonica 'Variegata'. GOLD-DUST PLANT.
AUSTRALIAN NUT: *Macadamia tetraphylla*.
Austrocedrus chilensis (D. Don) Florin & Boutelje. Syn.: *Libocedrus chilensis*. CHILEAN INCENSE CEDAR.
AVOCADO: *Persea americana*.
   MEXICAN: *Persea americana* var. *drymifolia*.
AZALEA, CHINESE: *Rhododendron molle*.
   FLAME: *Rhododendron calendulaceum*.
   KAEMPFER: *Rhododendron kaempferi*.
   KOREAN: *Rhododendron yedoense* var. *poukhanense*.
   KYUSHU: *Rhododendron kiusianum*.
   PINK-SHELL: *Rhododendron vaseyi*.
   PONTIC: *Rhododendron luteum*.
   ROYAL: *Rhododendron schlippenbachii*.
   TORCH: *Rhododendron kaempferi*.
   WESTERN: *Rhododendron occidentale*.
   YODOGAWA: *Rhododendron yedoense*.
AZARA, BOXLEAF: *Azara microphylla*.
   GOLDEN: *Azara petiolaris*.
   GOLDSPIRE: *Azara integrifolia*.
   LANCELEAF: *Azara lanceolata*.
Azara celastrina D. Don.
Azara dentata Ruiz & Pav.
Azara gilliesii Hook. & Arn. See *A. petiolaris*.
Azara integrifolia Ruiz & Pav. GOLDSPIRE AZARA.
Azara lanceolata Hook. f. LANCELEAF AZARA.
Azara microphylla Hook. f. BOXLEAF AZARA.
Azara microphylla 'Variegata'.
Azara petiolaris (D. Don) I. M. Johnst. Syn.: *A. gilliesii*. GOLDEN AZARA.

B

BABY'S BREATH: *Spiraea arguta*.
Baccharis pilularis DC. DWARF CHAPARRAL BROOM. DWARF COYOTE BRUSH.
Baccharis pilularis subsp. *consanguinea* (DC.) C. B. Wolf. CHAPARRAL BROOM. COYOTE BRUSH.
Baccharis pilularis 'Prostrata'. See *B. pilularis*.
Baccharis pilularis 'Twin Peaks'.
Baccharis sarothroides Gray.
Baccharis 'Twin Peaks'. See *B. pilularis* 'Twin Peaks'.
Baccharis viminea DC. MULE-FAT.
Bacularia monostachya (Mart.) F. J. Muell. ex Hook. f. See *Linospadix monostachya*.
Baeckia virgata Andr.
BALM-OF-GILEAD: *Populus* × *gileadensis*.
BAMBOO, ALPHONSE KARR: *Bambusa glaucescens* 'Alphonse Karr'.
   ARROW: *Sasa japonica*.
   BEECHEY: *Bambusa beecheyana*.
   BLACK: *Phyllostachys nigra*.
   BLUE: *Chimonobambusa falcata*.
   BLUE: *Sinarundinaria nitida*.
   BORY: *Phyllostachys nigra* 'Bory'.
   BUDDHA: *Bambusa ventricosa*.
   BUDDHA'S BELLY: *Bambusa ventricosa*.
   CALCUTTA: *Dendrocalamus strictus*.
   CHINESE GODDESS: *Bambusa glaucescens* var. *riviereorum*.
   COMMON: *Bambusa vulgaris*.
   DAVID BISSET: *Phyllostachys bissetii*.
   DWARF: *Arundinaria pygmaea*.
   DWARF FERN-LEAF: *Arundinaria disticha*.
   DWARF WHITE-STEM: *Arundinaria variegata*.
   FERN-LEAF HEDGE: *Bambusa glaucescens* 'Fernleaf'.
   FISH-POLE: *Phyllostachys aurea*.
   GIANT TIMBER: *Bambusa oldhamii*.
   GIANT TIMBER: *Phyllostachys bambusoides*.
   GOLDEN: *Phyllostachys aurea*.
   GOLDEN GODDESS HEDGE: *Bambusa glaucescens* 'Golden Goddess'.
   HEAVENLY: *Nandina domestica*.
   HEDGE: *Bambusa glaucescens*.
   HENON: *Phyllostachys nigra* 'Henon'.
   JAPANESE ARROW: *Sasa japonica*.
   JAPANESE TIMBER: *Phyllostachys bambusoides*.
   MAKINO: *Phyllostachys makinoi*.
   MALE: *Dendrocalamus strictus*.
   MARBLED: *Chimonobambusa marmorea*.
   MEXICAN: *Polygonum cuspidatum*.
   MEXICAN WEEPING: *Yushania aztecorum*.
   MEYER: *Phyllostachys meyeri*.
   MOSO: *Phyllostachys pubescens*.
   NARIHIRA: *Semiarundinaria fastuosa*.
   NINGALA: *Chimonobambusa falcata*.
   OLDHAM: *Bambusa oldhamii*.
   PALMATE: *Sasa palmata*.
   PIN-STRIPE: *Arundinaria humilis*.
   PUNTING-POLE: *Bambusa tuldoides*.
   PYGMY: *Arundinaria pygmaea*.
   RUSCUS-LEAVED: *Shibataea kumasasa*.

BAMBOO, continued
- SACRED: **Nandina domestica**.
- SICKLE: **Chimonobambusa falcata**.
- SILVERSTRIPE HEDGE: **Bambusa glaucescens 'Silverstripe'**.
- SIMON: **Arundinaria simonii**.
- SOFT SILVER-STRIPE: **Arundinaria argenteo-striata**.
- SQUARE-STEM: **Chimonobambusa quadrangularis**.
- STONE: **Phyllostachys angusta**.
- STRIPESTEM FERNLEAF HEDGE: **Bambusa glaucescens 'Stripestem Fernleaf'**.
- SUNRISE: **Arundinaria argenteo-striata 'Akebono'**.
- SWEET-SHOOT: **Phyllostachys dulcis**.
- TEXTILE: **Bambusa textilis**.
- WHITE: **Dendrocalamus membranaceus**.
- WRINKLED: **Phyllostachys bambusoides** var. **marliacea**.
- YELLOW-GROOVE: **Phyllostachys aureosulcata**.
- ZIG-ZAG: **Phyllostachys flexuosa**.

Bamburanta arnoldiana L. Linden. See **Hybophrynium braunianum** and **Ctenanthe pilosa**.
Bambusa argentea Hort. See **Bambusa glaucescens**.
Bambusa aurea Hort. See **Phyllostachys aurea**.
Bambusa beecheyana Munro. Syn.: *Sinocalamus beecheyanus*. BEECHEY BAMBOO.
Bambusa disticha Mitf. See **Arundinaria disticha**.
Bambusa falcata Vilmorin. See **Chimonobambusa falcata**.
Bambusa glaucescens (Willd.) Siebold ex Holtt. Syn.: *Bambusa argentea*. *B. multiplex*. *B. nana*. HEDGE BAMBOO.
Bambusa glaucescens 'Alphonse Karr'. ALPHONSE KARR BAMBOO.
Bambusa glaucescens 'Fernleaf'. FERN-LEAF HEDGE BAMBOO.
Bambusa glaucescens 'Golden Goddess'. GOLDEN GODDESS HEDGE BAMBOO.
Bambusa glaucescens var. *riviereorum* Maire. Syn.: *B. multiplex* var. *riviereorum*. CHINESE GODDESS BAMBOO.
Bambusa glaucescens 'Silverstripe'. SILVERSTRIPE HEDGE BAMBOO.
Bambusa glaucescens 'Stripestem Fernleaf'. STRIPESTEM FERNLEAF HEDGE BAMBOO.
Bambusa multiplex (Lour.) Raeuschel. See **B. glaucescens**.
Bambusa multiplex 'Alphonse Karr'. See **B. glaucescens 'Alphonse Karr'**.
Bambusa multiplex 'Fernleaf'. See **B. glaucescens 'Fernleaf'**.
Bambusa mulitplex 'Golden Goddess'. See **B. glaucescens 'Golden Goddess'**.
Bambusa multiplex 'Nana'?
Bambusa multiplex var. *riviereorum* Maire. See **B. glaucescens** var. *riviereorum*.
Bambusa multiplex 'Silverstripe'. See **B. glaucescens 'Silverstripe'**.
Bambusa multiplex 'Stripestem Fernleaf'. See **B. glaucescens 'Stripestem Fernleaf'**.
Bambusa mutabilis McClure.
Bambusa nana Roxb. See **B. glaucescens**.
Bambusa nana 'Alphonse Karr'. See **B. glaucescens 'Alphonse Karr'**.
Bambusa oldhamii Munro. Syn.: *Sinocalamus oldhamii*. OLDHAM BAMBOO. GIANT TIMBER BAMBOO. Plants called **Dendrocalamus latiflorus** may belong here.
Bambusa pygmaea Miq. See **Arundinaria pygmaea**.
Bambusa textilis McClure. TEXTILE BAMBOO.
Bambusa thouarsii Kunth. Probably not cult. Plants so called are **B. tuldoides**.
Bambusa tulda Roxb.
Bambusa tuldoides Munro. PUNTING-POLE BAMBOO. Plants called **B. thouarsii** belong here.
Bambusa variegata Siebold ex Miq. See **Arundinaria variegata**.
Bambusa ventricosa McClure. BUDDHA'S BELLY BAMBOO. BUDDHA BAMBOO.

Bambusa vulgaris Schrad. ex Wendl. COMMON BAMBOO.
Bambusa vulgaris 'Vittata'.
BANANA, ABYSSINIAN: **Ensete ventricosum**.
- CAVENDISH: **Musa acuminata 'Dwarf Cavendish'**.
- CHINESE DWARF: **Musa acuminata 'Dwarf Cavendish'**.
- DWARF: **Musa acuminata 'Dwarf Cavendish'**.
- DWARF: **Musa nana**.
- EDIBLE: **Musa acuminata**.
- EDIBLE: **Musa** × **paradisiaca**.
- FLOWERING: **Musa ornata**.
- INDIA: **Musa ornata**.
- LADYFINGER: **Musa acuminata 'Dwarf Cavendish'**.
- SUMATRA: **Musa sumatrana**.

BANANA SHRUB: **Michelia figo**.
Banksia ashbyi E. G. Baker.
Banksia baueri R. Br.
Banksia baxteri R. Br.
Banksia brownii Baxter.
Banksia burdettii E. G. Baker.
Banksia caleyi R. Br.
Banksia coccinea R. Br.
Banksia collina R. Br. See **B. spinulosa**.
Banksia ericifolia L. f.
Banksia grandis Willd.
Banksia hookerana Meissn.
Banksia integrifolia L. f.
Banksia lehmanniana Meissn.
Banksia marginata Cav.
Banksia media R. Br.
Banksia menziesii R. Br.
Banksia occidentalis R. Br.
Banksia praemorsa Andr.
Banksia prionotes Lindl.
Banksia serrata L. f.
Banksia serratifolia Salisb.
Banksia solandri R. Br.
Banksia speciosa R. Br.
Banksia spinulosa Sm. Syn.: *B. collina*.
Banksia victoriae Meissn.
BANYAN, AUSTRALIAN: **Ficus macrophylla**.
- WEEPING CHINESE: **Ficus benjamina**.

BARBADOS-PRIDE: **Caesalpinia pulcherrima**.
BARBERRY, BLACK: **Berberis gagnepainii**.
- CALIFORNIA: **Mahonia pinnata**.
- CHINESE: **Berberis asiatica**.
- CORAL: **Berberis** × **stenophylla 'Corallina Compacta'**.
- CREEPING: **Mahonia repens**.
- CRIMSON PYGMY: **Berberis thunbergii 'Atropurpurea Nana'**.
- DARWIN: **Berberis darwinii**.
- DESERT: **Mahonia fremontii**.
- HIGGINS: **Mahonia higginsiae**.
- HOLLY: **Berberis ilicifolia**.
- IRWIN: **Berberis** × **stenophylla 'Irwinii'**.
- JAPANESE: **Berberis thunbergii**.
- MAGELLAN: **Berberis buxifolia**.
- MENTOR: **Berberis** × **mentorensis**.
- RED: **Berberis thunbergii 'Atropurpurea'**.
- RED-LEAF JAPANESE: **Berberis thunbergii 'Atropurpurea'**.
- ROSEMARY: **Berberis** × **stenophylla**.
- SARGENT: **Berberis sargentiana**.
- THREE-SPINE: **Berberis gagnepainii**.
- TRUEHEDGE: **Berberis thunbergii 'Erecta'**.
- WILSON: **Berberis wilsoniae**.
- WINTERGREEN: **Berberis julianae**.
- WISLEY: **Berberis** × **wisleyensis**.

Barleria cristata L. PHILIPPINE VIOLET.

Barosma ovata (Thunb.) Bartl. & H. Wendl. See **Agathosma ovata**.
BARREL TREE: **Brachychiton rupestris**.
BARROOM PLANT: **Aspidistra elatior**.
BASSWOOD, AMERICAN: **Tilia americana**.
BAUHINIA, RED: **Bauhinia galpinii**.
    SWEETHEART: **Bauhinia tomentosa**.
Bauhinia 'Alba'. See **B. variegata 'Candida'**.
Bauhinia blakeana Dunn. HONG KONG ORCHID TREE.
Bauhinia candicans Benth. See **B. forficata**.
Bauhinia corniculata. Listed name for **B. forficata**.
Bauhinia corymbosa Roxb. ex DC.
Bauhinia forficata Link. Syn.: *B. candicans*.
Bauhinia galpinii N. E. Br. RED BAUHINIA.
Bauhinia grandiflora. Listed name for various species of **Bauhinia**.
Bauhinia purpurea L. ORCHID TREE. BUTTERFLY TREE.
Bauhinia purpurea 'Alba'. See **B. variegata 'Candida'**.
Bauhinia tomentosa L. SWEETHEART BAUHINIA.
Bauhinia variegata L. MOUNTAIN EBONY. PURPLE ORCHID TREE. ORCHID TREE.
Bauhinia variegata 'Alba'. See **B. variegata 'Candida'**.
Bauhinia variegata 'Candida'. WHITE ORCHID TREE.
Bauhinia variegata 'Rubra'. See **B. variegata**.
BAY, CALIFORNIA: **Umbellularia californica**.
    LOBLOLLY: **Gordonia lasianthus**.
    RED: **Persea borbonia**.
    SWAMP RED: **Persea borbonia**.
    SWEET: **Laurus nobilis**.
    SWEET: **Magnolia virginiana**.
BAY TREE, GRECIAN: **Laurus nobilis**.
BAYBERRY: **Myrica pensylvanica**.
BEAD TREE: **Melia azedarach**.
BEAM, WHITE: **Sorbus aria**.
BEAN, BLACK: **Castanospermum australe**.
    CASTOR: **Ricinus communis**.
    CHEROKEE: **Erythrina herbacea**.
    INDIAN: **Catalpa speciosa**.
    INDIAN: **Catalpa bignonioides**.
    KAFFIR: **Schotia afra**.
    MESCAL: **Sophora secundiflora**.
    SCREW: **Prosopis pubescens**.
    SNAIL: **Vigna caracalla**.
BEARBERRY: **Arctostaphylos uva-ursi**.
BEAR GRASS: **Xerophyllum tenax**.
BEAR'S-BREECH: **Acanthus mollis**.
Beaucarnea gracilis Lem.
Beaucarnea recurvata Lem. ELEPHANT-FOOT TREE. PONYTAIL. BOTTLE PALM.
Beaufortia sparsa R. Br. GRAVEL BOTTLEBRUSH.
Beaumontia grandiflora (Roxb.) Wallich. HERALD'S-TRUMPET. EASTER-LILY VINE.
BEAUTYBERRY: **Callicarpa bodinieri var. giraldii**.
BEAUTYBUSH: **Kolkwitzia amabilis**.
BEECH, AMERICAN: **Fagus grandifolia**.
    AUSTRALIAN: **Eucalyptus polyanthemos**.
    BLUE: **Carpinus caroliniana**.
    BRONZE: **Fagus sylvatica 'Atropunicea'**.
    COPPER: **Fagus sylvatica 'Cuprea'**.
    CUT-LEAF: **Fagus sylvatica 'Laciniata'**.
    EUROPEAN: **Fagus sylvatica**.
    FERN-LEAF: **Fagus sylvatica 'Asplenifolia'**.
    GOLDEN: **Fagus sylvatica 'Zlatia'**.
    PURPLE: **Fagus sylvatica 'Atropunicea'**.
    RIVERS PURPLE: **Fagus sylvatica 'Riversii'**.
    SOUTHERN HEMISPHERE: **Nothofagus moorei**.
    TRICOLOR: **Fagus sylvatica 'Tricolor'**.
    WEEPING: **Fagus sylvatica 'Pendula'**.
    WEEPING COPPER: **Fagus sylvatica 'Purpurea Pendula'**.
BEEFWOOD: **Stenocarpus salignus**.
    COAST: **Casuarina stricta**.
Begonia foliosa HBK.
Begonia foliosa var. *miniata* (Planch.) L. B. Sm. & Schubert. Syn.: *B. fuchsioides*.
Begonia fuchsioides Hook. See **B. foliosa var. miniata**.
Beilschmiedia miersii (Gay) Kostermans. CHILEAN LAUREL. Plants called *Cryptocarya miersii* belong here.
BELLFLOWER, CHILEAN: **Lapageria rosea**.
Beloperone californica Benth. See **Justicia californica**.
Beloperone guttata Brandegee. See **Justicia brandegeana**.
Beloperone lutea. Listed name, probably for **Justicia brandegeana 'Yellow Queen'**.
Beloperone tomentosa. Listed name, probably for **Justicia brandegeana**.
BENJAMINA: **Ficus benjamina**.
Berberis aquifolium Pursh. See **Mahonia aquifolium**.
Berberis asiatica Roxb. CHINESE BARBERRY.
Berberis 'Atropurpurea'. See **B. thunbergii 'Atropurpurea'**.
Berberis 'Atropurpurea Nana'. See **B. thunbergii 'Atropurpurea Nana'**.
Berberis buxifolia Lam. Syn.: *B. dulcis*. MAGELLAN BARBERRY.
Berberis buxifolia 'Nana'.
Berberis × *chenaultii* Ahrendt. [*B. gagnepainii* × *B. verruculosa*].
Berberis 'Corallina Compacta'. See **B. stenophylla 'Corallina Compacta'**.
Berberis 'Crimson Pygmy'. See **B. thunbergii 'Atropurpurea Nana'**.
Berberis darwinii Hook. DARWIN BARBERRY.
Berberis dulcis Paxt. See **B. buxifolia**.
Berberis dulcis 'Nana'. See **B. buxifolia 'Nana'**.
Berberis fremontii Torr. See **Mahonia fremontii**.
Berberis gagnepainii Schneid. BLACK BARBERRY. THREE-SPINE BARBERRY.
Berberis × *gladwynensis* E. Anderson. [*B. verruculosa* × *B. julianae*].
Berberis × *gladwynensis* 'William Penn'.
Berberis 'Gracilis'. See **B. × stenophylla 'Gracilis'**.
Berberis higginsiae Munz. See **Mahonia higginsiae**.
Berberis ilicifolia G. Forst. HOLLY BARBERRY. Some plants so called are × **Mahoberberis neubertii**.
Berberis × *irwinii* Byhouwer. See **B. × stenophylla 'Irwinii'**.
Berberis × *irwinii* 'Gracilis'. See **B. × stenophylla 'Gracilis'**.
Berberis julianae Schneid. WINTERGREEN BARBERRY. Some plants called **B. xanthoxylon** belong here.
Berberis juliana 'Nana'?
Berberis knightii (Lindl.) K. Koch. Plants so called are usually **B. manipurana**.
Berberis manipurana Ahrendt. Some plants called **B. knightii** and **B. xanthoxylon** belong here.
Berberis × *mentorensis* L. M. Ames. [*B. thunbergii* × *B. julianae*]. MENTOR BARBERRY.
Berberis nana. Listed name probably for **B. buxifolia 'Nana'**.
Berberis nervosa Pursh. See **Mahonia nervosa**.
Berberis nevinii Gray. See **Mahonia nevinii**.
Berberis pinnata Lag. See **Mahonia pinnata**.
Berberis piperana (Abrams) McMinn. See **Mahonia piperana**.
Berberis pruinosa Franch.
Berberis repens Lindl. See **Mahonia repens**.
Berberis sargentiana Schneid. SARGENT BARBERRY.
Berberis × *stenophylla* Lindl. [*B. darwinii* × *B. empetrifolia* Lam.]. ROSEMARY BARBERRY.

Berberis × *stenophylla* 'Compacta'.
Berberis × *stenophylla* 'Corallina Compacta'. CORAL BARBERRY.
Berberis × *stenophylla* 'Gracilis'.
Berberis × *stenophylla* 'Irwinii'. Syn.: *B.* × *irwinii*. IRWIN BARBERRY.
Berberis × *stenophylla* 'Nana Compacta'?
Berberis thunbergii DC. JAPANESE BARBERRY.
Berberis thunbergii 'Atropurpurea'. RED BARBERRY. RED-LEAF JAPANESE BARBERRY.
Berberis thunbergii 'Atropurpurea Nana'. CRIMSON PYGMY BARBERRY.
Berberis thunbergii 'Aurea'.
Berberis thunbergii 'Crimson Pygmy'. See **B. thunbergii 'Atropurpurea Nana'.**
Berberis thunbergii 'Erecta'. TRUEHEDGE BARBERRY.
Berberis thunbergii 'Kobold'.
Berberis thunbergii 'Minor'.
Berberis thunbergii 'Nana'.
Berberis thunbergii 'Rose Glow'.
Berberis triacanthophora Fedde. May not be cult. Plants so called may be **B.** × **wisleyensis**.
Berberis verruculosa Hemsl. & Wils.
Berberis vulgaris L. BARBERRY.
Berberis wilsoniae Hemsl. WILSON BARBERRY.
Berberis × **wisleyensis** Ahrendt. [*B. triacanthophora* × ? *B. gagnepainii*]. WISLEY BARBERRY. Some plants called **B. triacanthophora** belong here.
Berberis xanthoxylon Hassk. ex Schneid. May not be cult. Plants so called may be **B. julianae** or **B. manipurana**.
BERRYRUE: **Cneoridium dumosum**.
Beschorneria tubiflora Kunth.
Beschorneria yuccoides Hook.
Betula alba L. See **B. pendula**.
Betula alba fastigiata Clemenceau. See **B. pendula 'Fastigiata'**.
Betula alba laciniata Retzius. See **B. pendula 'Dalecarlica'**.
Betula alba 'Splendor Weeping'. See **B. pendula 'Splendor Weeping'**.
Betula 'Columnaris'.?
Betula fontinalis Sarg. See **B. occidentalis**.
Betula laciniata Wahlenb. See **B. pendula 'Dalecarlica'**.
Betula mandschurica (Regel) Nakai var. *japonica* (Miq.) Rehd. See **B. platyphylla** var. ***japonica***.
Betula maximowicziana Regel. MONARCH BIRCH.
Betula nigra L. RIVER BIRCH. BLACK BIRCH. RED BIRCH.
Betula occidentalis Hook. Syn.: *B. fontinalis*. WATER BIRCH. MOUNTAIN BIRCH.
Betula papyrifera Marsh. EASTERN WHITE BIRCH. PAPER BIRCH. CANOE BIRCH.
Betula pendula Roth. Syn.: *B. alba*. *B. verrucosa*. EUROPEAN WHITE BIRCH. WHITE BIRCH.
Betula pendula 'Dalecarlica'. Syn.: *B. laciniata*. *B. alba laciniata*. CUTLEAF WEEPING BIRCH.
Betula pendula 'Fastigiata'. PYRAMIDAL WHITE BIRCH.
Betula pendula 'Gracilis'.
Betula pendula 'Purpurea'. PURPLE BIRCH.
Betula pendula 'Splendor Weeping'.
Betula pendula 'Tristis'.
Betula pendula 'Youngii'. YOUNG'S WEEPING BIRCH.
Betula platyphylla Sukachev var. ***japonica*** (Miq.) Hara. Syn.: *B. mandschurica* var. *japonica*. JAPANESE WHITE BIRCH.
Betula populifolia Marsh. GRAY BIRCH.
Betula 'Pyramidalis'. See **B. pendula 'Fastigiata'**.
Betula verrucosa Ehrh. and listed cvs. See **B. pendula** and cvs.
Betula 'Youngii'. See **B. pendula 'Youngii'**.
BIG TREE: **Sequoiadendron giganteum**.
Bignonia callistegioides Cham. See **Clytostoma callistegioides**.
Bignonia capensis Thunb. See **Tecomaria capensis**.
Bignonia capreolata L. Syn.: *Anisostichus capreolatus*. *Campsis capreolata*. *Doxantha capreolata*. CROSS VINE. QUARTER VINE. TRUMPET FLOWER.
Bignonia chamberlaynii Sims. See **Anemopaegma chamberlaynii**.
Bignonia cherere Lindl. See **Distictis buccinatoria**.
Bignonia chinensis Lam. See **Campsis grandiflora**.
Bignonia grandiflora Thunb. See **Campsis grandiflora**.
Bignonia jasminoides Hort. See **Pandorea jasminoides**.
Bignonia purpurea Lodd. ex Hook. f. See **Clytostoma binatum**.
Bignonia radicans L. See **Campsis radicans**.
Bignonia speciosa R. Graham. See **Clytostoma callistegioides**.
Bignonia stans L. See **Tecoma stans**.
Bignonia tweediana Lindl. See **Macfadyena unguis-cati**.
Bignonia unguis-cati L. See **Macfadyena unguis-cati**.
Bignonia venusta Ker-Gawl. See **Pyrostegia venusta**.
Bignonia violacea Hort. See **Clytostoma callistegioides**.
Bilbergia nutans H. Wendl. FRIENDSHIP PLANT. QUEEN'S TEARS.
BIRCH, BLACK: **Betula nigra**.
  CANOE: **Betula papyrifera**.
  CUTLEAF WEEPING: **Betula pendula 'Dalecarlica'**.
  EASTERN WHITE: **Betula papyrifera**.
  EUROPEAN WHITE: **Betula pendula**.
  GRAY: **Betula populifolia**.
  JAPANESE WHITE: **Betula platyphylla** var. ***japonica***.
  MONARCH: **Betula maximowicziana**.
  MOUNTAIN: **Betula occidentalis**.
  PAPER: **Betula papyrifera**.
  PURPLE: **Betula pendula 'Purpurea'**.
  PYRAMIDAL WHITE: **Betula pendula 'Fastigiata'**.
  RED: **Betula nigra**.
  RIVER: **Betula nigra**.
  WATER: **Betula occidentalis**.
  WHITE: **Betula pendula**.
  YOUNG'S WEEPING: **Betula pendula 'Youngii'**.
BIRD-CATCHER TREE: **Pisonia umbellifera**.
BIRD-OF-PARADISE: **Strelitzia reginae**.
  GIANT: **Strelitzia nicolai**.
  QUEEN'S: **Strelitzia reginae**.
BIRD-OF-PARADISE BUSH: **Caesalpinia gilliesii**.
BIRD-OF-PARADISE FLOWER: **Caesalpinia gilliesii**.
BIRD-OF-PARADISE FLOWER: **Strelitzia reginae**.
BIRD-OF-PARADISE TREE: **Strelitzia nicolai**.
BIRD'S-EYE BUSH: **Ochna serrulata**.
Bischofia javanica Blume. Syn.: *B. trifoliata*.
Bischofia trifoliata (Roxb.) Hook. See **B. javanica**.
BITTERSWEET, AMERICAN: **Celastrus scandens**.
  CHINESE: **Celastrus rosthornianus**.
  ORIENTAL: **Celastrus orbiculatus**.
BLACKBOY: **Xanthorrhoea arborea**.
BLACKBOY: **Xanthorrhoea preissii**.
BLACKBOY: **Xanthorrhoea quadrangulata**.
BLACKBUTT: **Eucalyptus pilularis**.
  WOODWARD'S: **Eucalyptus woodwardii**.
BLADDERBUSH: **Isomeris arborea**.
BLADDER FLOWER, WHITE: **Araujia sericofera**.
BLADDERNUT, AMERICAN: **Staphylea trifolia**.
  EUROPEAN: **Staphylea pinnata**.
BLADDERPOD: **Isomeris arborea**.
Blechnum brasiliense Desv.
Blechnum brasiliense 'Crispum'.
Blechnum gibbum (Labill.) Mettenius. Some plants called **B. moorei** belong here.
Blechum moorei Christensen. May not be cult. Plants so called may be **B. gibbum**.

Blighia sapida Koenig. Not cult. in Calif. Plants so called are **Cupaniopsis anacardioides.**
BLOODWOOD, SWAMP: **Eucalyptus ptychocarpa.**
  YELLOW: **Eucalyptus eximia.**
BLUEBEARD: **Caryopteris incana.**
BLUEBELL, AUSTRALIAN: **Sollya heterophylla.**
BLUEBERRY, HIGHBUSH: **Vaccinium corymbosum.**
BLUEBERRY CLIMBER: **Ampelopsis brevipedunculata.**
BLUEBLOSSOM: **Ceanothus thyrsiflorus.**
  CREEPING: **Ceanothus thyrsiflorus** var. **repens.**
BLUE BRUSH: **Ceanothus thyrsiflorus.**
BLUEBUSH: **Eucalyptus macrocarpa.**
BLUE-CURLS, MOUNTAIN: **Trichostema parishii.**
  WOOLLY: **Trichostema lanatum.**
BLUEMIST: **Caryopteris × clandonensis.**
BLUSHING BRIDE: **Serruria florida.**
Bocconia arborea S. Wats.
Bocconia frutescens L.
Bolusanthus speciosus (Bolus) Harms. Syn.: *Lonchocarpus speciosus.* RHODESIAN WISTERIA. WISTERIA TREE.
Bombax ceiba L. Syn.: *B. malabaricum.* RED SILK-COTTON TREE.
Bombax ellipticum HBK. See **Pseudobombax ellipticum.**
Bombax malabaricum DC. See ***B. ceiba.***
BONE-BERRY, CHINESE: **Osteomeles schweriniae.**
BOOBYALLA: **Myoporum insulare.**
BOOJUM TREE: **Fouquieria columnaris.**
BORONIA, BROWN: **Boronia megastigma.**
  SCENTED: **Boronia megastigma.**
Boronia elatior Bartl.
Boronia megastigma Nees ex Bartl. BROWN BORONIA. SCENTED BORONIA.
BOTANICAL-WONDER: **× Fatshedera lizei.**
BO TREE: **Ficus religiosa.**
BOTTLEBRUSH, BRACELET: **Melaleuca armillaris.**
  BRILLIANT: **Callistemon citrinus 'Splendens'.**
  DWARF: **Callistemon subulatus.**
  FIERY: **Callistemon phoeniceus.**
  GRAVEL: **Beaufortia sparsa.**
  LEMON: **Callistemon citrinus.**
  LEMON: **Callistemon pallidus.**
  NARROW-LEAVED: **Callistemon linearis.**
  PINE-NEEDLE: **Callistemon pinifolius.**
  PRICKLY: **Callistemon brachyandrus.**
  RED: **Callistemon citrinus.**
  ROSE: **Melaleuca nesophila.**
  STIFF: **Callistemon rigidus.**
  WEEPING: **Callistemon viminalis.**
  WHITE: **Callistemon salignus.**
  WILLOW: **Callistemon salignus.**
BOTTLE TREE: **Brachychiton populneus.**
  FLAME: **Brachychiton acerifolius.**
  QUEENSLAND: **Brachychiton rupestris.**
  SCRUB: **Brachychiton discolor.**
***Bougainvillea* 'Afterglow'.**
Bougainvillea 'American Red'. See ***B.* 'San Diego Red'.**
***Bougainvillea* 'Barbara Karst'.**
***Bougainvillea* 'Betty Hendry'.**
Bougainvillea brasiliensis. Listed name for ***B. spectabilis*** and ***B. glabra.***
***Bougainvillea* 'Brilliant'.**
Bougainvillea × buttiana Holtt. & Standl. [?***B. glabra* × *B. peruviana*].**
***Bougainvillea × buttiana* 'Golden Glow'.**
***Bougainvillea × buttiana* 'Louis Wathen'.**
***Bougainvillea × buttiana* 'Mrs. Allan'.**
***Bougainvillea × buttiana* 'Mrs. Butt'.**
***Bougainvillea × buttiana* 'Orange King'.**
***Bougainvillea × buttiana* 'Praetoria'.**
***Bougainvillea × buttiana* 'Sunfire'.**
***Bougainvillea* 'California Glory'.**
***Bougainvillea* 'California Gold'.**
***Bougainvillea* 'Camarillo Fiesta'.**
***Bougainvillea* 'Carmencita'.**
***Bougainvillea* 'Convent'.**
***Bougainvillea* 'Convent Gardens'.**
***Bougainvillea* 'Crimson Jewel'.**
Bougainvillea 'Crimson Lake'. See ***B. × buttiana* 'Mrs. Butt'.**
***Bougainvillea* 'Easter Parade'.**
Bougainvillea glabra Choisy. PAPER-FLOWER. Some plants called *B. brasiliensis* belong here.
***Bougainvillea glabra* 'Magnifica'.**
***Bougainvillea glabra* 'Sanderana'.**
***Bougainvillea* 'Honolulu'.**
Bougainvillea 'Indian Maid'. See ***B.* 'Betty Hendry'.**
***Bougainvillea* 'Isabel Greensmith'.**
Bougainvillea 'Isabel Greenstreet'. See ***B.* 'Isabel Greensmith'.**
***Bougainvillea* 'Jamaica White'.**
***Bougainvillea* 'La Jolla'.**
Bougainvillea 'Lateritia'. See ***B. spectabilis* 'Lateritia'.**
***Bougainvillea* 'Logan Purple'.**
***Bougainvillea* 'Madonna'.**
Bougainvillea 'Magnifica'. See ***B. glabra* 'Magnifica'.**
***Bougainvillea* 'Manila Red'.**
***Bougainvillea* 'Marg Bacon'.**
***Bougainvillea* 'Monaco Elegance'.**
Bougainvillea 'Mrs. Butt'. See ***B. × buttiana* 'Mrs. Butt'.**
Bougainvillea 'Orange King'. See ***B. × buttiana* 'Orange King'.**
***Bougainvillea* 'Orange Red'.**
Bougainvillea 'Panama Queen'. See ***B.* 'Convent'.**
Bougainvillea peruviana Humb. & Bonpl.
***Bougainvillea* 'Pink Tiara'.**
***Bougainvillea* 'Rainbow Gold'.**
***Bougainvillea* 'Rosea'.**
***Bougainvillea* 'Rose Queen'.**
***Bougainvillea* 'Ruth Hendry'.**
Bougainvillea 'Sanderana'. See ***B. glabra* 'Sanderana'.**
Bougainvillea 'San Diego'. See ***B.* 'San Diego Red'.**
***Bougainvillea* 'San Diego Red'.**
***Bougainvillea* 'Scarlet O'Hara'.**
Bougainvillea spectabilis Willd. Some plants called *B. brasiliensis* belong here.
***Bougainvillea spectabilis* 'Lateritia'.**
***Bougainvillea* 'Sunset Yellow'.**
***Bougainvillea* 'Surprise'.**
***Bougainvillea* 'Tahitian Gold'.**
***Bougainvillea* 'Tahitian Maid'.**
***Bougainvillea* 'Temple Fire'.**
***Bougainvillea* 'Texas Dawn'.**
***Bougainvillea* 'Tropic Rainbow'.**
Boussingaultia basselloides HBK, not Hook. See **Anredera basselloides.** Plants called *B. basselloides* may be **Anredera cordifolia.**
Boussingaultia basselloides Hook., not HBK. See **Anredera cordifolia.**
Boussingaultia gracilis Miers. See **Anredera cordifolia.**
Bouvardia 'Albatross'. See ***B. longiflora* 'Albatross'.**
Bouvardia 'Fire Chief'. See ***B. leiantha* 'Fire Chief'.**
Bouvardia humboldtii Hort. See ***B. longiflora.***
Bouvardia humboldtii 'Albatross'. See ***B. longiflora* 'Albatross'.**
Bouvardia jacquinii HBK. See ***B. ternifolia.***
Bouvardia leiantha Benth.
***Bouvardia leiantha* 'Fire Chief'.**

Bouvardia longiflora (Cav.) HBK. Syn.: *B. humboldtii*.
Bouvardia longiflora **'Albatross'**.
Bouvardia ternifolia (Cav.) Schlechtend. Syn.: *B. jacquinii*.
Bowenia serrulata (Bull) Chamberlain.
BOWER PLANT: **Pandorea jasminoides**.
BOWER VINE: **Pandorea jasminoides**.
Bowkeria gerrardiana Harv. ex Hiern. CALCEOLARIA TREE.
BOWWOOD: **Maclura pomifera**.
BOX, BLUE: **Eucalyptus bauerana**.
  BRIMBLE: **Eucalyptus populnea**.
  BRISBANE: **Tristania conferta**.
  CRIMSON MALLEE: **Eucalyptus lansdowneana**.
  ENGLISH: **Buxus sempervirens**.
  FLOODED: **Eucalyptus microtheca**.
  JAPANESE: **Buxus microphylla** var. **japonica**.
  KIMBERLY GRAY: **Eucalyptus argillacea**.
  KOREAN: **Buxus harlandii**.
  KOREAN: **Buxus microphylla** var. **koreana**.
  RED: **Eucalyptus polyanthemos**.
  SILVER-LEAVED: **Eucalyptus pruinosa**.
  VICTORIAN: **Pittosporum undulatum**.
  WHITE: **Eucalyptus albens**.
  YELLOW: **Eucalyptus melliodora**.
BOX ELDER: **Acer negundo**.
  SILVER VARIEGATED: **Acer negundo 'Variegatum'**.
BOX THORN: **Bursaria spinosa**.
BOXWOOD: **Buxus sempervirens**.
  BROAD-LEAF ENGLISH: **Buxus sempervirens 'Rotundifolia'**.
  DWARF: **Buxus sempervirens 'Suffruticosa'**.
  EDGING: **Buxus sempervirens 'Suffruticosa'**.
  ENGLISH: **Buxus sempervirens**.
  JAPANESE: **Buxus microphylla** var. **japonica**.
  KOREAN: **Buxus harlandii**.
  KOREAN: **Buxus microphylla** var. **koreana**.
  OREGON: **Paxistima myrsinites**.
  VARIEGATED: **Buxus sempervirens 'Argentea'**.
Brachychiton acerifolius (A. Cunn.) F. J. Muell. Syn.: *Sterculia acerifolia*. FLAME TREE. FLAME BOTTLE TREE.
Brachychiton australis (Schott & Endl.) A. Terracciano. Syn.: *Sterculia trichosiphon*. Plants called *B. trichosiphon* belong here.
Brachychiton bidwillii Hook. Syn.: *Sterculia bidwillii*.
Brachychiton discolor F. J. Muell. Syn.: *B. luridus. Sterculia discolor. S. lurida.* WHITE KURRAJONG. QUEENSLAND LACEBARK. SCRUB BOTTLE TREE. PINK FLAME TREE.
Brachychiton diversifolius R. Br. Not cult. Plants so called are **B. populneus**.
Brachychiton gregori F. J. Muell. DESERT KURRAJONG. May not be cult. in Calif. Plants so called are usually **B. × hybridus** or **B. populneus**.
Brachychiton × *hybridus* Hort. [**B. acerifolius** × **B. populneus**]. Some plants called **B. gregori** or *Sterculia sextonii* belong here.
Brachychiton luridus C. Moore ex F. J. Muell. See **B. discolor**.
Brachychiton populneus (Schott & Endl.) R. Br. Syn.: *Sterculia diversifolia*. KURRAJONG. BOTTLE TREE. Some plants called **B. diversifolius** and **B. gregori** belong here.
Brachychiton rupestris (Lindl.) Schum. Syn.: *Sterculia rupestris*. BARREL TREE. QUEENSLAND BOTTLE TREE.
Brachychiton trichosiphon. Listed name for **B. australis**.
Brachyglottis rangiora Buchanan. See **B. repanda** var. **rangiora**.
Brachyglottis repanda J. R. Forst. & G. Forst.
Brachyglottis repanda var. *rangiora* (Buchanan) Allan. Syn.: *B. rangiora*.
Brachysema lanceolatum Meissn. SCIMITAR SHRUB. SWAN RIVER PEA SHRUB.

Brachystelma barberiae Harv. ex Hook. f.
Brahea armata S. Wats. Syn.: *Erythea armata. Glaucotheca armata.* BLUE FAN PALM. BLUE HESPER PALM. MEXICAN BLUE PALM.
Brahea bella L. H. Bailey.
Brahea brandegeei (C. Purpus) H. E. Moore. Syn.: *Erythea brandegeei*. SAN JOSE HESPER PALM.
Brahea dulcis (HBK) Mart. ROCK PALM.
Brahea edulis H. Wendl. ex S. Wats. Syn.: *Erythea edulis*. GUADALUPE PALM.
Brahea elegans (Franceschi ex Becc.) H. E. Moore. Syn.: *Erythea elegans*. FRANCESCHI PALM.
Brahea filamentosa (Fenzi) H. Wendl. ex Kuntze. See **Washingtonia filifera**.
Brassaia actinophylla Endl. Syn.: *Schefflera actinophylla*. AUSTRALIAN UMBRELLA TREE. QUEENSLAND UMBRELLA TREE. OCTOPUS TREE.
Brassaia arboricola. Listed name, presumably for **Schefflera arboricola**.
BREATH-OF-HEAVEN: **Coleonema album**.
BREATH-OF-HEAVEN: **Diosma ericoides**.
  PINK: **Coleonema pulchrum**.
  WHITE: **Coleonema album**.
BRIAR. See BRIER.
BRIDAL-WREATH: **Spiraea prunifolia**.
BRIDAL-WREATH: **Spiraea × vanhouttei**.
  DOUBLE: **Spiraea cantoniensis 'Lanceata'**.
  DWARF PINK: **Spiraea × bumalda 'Anthony Waterer'**.
BRIER, AUSTRIAN: **Rosa foetida**.
  AUSTRIAN COPPER: **Rosa foetida 'Bicolor'**.
  DOG: **Rosa canina**.
  SWEET: **Rosa rubiginosa**.
BRITTLEBUSH: **Encelia farinosa**.
BROOM, ATLAS: **Cytisus battandieri**.
  BRIDAL-VEIL: **Genista monosperma**.
  BURKWOOD: **Cytisus (Dallimore Hybrid) 'Burkwoodii'**.
  BUTCHER'S: **Ruscus aculeatus**.
  CANARY ISLAND: **Cytisus canariensis**.
  CHAPARRAL: **Baccharis pilularis** subsp. **consanguinea**.
  CLIMBING BUTCHER'S: **Semele androgyna**.
  DOROTHY WALPOLE: **Cytisus 'Dorothy Walpole'**.
  DWARF CHAPARRAL: **Baccharis pilularis**.
  EASTER: **Cytisus × spachianus**.
  FRENCH: **Cytisus monspessulanus**.
  KEW: **Cytisus × kewensis**.
  MONTPELIER: **Cytisus monspessulanus**.
  MOONLIGHT: **C. × praecox 'Warminster'**.
  MOONLIGHT: **C. scoparius 'Moonlight'**.
  MT. ETNA: **Genista aethnensis**.
  NORMANDY: **Cytisus scoparius 'Andreanus'**.
  PROSTRATE: **Cytisus decumbens**.
  PROVENCE: **Cytisus purgans**.
  PURPLE: **Cytisus purpureus**.
  SCOTCH: **Cytisus scoparius**.
  SCOTCH HYBRIDS: **Cytisus (Dallimore hybrids)**.
  SPANISH: **Genista hispanica**.
  SPANISH: **Spartium junceum**.
  SWEET: **Cytisus supranubius**.
  WARMINSTER: **Cytisus × praecox 'Warminster'**.
  WEAVER'S: **Spartium junceum**.
  WHITE SPANISH: **Cytisus multiflorus**.
Broussonetia papyrifera (L.) Vent. PAPER MULBERRY.
Bruckenthalia spiculifolia (Salisb.) Rchb. SPIKE HEATH.
Brugmansia arborea (L.) Lagerh. Syn.: *Datura arborea*. Some plants so called are **B. × candida**.

Brugmansia × *candida* Pers. [*B. aurea* Lagerh. × *B. versicolor* Lagerh.]. Syn.: *Datura* × *candida*. Some plants called *B. arborea* belong here.
Brugmansia × *insignis* (Barb.-Rodr.) Lockw. [*B. suaveolens* × *B. versicolor* Lagerh.]. Some plants called *B. suaveolens* belong here.
Brugmansia sanguinea (Ruiz & Pav.) D. Don. Syn.: *Datura sanguinea*. RED ANGEL'S-TRUMPET.
Brugmansia suaveolens (Humb. & Bonpl. ex Willd.) Bercht. & J. Presl. Syn.: *Datura suaveolens*. ANGEL'S-TRUMPET. WHITE ANGEL'S-TRUMPET. Some plants so called are *B.* × *insignis*.
Brunfelsia australis Benth. YESTERDAY-TODAY-AND-TOMORROW. MORNING-NOON-AND-NIGHT. Some plants called *B. latifolia* belong here.
Brunfelsia calycina Benth. See *B. pauciflora* var. *calycina*. Some plants called *B. calycina* are *B. pauciflora* 'Eximia' or *B. pauciflora* 'Floribunda'.
Brunfelsia calycina var. *eximia* (Scheidw.) L. H. Bailey & Raffill. See *B. pauciflora* 'Eximia'.
Brunfelsia calycina var. *floribunda* L. H. Bailey & Raffill. See *B. pauciflora* 'Floribunda'.
Brunfelsia calycina var. *macrantha* (Lem.) L. H. Bailey & Raffill. See *B. pauciflora* 'Macrantha'.
Brunfelsia eximia (Scheidw.) Bosse. See *B. pauciflora* 'Eximia'.
Brunfelsia floribunda Hort. See *B. pauciflora* 'Floribunda'.
Brunfelsia latifolia (Pohl) Benth. Some plants so called are *B. australis*.
Brunfelsia macrantha Lem. See *B. pauciflora* 'Macrantha'.
Brunfelsia pauciflora (Cham. & Schlechtend.) Benth.
Brunfelsia pauciflora var. *calycina* (Benth.) J. A. Schmidt. Syn.: *B. calycina*. BRAZIL RAIN TREE.
Brunfelsia pauciflora 'Eximia'. Syn:. *B. eximia. B. calycina* var. *eximia*. MORNING-NOON-AND-NIGHT. YESTERDAY-TODAY-AND-TOMORROW. YESTERDAY-AND-TODAY. Some plants called *B. calycina* belong here.
Brunfelsia pauciflora 'Floribunda'. Syn.: *B. calycina* var. *floribunda*. *B. floribunda*. YESTERDAY-TODAY-AND-TOMORROW. Some plants called *B. calycina* belong here.
Brunfelsia pauciflora 'Macrantha'. Syn.: *B. calycina* var. *macrantha*.
BUCHU: **Diosma ericoides**.
BUCKBRUSH: **Ceanothus cuneatus**.
BUCKBRUSH: **Ceanothus velutinus**.
BUCKEYE, CALIFORNIA: **Aesculus californica**.
　OHIO: **Aesculus glabra**.
　RED: **Aesculus pavia**.
　YELLOW: **Aesculus octandra**.
Bucklandia populnea R. Br. See *Exbucklandia populnea*.
BUCKTHORN: **Rhamnus crocea**.
　ALDER: **Rhamnus frangula**.
　EUROPEAN: **Rhamnus cathartica**.
　ITALIAN: **Rhamnus alaternus**.
　VARIEGATED ITALIAN: **Rhamnus alaternus 'Argenteo-variegata'**.
BUCKWHEAT, ASHYLEAF: **Eriogonum cinereum**.
　CALIFORNIA: **Eriogonum fasciculatum**.
　GRANITE: **Eriogonum lobbii**.
　PINK: **Eriogonum grande**.
　RED: **Eriogonum grande** var. **rubescens**.
　SAFFRON: **Eriogonum crocatum**.
　SANTA CRUZ ISLAND: **Eriogonum arborescens**.
　SEA CLIFF: **Eriogonum parvifolium**.
　WILD: **Eriogonum fasciculatum**.
　WRIGHT: **Eriogonum wrightii** subsp. **subscaposum**.
　YELLOW: **Eriogonum crocatum**.
BUDDLEIA, ALTERNATE-LEAF: **Buddleia alternifolia**.
　WEYER: **Buddleia** × **weyerana**.
Buddleia alternifolia Maxim. FOUNTAIN BUTTERFLYBUSH. ALTERNATE-LEAF BUDDLEIA.
Buddleia asiatica Lour. WHITE BUTTERFLYBUSH.
Buddleia davidii Franch. BUTTERFLYBUSH. SUMMER LILAC.
Buddleia davidii 'Black Knight'.
Buddleia davidii 'Charming'.
Buddleia davidii 'Dubonnet'.
Buddleia davidii 'Empire Blue'.
Buddleia davidii 'Fascinating'.
Buddleia davidii 'Fortune'.
Buddleia davidii 'Ile de France'.
Buddleia davidii 'Peace'.
Buddleia davidii 'Purple Prince'.
Buddleia davidii 'Royal Red'.
Buddleia davidii 'White Bouquet'.
Buddleia davidii 'White Cloud'.
Buddleia davidii 'White Profusion'.
Buddleia globosa Hope.
Buddleia hartwegii. Listed name for *B. davidii*.
Buddleia saligna Willd. Syn.: *Chilianthus arboreus*.
Buddleia salviifolia (L.) Lam.
Buddleia × *weyerana* Weyer. [*B. davidii* var. *magnifica* (Wils.) Rehd. & Wils. × *B. globosa*]. WEYER BUDDLEIA.
BUFFALO BERRY: **Shepherdia argentea**.
BUFFALO-HORN: **Burchellia bubalina**.
BULL BAY: **Magnolia grandiflora**.
BUNCHBERRY: **Cornus canadensis**.
BUNYA-BUNYA: **Araucaria bidwillii**.
Bupleurum fruticosum L.
Burchellia bubalina (L. f.) Sims. Syn.: *B. capensis*. BUFFALO-HORN.
Burchellia capensis R. Br. See *B. bubalina*.
BURNING BUSH, WINGED: **Euonymus alata**.
BURRO-FAT: **Isomeris arborea**.
BURRO-TAIL: **Sedum morganianum**.
Bursaria spinosa Cav. BOX THORN.
Bursera fagaroides (HBK) Engler.
BUSHMAN'S POISON: **Acokanthera oppositifolia**.
BUSHRUE: **Cneoridium dumosum**.
Butia capitata (Mart.) Becc. SOUTH AMERICAN JELLY PALM. PINDO PALM. HARDY BLUE COCOS PALM. Some plants called *Cocos australis* belong here.
Butia eriospatha (Mart. ex Drude) Becc. WOOLLY BUTIA PALM.
BUTTERFLY TREE: **Bauhinia purpurea**.
BUTTERFLYBUSH: **Buddleia davidii**.
　FOUNTAIN: **Buddleia alternifolia**.
　WHITE: **Buddleia asiatica**.
BUTTERNUT: **Juglans cinerea**.
BUTTONWILLOW: **Cephalanthus occidentalis**.
BUTTONWOOD: **Platanus occidentalis**.
BUTTONWOOD: **Platanus racemosa**.
Buxus arborescens Mill. See *B. sempervirens*.
Buxus balearica Lam.
Buxus 'Green Beauty'. See *B. microphylla* var. *japonica* 'Green Beauty'.
Buxus harlandii Hance. KOREAN BOXWOOD. KOREAN BOX.
Buxus hildebrandtii Baill.
Buxus japonica Muell.-Arg. See *B. microphylla* var. *japonica*.
Buxus microphylla Siebold & Zucc.
Buxus microphylla 'Compacta'.
Buxus microphylla 'Green Beauty'. See *B. microphylla* var. *japonica* 'Green Beauty'.
Buxus microphylla var. *japonica* (Muell.-Arg.) Rehd. & Wils. Syn.: *B. japonica*. JAPANESE BOX. JAPANESE BOXWOOD.
Buxus microphylla var. *japonica* 'Green Beauty'.

Buxus microphylla var. *japonica* 'Richardii'.
Buxus microphylla var. *japonica* 'Rotundifolia'.
Buxus microphylla var. *koreana* Nakai. KOREAN BOX. KOREAN BOXWOOD.
Buxus microphylla 'Richardii'. See **B. microphylla** var. *japonica* 'Richardii'.
Buxus microphylla 'Rotundifolia'. See **B. microphylla** var. *japonica* 'Rotundifolia'.
Buxus 'Richardii'. See **B. microphylla** var. *japonica* 'Richardii'.
Buxus 'Rotundifolia'. See **B. microphylla** var. *japonica* 'Rotundifolia' or **B. sempervirens** 'Rotundifolia'.
Buxus sempervirens L. Syn.: *B. arborescens. B. sempervirens* var. *arborescens*. ENGLISH BOX. BOXWOOD. ENGLISH BOXWOOD.
Buxus sempervirens var. *arborescens* L. See **B. sempervirens**.
Buxus sempervirens 'Arborescens'. See **B. sempervirens**.
Buxus sempervirens 'Argentea'. VARIEGATED BOXWOOD.
Buxus sempervirens 'Argenteo-variegata'.
Buxus sempervirens 'Aureo-variegata'.
Buxus sempervirens 'Fastigiata'.
Buxus sempervirens 'Handsworthiensis'.
Buxus sempervirens 'Handsworthii'. See *Buxus sempervirens* 'Handsworthiensis'.
Buxus sempervirens 'Inglis'?
Buxus sempervirens 'Rotundifolia'. BROAD-LEAF ENGLISH BOXWOOD.
Buxus sempervirens 'Suffruticosa'. Syn.: *B. suffruticosa*. DWARF BOXWOOD. EDGING BOXWOOD.
Buxus sempervirens 'Trustee'.?
Buxus sempervirens 'Vardar Valley'.
Buxus sempervirens 'Welleri'.
Buxus suffruticosa Weston. See **B. sempervirens** 'Suffruticosa'.

CABBAGE TREE: **Cordyline australis.**
CABBAGE TREE: **Cussonia spicata.**
  SPIKED: **Cussonia spicata.**
CACTUS, APPLE: **Cereus peruvianus.**
  GIANT: **Carnegiea gigantea.**
  SPINELESS: **Opuntia ficus-indica.**
  TOAD: **Stapelia variegata.**
Caesalpinia gilliesii (Wallich ex Hook.) D. Dietrich. Syn.: *Poinciana gilliesii*. BIRD-OF-PARADISE BUSH. BIRD-OF-PARADISE FLOWER. PARADISE POINCIANA.
Caesalpinia pulcherrima (L.) Swartz. Syn.: *Poinciana pulcherrima*. BARBADOS-PRIDE. DWARF POINCIANA.
Caesalpinia spinosa (Molina) Kuntze. TARA.
CAJEPUT TREE: **Melaleuca quinquenervia.**
Caladium esculentum Hort. See *Colocasia esculenta.*
CALAMONDIN: × **Citrofortunella mitis.**
Calceolaria multiflora Cav.
CALCEOLARIA TREE: **Bowkeria gerrardiana.**
Calibanus hookeri (Lem.) Trel.
CALICO BUSH: **Kalmia latifolia.**
Calliandra californica Benth.
Calliandra costaricensis (Britt. & Rose) Standl.
Calliandra eriophylla Benth. FAIRY-DUSTER. FALSE MESQUITE.
Calliandra guildingii Benth. Plants so called are usually **C. tweedii.**
Calliandra haematocephala Hassk. Syn.: *C. inaequilatera*. PINK POWDER PUFF.
Calliandra haematocephala 'Snowball'.

Calliandra houstoniana (Mill.) Standl.
Calliandra inaequilatera Rusby. See **C. haematocephala.**
Calliandra portoricensis (Jacq.) Benth.
Calliandra tweedii Benth. BRAZILIAN FLAMEBUSH. TRINIDAD FLAMEBUSH. MEXICAN FLAMEBUSH. Some plants called **C. guildingii** belong here.
Callicarpa bodinieri Lev. var. *giraldii* (Hesse ex Rehd.) Rehd. Syn.: *C. giraldiana*. BEAUTYBERRY.
Callicarpa dichotoma (Lour.) K. Koch.
Callicarpa giraldiana Hesse. See **C. bodinieri** var. *giraldii.*
Callicarpa rubella Lindl.
Callistemon acuminatus Cheel.
Callistemon brachyandrus Lindl. PRICKLY BOTTLEBUSH.
Callistemon citrinus (Curtis) Skeels. Syn.: *C. lanceolatus*. LEMON BOTTLEBRUSH. RED BOTTLEBRUSH.
Callistemon citrinus 'Chico Red'.
Callistemon citrinus 'Compactus'.
Callistemon citrinus 'McCaskillii'. See **C. viminalis** 'McCaskillii'.
Callistemon citrinus 'Splendens'. BRILLIANT BOTTLEBRUSH.
Callistemon coccineus F. J. Muell. See **C. macropunctatus.**
Callistemon cupressifolius?
Callistemon 'Jeffers'. Some plants called *C. violaceus* may belong here.
Callistemon lanceolatus (Sm.) DC. See **C. citrinus.**
Callistemon lanceolatus 'Compactus'. See **C. citrinus** 'Compactus'.
Callistemon lilacinus Cheel. Name of uncertain application.
Callistemon linearis (Schrad. & Wendl.) DC. NARROW-LEAVED BOTTLEBRUSH.
Callistemon linearis pumilus?
Callistemon macropunctatus (Dum.-Cours.) Court. Syn.: *C. coccineus. C. rugulosus.*
Callistemon 'McCaskillii'. See **C. viminalis** 'McCaskillii'.
Callistemon pachyphyllus Cheel.
Callistemon pallidus (Bonpl.) DC. LEMON BOTTLEBRUSH.
Callistemon phoeniceus Lindl. FIERY BOTTLEBRUSH.
Callistemon pinifolius (Wendl.) DC. PINE-NEEDLE BOTTLEBRUSH.
Callistemon 'Red Cascade'.
Callistemon rigidus B. Br. STIFF BOTTLEBRUSH.
Callistemon roseus (Guilfoyle) Cheel. Syn.: *C. salignus* var. *roseus*.
Callistemon rugulosus (Willd. ex Link) DC. See **C. macropunctatus.**
Callistemon salignus (Sm.) DC. WHITE BOTTLEBRUSH. WILLOW BOTTLEBRUSH. PINK-TIPS.
Callistemon salignus var. *roseus* Guilfoyle. See **C. roseus.**
Callistemon sieberi DC.
Callistemon speciosus (Sims) DC.
Callistemon subulatus Cheel. DWARF BOTTLEBRUSH.
Callistemon validus?
Callistemon viminalis (Soland. ex Gaertner) Cheel. WEEPING BOTTLEBRUSH.
Callistemon viminalis 'McCaskillii'.
Callistemon violaceus. Listed name of uncertain application. Some plants so called may be **C.** 'Jeffers'.
Callistemon viridiflorus (Sims) DC.
Callitris calcarata R. Br. ex F. M. Bailey. See **C. endlicheri.**
Callitris columellaris F. J. Muell. Syn.: *C. glauca*.
Callitris endlicheri (Parl.) F. M. Bailey. Syn.: *C. calcarata*.
Callitris glauca R. Br. ex R. T. Baker & H. G. Sm. See **C. columellaris.**
Callistris preissii Miq. Syn.: *C. robusta. C. verrucosa.* CYPRESS PINE.
Callitris robusta (A. Cunn. ex Parl.) R. Br. ex F. M. Bailey. See **C. preissii.**
Callitris verrucosa (A. Cunn. ex Endl.) F. J. Muell. See **C. preissii.**

Calluna vulgaris (L.) Hull. Syn.: *Erica vulgaris*. SCOTCH HEATHER.
Calluna vulgaris 'Aberdeen'.
Calluna vulgaris 'Alba'. WHITE SCOTCH HEATHER.
Calluna vulgaris 'Alba Plena'.
Calluna vulgaris 'Alportii'.
Calluna vulgaris 'Aurea'. GOLDEN HEATHER.
Calluna vulgaris 'Aureifolia'. See **C. vulgaris 'Hammondii Aureifolia'.**
Calluna vulgaris 'Autumn Glow'.
Calluna vulgaris 'Barnett Anley'.
Calluna vulgaris 'Carlton'.
Calluna vulgaris 'County Wicklow'. COUNTY WICKLOW HEATHER.
Calluna vulgaris 'Crispa'.
Calluna vulgaris 'Cuprea'.
Calluna vulgaris 'C. W. Nix'.
Calluna vulgaris 'Cyndy'.
Calluna vulgaris 'Dainty Bess'.
Calluna vulgaris 'Darleyensis'.
Calluna vulgaris 'David Eason'.
Calluna vulgaris 'Durfordii'.
Calluna vulgaris 'Elata'.
Calluna vulgaris 'Elegantissima'.
Calluna vulgaris 'Else Frye'.
Calluna vulgaris 'Elsie Purnell'.
Calluna vulgaris 'Flore Pleno'.
Calluna vulgaris 'Foxhollow Wanderer'.
Calluna vulgaris 'Foxii'.
Calluna vulgaris 'Foxii Nana'.
Calluna vulgaris 'Goldsworth Crimson'.
Calluna vulgaris 'Hammondii Aureifolia'.
Calluna vulgaris 'H. E. Beale'. PINK SCOTCH HEATHER.
Calluna vulgaris 'Hibernica'.
Calluna vulgaris 'Hiemalis'. JOHNSON HEATHER.
Calluna vulgaris 'Hirsuta'.
Calluna vulgaris 'Hirsuta Albiflora'.
Calluna vulgaris 'J. H. Hamilton'.
Calluna vulgaris 'Joan Sparkes'.
Calluna vulgaris 'Johnson'. See **C. vulgaris 'Hiemalis'.**
Calluna vulgaris 'Kuphaldtii'.
Calluna vulgaris 'Mair's Variety'.
Calluna vulgaris 'Mrs. Pat'.
Calluna vulgaris 'Mrs. Ronald Gray'.
Calluna vulgaris 'Mullion'.
Calluna vulgaris 'Nana'.
Calluna vulgaris 'Nana Compacta'.
Calluna vulgaris 'Penhale'.
Calluna vulgaris 'Purpurea'. See **C. vulgaris 'Serlei Purpurea'.**
Calluna vulgaris 'Pygmaea'.
Calluna vulgaris 'Rainbow'.
Calluna vulgaris 'Rigida'.
Calluna vulgaris 'Robert Chapman'.
Calluna vulgaris 'Roma'.
Calluna vulgaris 'Rubra'. See **C. vulgaris 'Serlei Rubra'.**
Calluna vulgaris 'Ruth Sparkes'.
Calluna vulgaris 'Serlei'.
Calluna vulgaris 'Serlei Aurea'.
Calluna vulgaris 'Serlei Purpurea'. LAVENDER SCOTCH HEATHER.
Calluna vulgaris 'Serlei Rubra'.
Calluna vulgaris 'Sister Anne'.
Calluna vulgaris 'Smith's Minima'.
Calluna vulgaris 'Spitfire'.
Calluna vulgaris 'St. Nick'.
Calluna vulgaris 'Tenuis'.
Calluna vulgaris 'Tib'.
Calluna vulgaris 'Tom Thumb'.
Calluna vulgaris 'Tomentosa'.
Calocedrus decurrens (Torr.) Florin. Syn.: *Libocedrus decurrens*. INCENSE CEDAR.
Calocedrus decurrens 'Compacta'.
Calocephalus brownii (Cass.) F. J. Muell. CUSHIONBUSH.
Calodendrum capense Thunb. CAPE CHESTNUT.
Calothamnus gilesii F. J. Muell. GILES NETBUSH.
Calothamnus quadrifidus R. Br.
Calothamnus rupestris Schauer.
Calothamnus villosus R. Br. HAIRY NETBUSH.
Calpurnia aurea (Ait.) Benth. EAST AFRICAN LABURNUM.
Calpurnia villosa Harv.
Calycanthus floridus L. SWEET SHRUB. CAROLINA ALLSPICE.
Calycanthus occidentalis Hook. & Arn. WESTERN SPICEBUSH.
CAMELLIA, JAPANESE: **Camellia japonica.**
  SASANQUA: **Camellia sasanqua.**
  SUN: **Camellia sasanqua.**
Camellia 'Fragrant Pink'. [**C. japonica** subsp. *rusticana* (Honda) Kitamura × **C. lutchuensis** Ito].
Camellia granthamiana Sealy.
Camellia hiemalis Nakai.
Camellia 'Howard Asper'. [**C. reticulata** × **C. japonica**].
Camellia japonica L. JAPANESE CAMELLIA.
Camellia japonica 'Leucocarpa'?
Camellia reticulata Lindl.
Camellia reticulata 'Captain Rawes'.
Camellia saluenensis Stapf ex Bean.
Camellia sasanqua Thunb. SUN CAMELLIA. SASANQUA CAMELLIA.
Camellia sinensis (L.) Kuntze. Syn:. *Thea sinensis*. TEA PLANT.
Camellia vernalis (Makino) Makino.
Camellia × williamsii W. W. Sm. [**C. japonica** × **C. saluenensis**].
Camellia × williamsii 'Donation'.
Camellia × williamsii 'J. C. Williams'.
Camellia × williamsii 'Mary Christian'.
Campanula vidalii H. C. Wats.
CAMPHOR TREE: **Cinnamomum camphora.**
  NEPAL: **Cinnamomum glanduliferum.**
Camphora officinalis Nees ex Steud. See **Cinnamomum camphora.**
Camphora officinarum Nees. See **Cinnamomum camphora.**
Campsis capreolata Hort. See **Bignonia capreolata.**
Campsis 'Golden Trumpet'?
Campsis grandiflora (Thunb.) K. Schum. Syn.: *Bignonia chinensis. B. grandiflora*. CHINESE TRUMPET CREEPER.
Campsis 'Madame Galen'. See **C. × tagliabuana 'Madame Galen'.**
Campsis radicans (L.) Seem. ex Bur. Syn.: *Bignonia radicans*. TRUMPET CREEPER.
Campsis × tagliabuana (Vis.) Rehd. [**C. grandiflora** × **C. radicans**].
Campsis × tagliabuana 'Madame Galen'.
Camptotheca acuminata Decne.
CANARY-BIRD BUSH: **Crotalaria agatiflora.**
CANDLE BUSH: **Cassia alata.**
CANDLEBERRY: **Myrica cerifera.**
Candollea cuneiformis Labill. See **Hibbertia cuneiformis.**
CANDYTUFT: **Iberis semperflorens.**
  EDGING: **Iberis sempervirens.**
CANE, DUMB: **Dieffenbachia seguine.**
  SPOTTED DUMB: **Dieffenbachia maculata.**
Cantua bicolor Lam.
Cantua buxifolia Juss. ex Lam. MAGIC FLOWER. SACRED-FLOWER-OF-THE-INCAS.
CARACOL: **Vigna caracalla.**
Caragana arborescens Lam. SIBERIAN PEA SHRUB.

CARDINAL-SPEAR: **Erythrina herbacea.**
Carica candamarcensis Hook. f. See **C. pubescens.**
Carica papaya L. PAPAYA.
Carica pubescens Lenné & K. Koch. Syn.: *C. candamarcensis.* MOUNTAIN PAPAYA.
Carissa 'Boxwood Beauty'. See **C. macrocarpa 'Boxwood Beauty'.**
Carissa carandas L. KARANDA.
Carissa edulis (Forssk.) Vahl.
Carissa 'Fancy'. See **C. macrocarpa 'Fancy'.**
Carissa grandiflora (E. H. Mey.) A. DC. and all listed cvs. See **C. macrocarpa** and cvs.
Carissa 'Green Carpet'. See **C. macrocarpa 'Green Carpet'.**
Carissa macrocarpa (Ecklon) A. DC. Syn.: *C. grandiflora.* NATAL PLUM.
Carissa macrocarpa 'Alles'.
Carissa macrocarpa 'Boxwood Beauty'.
Carissa macrocarpa 'Bronze'.
Carissa macrocarpa 'Coolidge'.
Carissa macrocarpa 'Fancy'.
Carissa macrocarpa 'Green Carpet'.
Carissa macrocarpa 'Kenya'.
Carissa macrocarpa 'Lowrey'.
Carissa macrocarpa 'Minima'.
Carissa macrocarpa 'Pygmy'.
Carissa macrocarpa 'Prostrata'.
Carissa macrocarpa 'Ruby Point'.
Carissa macrocarpa 'Tomlinson'.
Carissa macrocarpa 'Tuttle'.
Carissa minima. Listed name probably for **C. macrocarpa 'Minima'.**
Carissa 'Prostrata'. See **C. macrocarpa 'Prostrata'.**
Carissa 'Prostrata Coolidgei'. See **C. macrocarpa 'Coolidge'.**
Carissa 'Prostrata Tuttlei'. See **C. macrocarpa 'Tuttle'.**
Carissa 'Tuttlei'. See **C. macrocarpa 'Tuttle'.**
CARMEL CREEPER: **Ceanothus griseus** var. **horizontalis.**
Carmichaelia arborea (G. Forst.) Druce.
Carmichaelia australis of authors, probably not R. Br. See **C. arborea.**
Carmichaelia enysii T. Kirk.
Carnegiea gigantea (Engelm.) Britt. & Rose. SAGUARO. SAHUARA. GIANT CACTUS.
CAROB: **Ceratonia siliqua.**
Carpenteria californica Torr. BUSH ANEMONE.
Carpinus betulus L. EUROPEAN HORNBEAM.
Carpinus betulus 'Fastigiata'.
Carpinus betulus 'Quercifolia'.
Carpinus caroliniana Walter. AMERICAN HORNBEAM. BLUE BEECH.
Carpinus japonica Blume. JAPANESE HORNBEAM.
Carpobrotus acinaciformis (L.) L. Bolus.
CARRION FLOWER: **Stapelia gigantea.**
CARROTWOOD: **Cupaniopsis anacardioides.**
Carya illinoensis (Wangenh.) K. Koch. (Not *C.* "*illinoinensis*".) Syn.: *C. pecan.* PECAN.
Carya pecan (Marsh.) Engler & Graebn. See **C. illinoensis.**
Caryopteris × clandonensis A. Simmonds ex Rehd. [**C. incana × C. mongholica**]. BLUE MIST.
Caryopteris incana (Thunb.) Miq. Syn.: *C. mastacanthus. C. tangutica.* BLUEBEARD. BLUE SPIRAEA.
Caryopteris mastacanthus Schauer. See **C. incana.**
Caryopteris mongholica Bunge.
Caryopteris tangutica Maxim. See **C. incana.**
Caryota mitis Lour. CLUSTERED FISHTAIL PALM.
Caryota ochlandra Hance. CANTON FISHTAIL PALM.
Caryota urens L. FISHTAIL WINE PALM.
CASCARA: **Rhamnus purshiana.**

CASCARA SAGRADA: **Rhamnus purshiana.**
Casimiroa edulis Llave & Lex. WHITE SAPOTE. MEXICAN APPLE.
Casimiroa edulis 'Coleman'.
Casimiroa edulis 'Pike'.
Casimiroa edulis 'Suebelle'.
Casimiroa edulis 'Wilson'.
CASSIA, FEATHERY: **Cassia artemisioides.**
SILVER: **Cassia artemisioides.**
Cassia alata L. CANDLE BUSH.
Cassia armata S. Wats.
Cassia artemisioides Gaud.-Beaup. ex A. DC. SILVER CASSIA. FEATHERY CASSIA.
Cassia bicapsularis L.
Cassia brewsteri (F. J. Muell.) Benth.
Cassia candolleana Vogel. See **C. coluteoides.**
Cassia carnaval Spegazzini. See **C. leptophylla.**
Cassia coluteoides Colladon.
Cassia corymbosa Lam. FLOWERY SENNA.
Cassia desolata F. J. Muell. var. *involucrata* J. M. Black. See **C. helmsii.**
Cassia didymobotrya Fresenius. Syn.: *C. nairobensis.*
Cassia eremophila A. Cunn. ex Vogel.
Cassia excelsa Hort. See **C. spectabilis.**
Cassia fistula L. GOLDEN-SHOWER TREE.
Cassia floribunda Cav. See **C. laevigata.**
Cassia glauca Lam. See **C. surattensis.**
Cassia goldmanii Rose.
Cassia helmsii Symon. Syn.: *C. desolata* var. *involucrata.*
Cassia laevigata Willd. Syn.: *C. floribunda.*
Cassia latopetiolata Dombey ex Vogel.
Cassia leptocarpa Benth.
Cassia leptophylla Vogel. Syn.: *C. carnaval.* GOLD MEDALLION TREE.
Cassia liebmannii Benth.
Cassia multijuga L. Rich.
Cassia nairobensis Hort. See **C. didymobotrya.**
Cassia sophera L.
Cassia spectabilis A. DC. Syn.: *C. excelsa.* CROWN-OF-GOLD TREE.
Cassia splendida Vogel. GOLDEN-WONDER SENNA.
Cassia surattensis Burm. f. Syn.: *C. glauca.*
Cassia tomentosa L. f. WOOLLY SENNA.
CASSINE: **Ilex vomitoria.**
Cassine orientalis (Jacq.) Kuntze. Syn.: *Elaeodendron orientale.* FALSE OLIVE.
Cassiope lycopodioides (Pall.) D. Don.
Cassiope mertensiana (Bong.) D. Don.
Cassiope mertensiana var. **gracilis** (Piper) C. L. Hitchc.
Cassiope 'Muirhead'. [**C. lycopodioides × C. wardii**].
Cassiope wardii Marquand.
Castanea americana Raf. See **C. dentata.**
Castanea dentata (Marsh.) Borkh. Syn:. *C. americana.* AMERICAN CHESTNUT.
Castanea mollissima Blume. CHINESE CHESTNUT.
Castanea sativa Mill. SPANISH CHESTNUT.
Castanopsis chrysophylla (Dougl. ex Hook.) A. DC. See **Chrysolepis chrysophylla.**
Castanopsis chrysophylla var. *minor* (Benth.) A. DC. See **Chrysolepis chrysophylla** var. **minor.**
Castanopsis sempervirens (Kell.) Dudl. See **Chrysolepis sempervirens.**
Castanospermum australe A. Cunn. & Fraser ex Hook. MORETON BAY CHESTNUT. BLACK BEAN.
CAST-IRON PLANT: **Aspidistra elatior.**
CASTOR-OIL PLANT: **Ricinus communis.**

Casuarina cunninghamiana Miq. HORSETAIL TREE. RIVER SHE OAK. Plants in Calif. called **C. equisetifolia** belong here.
Casuarina equisetifolia J. R. Forst. & G. Forst. May not be cult. in Calif. Plants so called are **C. cunninghamiana**.
Casuarina glauca Sieber ex K. Spreng.
Casuarina littoralis Salisb. Syn.: *C. suberosa*.
Casuarina nana Sieber ex K. Spreng. STUNTED SHE OAK.
Casuarina stricta Dryander. MOUNTAIN SHE OAK. DROOPING SHE OAK. COAST BEEFWOOD.. HORSETAIL TREE.
Casuarina suberosa Otto & A. Dietrich. See **C. littoralis**.
CATALINA PERFUME.: **Ribes viburnifolium**.
CATALPA: **Catalpa bignonioides**.
    UMBRELLA: **Catalpa bignonioides 'Nana'**.
    WESTERN: **Catalpa speciosa**.
Catalpa bignonioides Walter. INDIAN BEAN. CATALPA.
Catalpa bignonioides 'Aurea'.
Catalpa bignonioides 'Nana'. UMBRELLA CATALPA.
Catalpa bungei C. A. Mey.
Catalpa speciosa Warder ex Engelm. CIGAR TREE. INDIAN BEAN. WESTERN CATALPA.
Catha edulis (Vahl) Forssk. ex Endl. KHAT. ARABIAN TEA.
CAT'S-CLAW: **Acacia greggii**.
CAT'S-CLAW: **Macfadyena unguis-cati**.
CAULIFLOWER EARS: **Crassula ovata**.
CEANOTHUS, BIG POD: **Ceanothus megacarpus**.
    BOLINAS RIDGE: **Ceanothus masonii**.
    CALISTOGA: **Ceanothus divergens**.
    CARMEL: **Ceanothus griseus**.
    COAST: **Ceanothus ramulosus**.
    CUPLEAF: **Ceanothus greggii** var. **perplexans**.
    FELTLEAF: **Ceanothus arboreus**.
    FRENCH HYBRID: **Ceanothus × delilianus**.
    GREEN BARK: **Ceanothus spinosus**.
    HEARST: **Ceanothus hearstiorum**.
    HOARYLEAF: **Ceanothus crassifolius**.
    HOLLYLEAF: **Ceanothus purpureus**.
    LOMPOC: **Ceanothus ramulosus** var. **fascicularis**.
    MARITIME: **Ceanothus maritimus**.
    MONTEREY: **Ceanothus rigidus**.
    MOUNT ST. HELENA: **Ceanothus divergens**.
    MOUNT TRANQUILLON: **Ceanothus papillosus** var. **roweanus**.
    MOUNT VISION: **Ceanothus gloriosus** var. **porrectus**.
    NAVARRO: **Ceanothus gloriosus** var. **exaltatus**.
    NIPOMO: **Ceanothus impressus** var. **nipomensis**.
    PALMER: **Ceanothus palmeri**.
    POINT REYES: **Ceanothus gloriosus**.
    ROWE: **Ceanothus papillosus** var. **roweanus**.
    SAN DIEGO: **Ceanothus cyaneus**.
    SANTA BARBARA: **Ceanothus impressus**.
    VARNISH-LEAF: **Ceanothus velutinus**.
    VEITCH: **Ceanothus × veitchianus**.
    VINE HILL: **Ceanothus foliosus** var. **vineatus**.
    WART-LEAF: **Ceanothus papillosus**.
    WARTSTEM: **Ceanothus verrucosus**.
    WAVY-LEAF: **Ceanothus foliosus**.
    WHITE MONTEREY: **Ceanothus rigidus 'Albus'**.
    WOOLLY-LEAF: **Ceanothus tomentosus** var. **olivaceus**.
Ceanothus americanus L. NEW JERSEY TEA.
Ceanothus arboreus Greene. FELTLEAF CEANOTHUS. CATALINA MOUNTAIN LILAC.
Ceanothus arboreus 'Owlswood Blue'. See **C. 'Owlswood Blue'**.
Ceanothus arboreus 'Ray Hartman'. See **C. 'Ray Hartman'**.
Ceanothus 'Bamico'.
Ceanothus 'Blue Buttons'.
Ceanothus 'Blue Cloud'.
Ceanothus 'Blue Cushion'.

Ceanothus 'Blue Sky'.
Ceanothus 'Blue Wisp'.
Ceanothus 'Burtonensis'. [**C. impressus** × **C. thyrsiflorus**].
Ceanothus 'Concha'.
Ceanothus confusus J. T. Howell. See **C. divergens** subsp. **confusus**.
Ceanothus 'Consuelo'.
Ceanothus cordulatus Kell. SNOWBUSH. MOUNTAIN WHITEHORN.
Ceanothus crassifolius Torr. HOARYLEAF CEANOTHUS.
Ceanothus cuneatus (Hook.) Nutt. BUCKBRUSH.
Ceanothus cyaneus Eastw. SAN DIEGO CEANOTHUS.
Ceanothus cyaneus 'Sierra Blue'. See **C. 'Sierra Blue'**.
Ceanothus 'Dark Star'.
Ceanothus × delilianus Spach. [**C. americanus** × **C. coeruleus** Lag.]. FRENCH HYBRID CEANOTHUS.
Ceanothus dentatus Torr. & Gray.
Ceanothus divergens Parry. CALISTOGA CEANOTHUS. MOUNT ST. HELENA CEANOTHUS.
Ceanothus divergens subsp. **confusus** (J. T. Howell) Abrams. Syn.: *C. confusus*.
Ceanothus diversifolius Kell. PINE MAT.
Ceanothus 'Eleanor Taylor'.
Ceanothus 'Emily Brown'.
Ceanothus 'Far Horizons'.
Ceanothus fendleri Gray.
Ceanothus foliosus Parry. WAVY-LEAF CEANOTHUS.
Ceanothus foliosus var. **vineatus** McMinn. VINE HILL CEANOTHUS.
Ceanothus fresnensis Dudl. ex Abrams.
Ceanothus 'Frosty Blue'.
Ceanothus 'Gentian Plume'.
Ceanothus 'Gloire de Versailles'.
Ceanothus gloriosus J. T. Howell. POINT REYES CREEPER. POINT REYES CEANOTHUS.
Ceanothus gloriosus 'Anchor Bay'.
Ceanothus gloriosus var. **exaltatus** J. T. Howell. NAVARRO CEANOTHUS.
Ceanothus gloriosus var. *exaltatus* 'Emily Brown'. See **C. 'Emily Brown'**.
Ceanothus gloriosus var. **porrectus** J. T. Howell. MOUNT VISION CEANOTHUS.
Ceanothus gloriosus 'Santa Ana'. See **C. 'Santa Ana'**.
Ceanothus gloriosus 'Tuttle'.
Ceanothus greggii Gray var. **perplexans** (Trel.) Jeps. CUPLEAF CEANOTHUS.
Ceanothus griseus (Trel.) McMinn. CARMEL CEANOTHUS.
Ceanothus griseus 'Bamico'. See **C. 'Bamico'**.
Ceanothus griseus var. **horizontalis** McMinn. CARMEL CREEPER.
Ceanothus griseus var. **horizontalis** 'Compacta'?
Ceanothus griseus var. **horizontalis** 'Hurricane Point'.
Ceanothus griseus var. **horizontalis** 'Yankee Point'.
Ceanothus griseus 'Louis Edmunds'.
Ceanothus griseus 'Santa Ana'. See **C. 'Santa Ana'**.
Ceanothus hearstiorum Hoover & Roof. HEARST CEANOTHUS.
Ceanothus horizontalis. Listed name for **C. griseus** var. **horizontalis**.
Ceanothus impressus Trel. SANTA BARBARA CEANOTHUS.
Ceanothus impressus 'Mountain Haze'. See **C. 'Mountain Haze'**.
Ceanothus impressus var. **nipomensis** McMinn. NIPOMO CEANOTHUS.
Ceanothus impressus 'Puget Blue'. See **C. 'Puget Blue'**.
Ceanothus incanus Torr. & Gray. COAST WHITETHORN.
Ceanothus insularis Eastw.
Ceanothus integerrimus Hook. & Arn. DEERBRUSH.
Ceanothus 'J. K. Lawrence'.
Ceanothus 'Joyce Coulter'.
Ceanothus 'Julia Phelps'.

Ceanothus 'Junior'.
Ceanothus leucodermis Greene. CHAPARRAL WHITETHORN.
Ceanothus 'Louis Edmunds'. See **C. griseus 'Louis Edmunds'**.
Ceanothus maritimus Hoover. MARITIME CEANOTHUS.
Ceanothus 'Mary Lake'.
Ceanothus 'Mary Simpson'.
Ceanothus masonii McMinn. BOLINAS RIDGE CEANOTHUS.
Ceanothus megacarpus Nutt. BIG POD CEANOTHUS.
Ceanothus 'Mills Glory'.
Ceanothus 'Mountain Haze'.
Ceanothus nipomensis. Listed name for **C. impressus** var. **nipomensis**.
Ceanothus oliganthus Nutt.
Ceanothus 'Owlswood Blue'. [**C. arboreus** × **C.** sp.].
Ceanothus palmeri Trel. PALMER CEANOTHUS.
Ceanothus papillosus Torr. & Gray. WART-LEAF CEANOTHUS.
Ceanothus papillosus var. *roweanus* McMinn. MOUNT TRANQUILLON CEANOTHUS. ROWE CEANOTHUS.
Ceanothus parryi Trel.
Ceanothus prostratus Benth. SQUAW CARPET. MAHALA MATS.
Ceanothus prostratus var. *occidentalis* McMinn.
Ceanothus 'Puget Blue'.
Ceanothus pumilus Greene. SISKIYOU-MAT.
Ceanothus pumilus 'French Hill'?
Ceanothus purpureus Jeps. HOLLYLEAF CEANOTHUS.
Ceanothus ramulosus (Greene) McMinn. COAST CEANOTHUS.
Ceanothus ramulosus var. *fascicularis* McMinn. LOMPOC CEANOTHUS.
Ceanothus 'Ray Hartman'. [?**C. arboreus** × **C. griseus**].
Ceanothus repens. Listed name for **C. thyrsiflorus** var. **repens**.
Ceanothus rigidus Nutt. MONTEREY CEANOTHUS.
Ceanothus rigidus 'Albus'. WHITE MONTEREY CEANOTHUS.
Ceanothus rigidus 'Snowball'.
Ceanothus roweanus. Listed name for **C. papillosus** var. **roweanus**.
Ceanothus 'Royal Blue'.
Ceanothus 'Santa Ana'.
Ceanothus 'Sierra Blue'.
Ceanothus sorediatus Hook. & Arn. JIM-BRUSH.
Ceanothus spinosus Nutt. RED-HEART. GREEN BARK CEANOTHUS.
Ceanothus 'Theodore Payne'.
Ceanothus thyrsiflorus Eschsch. BLUE BRUSH. BLUEBLOSSOM.
Ceanothus thyrsiflorus 'Compactus'.
Ceanothus thyrsiflorus 'Millerton Point'.
Ceanothus thyrsiflorus var. *repens* McMinn. CREEPING BLUEBLOSSOM.
Ceanothus thyrsiflorus 'Skylark'.
Ceanothus tomentosus Parry var. *olivaceus* Jeps. WOOLLY-LEAF CEANOTHUS.
Ceanothus × *veitchianus* Hook. [**C. griseus** × **C. rigidus**]. VEITCH CEANOTHUS.
Ceanothus velutinus Dougl. ex Hook. TOBACCO BRUSH. VARNISH-LEAF CEANOTHUS. BUCKBRUSH.
Ceanothus velutinus var. *laevigatus* (Hook.) Torr. & Gray.
Ceanothus verrucosus Nutt. WARTSTEM CEANOTHUS.
Ceanothus vineatus. Listed name for **C. foliosus** var. **vineatus**.
Ceanothus 'Yankee Point'. See **C. griseus** var. **horizontalis** 'Yankee Point'.
CEDAR, AFRICAN: **Widdringtonia schwarzii**.
  ALASKA: **Chamaecyparis nootkatensis**.
  ATLAS MOUNTAIN: **Cedrus atlantica**.
  BLUE ATLAS: **Cedrus atlantica 'Glauca'**.
  BURK EASTERN RED: **Juniperus virginiana 'Burkii'**.
  CANOE: **Thuja plicata**.
  CHILEAN INCENSE: **Austrocedrus chilensis**.
  CHINESE: **Cedrela sinensis**.
  COMPACT DEODAR: **Cedrus deodara 'Compacta'**.
  CREEPING DEODAR: **Cedrus deodara 'Prostrata'**.
  CYPRUS: **Cedrus brevifolia**.
  DEERHORN: **Thujopsis dolabrata**.
  DEODAR: **Cedrus deodara**.
  DWARF ALASKA: **Chamaecyparis nootkatensis 'Compacta'**.
  EASTERN RED: **Juniperus virginiana**.
  GIANT: **Thuja plicata**.
  GOLDEN ATLAS: **Cedrus atlantica 'Aurea'**.
  HIBA: **Thujopsis dolabrata**.
  HIMALAYAN: **Cedrus deodara**.
  INCENSE: **Calocedrus decurrens**.
  JAPANESE: **Crytomeria japonica**.
  NORTHERN WHITE: **Thuja occidentalis**.
  PINK: **Acrocarpus fraxinifolius**.
  PLUME: **Cryptomeria japonica 'Elegans'**.
  PORT ORFORD: **Chamaecyparis lawsoniana**.
  RED: **Juniperus virginiana**.
  RED: **Thuja plicata**.
  SILVER RED: **Juniperus virginiana 'Glauca'**.
  WEEPING DEODAR: **Cedrus deodara 'Pendula'**.
  WESTERN RED: **Juniperus occidentalis**.
  WESTERN RED: **Thuja plicata**.
  WHITE: **Chamaecyparis thyoides**.
CEDAR-OF-LEBANON: **Cedrus libani**.
Cedrela fissilis Vell.
Cedrela sinensis Juss. CHINESE CEDAR.
Cedrus atlantica (Endl.) Manetti ex Carrière. ATLAS MOUNTAIN CEDAR.
Cedrus atlantica 'Argentea'.
Cedrus atlantica 'Aurea'. GOLDEN ATLAS CEDAR.
Cedrus atlantica 'Fastigiata'.
Cedrus atlantica 'Glauca'. BLUE ATLAS CEDAR.
Cedrus atlantica 'Glauca Pendula'.
Cedrus atlantica 'Pendula'.
Cedrus brevifolia (Hook. f.) Dode. CYPRUS CEDAR.
Cedrus deodara (D. Don) G. Don. HIMALAYAN CEDAR. CALIFORNIA CHRISTMAS TREE. DEODAR CEDAR.
Cedrus deodara 'Aurea'.
Cedrus deodara 'Aurea Wells'.
Cedrus deodara 'Compacta'. COMPACT DEODAR CEDAR.
Cedrus deodara 'Descanso Dwarf'.
Cedrus deodara 'Pendula'. WEEPING DEODAR CEDAR.
Cedrus deodara 'Prostrata'. CREEPING DEODAR CEDAR.
Cedrus libanensis Juss. ex Mirb. See **C. libani**.
Cedrus libani A. Rich. Syn.: *C. libanotica. C. libanensis.* CEDAR-OF-LEBANON.
Cedrus libani 'Nana'.
Cedrus libani 'Pendula'.
Cedrus libani 'Sargentii'.
Cedrus libani subsp. *stenocoma* (O. Schwarz) Davis.
Cedrus libanotica Link. See **C. libani**.
Celastrus angulatus Maxim.
Celastrus loeseneri Rehd. & Wils. See **C. rosthornianus**.
Celastrus orbiculatus Thunb. ORIENTAL BITTERSWEET.
Celastrus rosthornianus Loes. Syn.: *C. loeseneri.* CHINESE BITTERSWEET.
Celastrus scandens L. AMERICAN BITTERSWEET.
Celtis australis L. EUROPEAN HACKBERRY. MEDITERRANEAN HACKBERRY.
Celtis douglasii Planch. See **C. reticulata**.
Celtis laevigata Willd. MISSISSIPPI HACKBERRY. SUGARBERRY.
Celtis occidentalis L. HACKBERRY.
Celtis reticulata Torr. Syn.: *C. douglasii.* WESTERN HACKBERRY.
Celtis sinensis Pers. CHINESE HACKBERRY.
CENIZA: **Leucophyllum frutescens**.

CENTAUREA, VELVET: **Centaurea gymnocarpa.**
Centaurea cineraria L. DUSTY-MILLER.
Centaurea gymnocarpa Moris & De Notaris. VELVET CENTAUREA.
CENTIPEDE PLANT: **Homalocladium platycladum.**
CENTURY PLANT: **Agave americana.**
Cephalanthus occidentalis L. BUTTONWILLOW.
Cephalotaxus drupacea Siebold & Zucc. See **C. harringtonia** var. **drupacea.**
Cephalotaxus fortunei Hook. CHINESE PLUM YEW.
Cephalotaxus harringtonia (D. Don) K. Koch. HARRINGTON PLUM YEW. KOREAN YEW. KOREAN PLUM YEW.
Cephalotaxus harringtonia var. *drupacea* (Siebold & Zucc.) Koidz. Syn.: *C. drupacea.* JAPANESE PLUM YEW.
Cephalotaxus harringtonia 'Fastigiata'.
Ceratonia siliqua L. ST. JOHN'S-BREAD. CAROB. ALGARROBA.
Ceratostigma griffithii C. B. Clarke. BURMESE PLUMBAGO.
Ceratostigma larpentiae. Listed name for **C. plumbaginoides.**
Ceratostigma plumbaginoides Bunge. Syn.: *Plumbago larpentiae.* DWARF PLUMBAGO.
Ceratostigma willmottianum Stapf. CHINESE PLUMBAGO.
Ceratozamia latifolia Miq. See **C. mexicana** var. **latifolia.**
Ceratozamia longifolia Miq. See **C. mexicana** var. **longifolia.**
Ceratozamia mexicana Brongn. MEXICAN HORNCONE.
Ceratozamia mexicana var. **latifolia** (Miq.) J. Schuster. Syn.: *C. latifolia.*
Ceratozamia mexicana var. **longifolia** (Miq.) J. Schuster. Syn.: *C. longifolia.*
Cercidiphyllum japonicum Siebold & Zucc. ex J. Hoffm. & H. Schult. KATSURA TREE.
Cercidiphyllum japonicum var. *magnificum* Nakai. See **C. magnificum.**
Cercidiphyllum magnificum (Nakai) Nakai. Syn.: *C. japonicum* var. *magnificum.*
Cercidium floridum Benth. ex Gray. BLUE PALO VERDE. PALO VERDE.
Cercidium microphyllum (Torr.) Rose & I. M. Johnst. LITTLELEAF PALO VERDE.
Cercis canadensis L. EASTERN REDBUD.
Cercis canadensis 'Alba'.
Cercis canadensis 'Flame'.
Cercis canadensis 'Forest Pansy'.
Cercis canadensis 'Oklahoma'. See **C. reniformis 'Oklahoma'.**
Cercis canadensis 'Plena'.
Cercis canadensis 'Ruby Atkinson'.
Cercis chinensis Bunge. CHINESE REDBUD.
Cercis occidentalis Torr. ex Gray. CALIFORNIA REDBUD. WESTERN REDBUD. JUDAS TREE.
Cercis 'Oklahoma'. See **C. reniformis 'Oklahoma'.**
Cercis reniformis Engelm. ex Coult.
Cercis reniformis 'Oklahoma'. OKLAHOMA REDBUD.
Cercis siliquastrum L. JUDAS TREE.
Cercis siliquastrum 'Alba'. WHITE JUDAS TREE.
Cercocarpus alnifolius Rydb. See **C. betuloides** var. **blancheae.**
Cercocarpus betuloides Nutt. ex Torr. & Gray. Syn.: *C. betuloides* var. *multiflorus.* MOUNTAIN MAHOGANY. HARDTACK.
Cercocarpus betuloides var. **blancheae** (Schneid.) Little. Syn.: *C. alnifolius.* ALDER-LEAF MOUNTAIN MAHOGANY.
Cercocarpus betuloides var. *multiflorus* Jeps. See **C. betuloides.**
Cercocarpus betuloides var. **traskiae** (Eastw.) Dunkle. Syn.: *C. traskiae.* CATALINA MOUNTAIN MAHOGANY.
Cercocarpus ledifolius Nutt. DESERT MOUNTAIN MAHOGANY. CURL-LEAF MOUNTAIN MAHOGANY.
Cercocarpus minutiflorus Abrams. SAN DIEGO MOUNTAIN MAHOGANY.
Cercocarpus traskiae Eastw. See **C. betuloides** var. **traskiae.**

Cereus peruvianus (L.) Mill. PERUVIAN APPLE. APPLE CACTUS.
Cereus peruvianus 'Monstrosus'. GIANT-CLUB. CURIOSITY PLANT.
CERIMAN: **Monstera deliciosa.**
Ceropegia woodii Schlechter. ROSARY VINE.
CESTRUM, ORANGE: **Cestrum aurantiacum.**
Cestrum aurantiacum Lindl. ORANGE CESTRUM.
Cestrum diurnum L. DAY-BLOOMING JESSAMINE.
Cestrum elegans (Brongn.) Schlechtend. Syn.: *C. purpureum.*
Cestrum elegans 'Smithii'.
Cestrum fasciculatum (Schlechtend.) Miers.
Cestrum 'Newellii'.
Cestrum nocturnum L. NIGHT-BLOOMING JESSAMINE.
Cestrum parqui L'Her. WILLOW-LEAVED JESSAMINE.
Cestrum purpureum (Lindl.) Standl. See **C. elegans.**
Chaenomeles 'Afterglow'. See **C. × vilminiana 'Afterglow'.**
Chaenomeles 'Alba'?
Chaenomeles 'Apple Blossom'. See **C. speciosa 'Apple Blossom'.**
Chaenomeles 'Aurora'. See **C. speciosa 'Aurora'.**
Chaenomeles 'Blood Red'. See **C. speciosa 'Blood Red'.**
Chaenomeles 'Bonfire'. See **C. speciosa 'Bonfire'.**
Chaenomeles × californica W. B. Clarke ex C. Weber [**C. cathayensis** × **C. × superba**].
Chaenomeles × californica 'Clarke's Giant Red'.
Chaenomeles × californica 'Enchantress'.
Chaenomeles × californica 'Flamingo'.
Chaenomeles × californica 'Pink Beauty'.
Chaenomeles × californica 'Rosy Morn'.
Chaenomeles 'Cameo'. See **C. × superba 'Cameo'.**
Chaenomeles 'Candida'. See **C. speciosa 'Candida'.**
Chaenomeles 'Cardinalis'. See **C. speciosa 'Cardinalis'.**
Chaenomeles cathayensis (Hemsl.) Schneid.
Chaenomeles 'Charming'. See **C. × superba 'Charming'.**
Chaenomeles 'Clarke's Giant'. See **C. japonica 'Clarke's Giant'.**
Chaenomeles 'Clarke's Giant Red'. See **C. × californica 'Clarke's Giant Red'.**
Chaenomeles × clarkiana C. Weber. [**C. cathayensis** × **C. japonica**].
Chaenomeles × clarkiana 'Minerva'.
Chaenomeles 'Contorta'. See **C. speciosa 'Contorta'.**
Chaenomeles 'Coral Beauty'. See **C. × superba 'Coral Beauty'.**
Chaenomeles 'Coral Glow'. See **C. × superba 'Corallina'.**
Chaenomeles 'Coral Sea'. See **C. × superba 'Coral Sea'.**
Chaenomeles 'Corallina'. See **C. × superba 'Corallina'.**
Chaenomeles 'Crimson and Gold'. See **C. × superba 'Crimson and Gold'.**
Chaenomeles 'Enchantress'. See **C. × californica 'Enchantress'.**
Chaenomeles 'Falconnet Charlet'. See **C. speciosa 'Falconnet Charlet'.**
Chaenomeles 'Flamingo'. See **C. × californica 'Flamingo'.**
Chaenomeles 'Glowing-Ember'. See **C. × superba 'Glowing-Ember'.**
Chaenomeles 'Hollandia'. See **C. × superba 'Hollandia'.**
Chaenomeles japonica (Thunb.) Lindl. ex Spach. Syn.: *Cydonia japonica. Pyrus japonica.* JAPANESE FLOWERING QUINCE.
Chaenomeles japonica 'Clarke's Giant'.
Chaenomeles japonica 'Minerva'. See **C. × clarkiana 'Minerva'.**
Chaenomeles 'Jet Trail'. See **C. × superba 'Jet Trail'.**
Chaenomeles 'Knap Hill'. See **C. × superba 'Knap Hill Scarlet'.**
Chaenomeles lagenaria (Loisel.) G. Koidz. See **C. speciosa.**
Chaenomeles lagenaria 'Rubra'. See **C. speciosa 'Rubra'.**
Chaenomeles 'Minerva'. See **C. × clarkiana 'Minerva'.**
Chaenomeles 'Mount Shasta'. See **C. × superba 'Mount Shasta'.**
Chaenomeles 'Nivalis'. See **C. speciosa 'Nivalis'.**
Chaenomeles 'Pink Beauty'. See **C. × californica 'Pink Beauty'.**
Chaenomeles 'Pink Lady'. See **C. × superba 'Pink Lady'.**

Chaenomeles 'Red Chief'. See **C.** × ***superba* 'Red Chief'**.
Chaenomeles 'Red Ripples'. See **C.** ***speciosa* 'Red Ruffles'**.
Chaenomeles 'Red Ruffles'. See **C.** ***speciosa* 'Red Ruffles'**.
Chaenomeles 'Rosy Morn'. See **C.** × ***californica* 'Rosy Morn'**.
Chaenomeles 'Rowallane'. See **C.** × ***superba* 'Rowallane'**.
Chaenomeles 'Roxana Foster'. See **C.** × ***superba* 'Roxana Foster'**.
Chaenomeles 'Rubra'. See **C.** ***speciosa* 'Rubra'**.
Chaenomeles 'Simonii'. See **C.** ***speciosa* 'Simonii'**.
Chaenomeles sinensis (Dum.-Cours.) Schneid. See **Cydonia sinensis.**
Chaenomeles 'Snow'. See **C.** ***speciosa* 'Snow'**.
Chaenomeles speciosa (Sweet) Nakai. Syn.: *C. lagenaria*. FLOWERING QUINCE. JAPANESE QUINCE.
Chaenomeles speciosa 'Apple Blossom'.
Chaenomeles speciosa 'Aurora'.
Chaenomeles speciosa 'Blood Red'.
Chaenomeles speciosa 'Bonfire'.
Chaenomeles speciosa 'Cameo'. See **C.** × ***superba* 'Cameo'**.
Chaenomeles speciosa 'Candida'.
Chaenomeles speciosa 'Candidissima'.
Chaenomeles speciosa 'Cardinalis'.
Chaenomeles speciosa 'Contorta'.
Chaenomeles speciosa 'Crimson Beauty'. See **C.** × ***superba* 'Crimson Beauty'**.
Chaenomeles speciosa 'Falconnet Charlet'.
Chaenomeles speciosa 'Nivalis'.
Chaenomeles speciosa 'Red Ruffles'.
Chaenomeles speciosa 'Rubra'.
Chaenomeles speciosa 'Simonii'.
Chaenomeles speciosa 'Snow'.
Chaenomeles speciosa 'Toyo-Nishiki'.
Chaenomeles 'Stanford Red'. See **C.** × ***superba* 'Stanford Red'**.
Chaenomeles × **superba** (Frahm) Rehd. [*C. japonica* × *C. speciosa*].
Chaenomeles × ***superba* 'Cameo'.**
Chaenomeles × ***superba* 'Charming'.**
Chaenomeles × ***superba* 'Coral Beauty'.**
Chaenomeles × ***superba* 'Coral Sea'.**
Chaenomeles × ***superba* 'Corallina'.**
Chaenomeles × ***superba* 'Crimson and Gold'.**
Chaenomeles × ***superba* 'Crimson Beauty'.**
Chaenomeles × ***superba* 'Glowing-Ember'.**
Chaenomeles × ***superba* 'Hollandia'.**
Chaenomeles × ***superba* 'Jet Trail'.**
Chaenomeles × ***superba* 'Knap Hill Scarlet'.**
Chaenomeles × ***superba* 'Mount Shasta'.**
Chaenomeles × ***superba* 'Pink Lady'.**
Chaenomeles × ***superba* 'Red Chief'.**
Chaenomeles × ***superba* 'Rowallane'.**
Chaenomeles × ***superba* 'Roxana Foster'.**
Chaenomeles × ***superba* 'Stanford Red'.**
Chaenomeles × ***superba* 'Texas Scarlet'.**
Chaenomeles 'Texas Scarlet'. See **C.** × ***superba* 'Texas Scarlet'**.
Chaenomeles 'Thornless Pink'. See **C.** × ***superba* 'Pink Lady'**.
Chaenomeles 'Toyo-Nishiki'. See **C.** ***speciosa* 'Toyo-Nishiki'**.
Chaenomeles × **vilmoriniana** C. Weber. [*C. cathayensis* × *C. speciosa*].
Chaenomeles × ***vilmoriniana* 'Afterglow'.**
CHALICE VINE: **Solandra maxima.**
Chamaecyparis 'Allumii'. See **C.** ***lawsoniana* 'Allumii'**.
Chamaecyparis 'Azurea'. See **C.** ***lawsoniana* 'Azurea'**.
Chamaecyparis 'Blue Haven'?
Chamaecyparis 'Columnaris Glauca'. See **C.** ***lawsoniana* 'Columnaris'**.
Chamaecyparis 'Cyanoviridis'. See **C.** ***pisifera* 'Boulevard'**.
Chamaecyparis 'Ellwoodii'. See **C.** ***lawsoniana* 'Ellwoodii'**.
Chamaecyparis 'Ericoides'. See **C.** ***lawsoniana* 'Ericoides'** or **C.** ***obtusa* 'Ericoides'** or **C.** ***pisifera* 'Ericoides'** or **C.** ***nootkatensis* 'Ericoides'**.
Chamaecyparis 'Filifera Aurea'. See **C.** ***pisifera* 'Filifera Aurea'**.
Chamaecyparis 'Filifera Aurea Nana'. See **C.** ***pisifera* 'Filifera Aurea Nana'**.
Chamaecyparis 'Fraseri'. See **C.** ***lawsoniana* 'Fraseri'**.
Chamaecyparis funebris (Endl.) Franco. Syn.: *Cupressus funebris*. MOURNING CYPRESS.
Chamaecyparis 'Knowefieldensis'. See **C.** ***lawsoniana* 'Knowefieldensis'**.
Chamaecyparis lawsoniana (A. Murr.) Parl. LAWSON CYPRESS. PORT ORFORD CEDAR.
Chamaecyparis lawsoniana 'Allumii'. PYRAMIDAL BLUE CYPRESS. BLUE LAWSON CYPRESS. SCARAB CYPRESS.
Chamaecyparis lawsoniana 'Azurea'. AZURE CYPRESS.
Chamaecyparis lawsoniana 'Columnaris'.
Chamaecyparis lawsoniana 'Columnaris Glauca'. See **C.** ***lawsoniana* 'Columnaris'**.
Chamaecyparis lawsoniana 'Compacta'.
Chamaecyparis lawsoniana 'Crippsii Aurea'. See **C.** ***obtusa* 'Crippsii'**.
Chamaecyparis lawsoniana 'Elegantissima'.
Chamaecyparis lawsoniana 'Ellwoodii'. ELLWOOD CYPRESS.
Chamaecyparis lawsoniana 'Ericoides'.
Chamaecyparis lawsoniana 'Fletcheri'. FLETCHER CYPRESS.
Chamaecyparis lawsoniana 'Forsteckensis'. FORSTECK CYPRESS.
Chamaecyparis lawsoniana 'Forsteckiana'. See **C.** ***lawsoniana* 'Forsteckensis'**.
Chamaecyparis lawsoniana 'Fraseri'.
Chamaecyparis lawsoniana 'Grandi'.
Chamaecyparis lawsoniana 'Knowefieldensis'. KNOWEFIELD CYPRESS.
Chamaecyparis lawsoniana 'Lutea'. GOLDEN LAWSON CYPRESS.
Chamaecyparis lawsoniana 'Lycopodioides'.
Chamaecyparis lawsoniana 'Minima'. DWARF LAWSON CYPRESS.
Chamaecyparis lawsoniana 'Minima Glauca'. LITTLE BLUE CYPRESS.
Chamaecyparis lawsoniana 'Moerheimii'.
Chamaecyparis lawsoniana 'Nana Argentea'.
Chamaecyparis lawsoniana 'Nestoides'. NEST CYPRESS.
Chamaecyparis lawsoniana 'Nidiformis'. Syn.: *C.* × *nidifera*. NEST CYPRESS.
Chamaecyparis lawsoniana 'Pendula'. WEEPING CYPRESS.
Chamaecyparis lawsoniana 'Silver Ball'?
Chamaecyparis lawsoniana 'Silver Queen'. SILVER QUEEN CYPRESS.
Chamaecyparis lawsoniana 'Stewartii'. STEWART GOLDEN CYPRESS.
Chamaecyparis lawsoniana 'Westermannii'. WESTERMANN CYPRESS.
Chamaecyparis lawsoniana 'Wisselii'. WISSEL CYPRESS.
Chamaecyparis lawsoniana 'Yellow Transparent'.
Chamaecyparis 'Nestoides'. See **C.** ***lawsoniana* 'Nestoides'**.
Chamaecyparis × **nidifera** (Nichols.) Hornibr. See **C.** ***lawsoniana* 'Nidiformis'**.
Chamaecyparis 'Nidiformis'. See **C.** ***lawsoniana* 'Nidiformis'**.
Chamaecyparis nootkatensis (D. Don) Spach. ALASKA CEDAR. ALASKA CYPRESS. NOOTKA CYPRESS.
Chamaecyparis nootkatensis 'Compacta'. DWARF NOOTKA CYPRESS. DWARF ALASKA CEDAR. DWARF ALASKA CYPRESS.
Chamaecyparis nootkatensis 'Compacta Glauca'.
Chamaecyparis nootkatensis 'Ericoides'.
Chamaecyparis nootkatensis 'Pendula'. See **C.** ***nootkatensis***.
Chamaecyparis obtusa (Siebold & Zucc.) Endl. Syn.: *Cupressus obtusa*. HINOKI. HINOKI CYPRESS.

Chamaecyparis obtusa **'Aurea'**. GOLDEN HINOKI CYPRESS. GOLDEN HINOKI.
Chamaecyparis obtusa 'Aurea Crippsii'. See **C. obtusa 'Crippsii'**.
Chamaecyparis obtusa 'Aurea Nana'. See **C. obtusa 'Nana Aurea'**.
***Chamaecyparis obtusa* 'Compacta'**.
***Chamaecyparis obtusa* 'Coralliformis'**.
***Chamaecyparis obtusa* 'Crippsii'**. CRIPPS GOLDEN CYPRESS.
***Chamaecyparis obtusa* 'Ericoides'**.
***Chamaecyparis obtusa* 'Filicoides'**. FERNSPRAY CYPRESS.
Chamaecyparis obtusa* var. *formosana (Hayata) Rehd.
***Chamaecyparis obtusa* 'Gracilis'**. SLENDER HINOKI CYPRESS.
Chamaecyparis obtusa 'Gracilis Nana'. See **C. obtusa 'Nana Gracilis'**.
***Chamaecyparis obtusa* 'Graciosa'**.
***Chamaecyparis obtusa* 'Juniperoides'**.
***Chamaecyparis obtusa* 'Kosteri'**.
***Chamaecyparis obtusa* 'Lycopodioides'**. CLUB-MOSS CYPRESS.
Chamaecyparis obtusa 'Lycopodioides Coralliformis'. See **C. obtusa 'Coralliformis'**.
***Chamaecyparis obtusa* 'Nana'**. DWARF HINOKI CYPRESS.
***Chamaecyparis obtusa* 'Nana Aurea'**. DWARF GOLDEN HINOKI CYPRESS. DWARF GOLDEN HINOKI.
***Chamaecyparis obtusa* 'Nana Gracilis'**.
Chamaecyparis obtusa 'Nana Kosteri'. See **C. obtusa 'Kosteri'**.
***Chamaecyparis obtusa* 'Prostrata'**.
***Chamaecyparis obtusa* 'Sanderi'**.
***Chamaecyparis obtusa* 'Tetragona'**.
Chamaecyparis obtusa 'Torulosa'. See **C. obtusa 'Coralliformis'**.
Chamaecyparis pisifera (Siebold & Zucc.) Endl. SAWARA RETINOSPORA. SAWARA CYPRESS.
Chamaecyparis pisifera 'Aurea Compacta Nana'?
***Chamaecyparis pisifera* 'Aurea Nana'**.
***Chamaecyparis pisifera* 'Boulevard'**. BOULEVARD CYPRESS. Plants called C. pisifera 'Cyanoviridis' belong here.
***Chamaecyparis pisifera* 'Compacta'**.
Chamaecyparis pisifera 'Compacta Nana'?
***Chamaecyparis pisifera* 'Compacta Variegata'**.
Chamaecyparis pisifera 'Compressa'. Plants so called may be **C. pisifera 'Plumosa Compressa'**.
Chamaecyparis pisifera 'Cyanoviridis'. See **C. pisifera 'Boulevard'**.
***Chamaecyparis pisifera* 'Ericoides'**.
***Chamaecyparis pisifera* 'Filifera'**. THREAD-BRANCH RETINOSPORA. THREAD-BRANCH CYPRESS.
***Chamaecyparis pisifera* 'Filifera Aurea'**. GOLDEN THREAD-BRANCH CYPRESS. GOLDEN THREAD-BRANCH RETINOSPORA.
***Chamaecyparis pisifera* 'Filifera Aurea Nana'**. DWARF GOLDEN THREAD-BRANCH CYPRESS. DWARF GOLDEN THREAD-BRANCH RETINOSPORA.
Chamaecyparis pisifera 'Glauca Compacta Nana'?
Chamaecyparis pisifera 'Juniperoides'?
Chamaecyparis pisifera 'Minima'. Listed name. Some plants so called may be **C. pisifera 'Squarrosa Minima'**.
Chamaecyparis pisifera 'Minima Aurea'?
Chamaecyparis pisifera 'Minima Variegata'?
***Chamaecyparis pisifera* 'Nana'**.
***Chamaecyparis pisifera* 'Nana Aureovariegata'**.
***Chamaecyparis pisifera* 'Nana Variegata'**.
***Chamaecyparis pisifera* 'Plumosa'**. PLUME CYPRESS.
***Chamaecyparis pisifera* 'Plumosa Aurea'**. GOLDEN PLUME CYPRESS.
***Chamaecyparis pisifera* 'Plumosa Compressa'**.
Chamaecyparis pisifera 'Plumosa Minima'? Some plants so called may be **C. pisifera 'Plumosa Nana'** or **C. pisifera 'Squarrosa Minima'**.

Chamaecyparis pisifera 'Plumosa Minima Aurea'. Plants so called may be **C. pisifera 'Plumosa Nana Aurea'**.
***Chamaecyparis pisifera* 'Plumosa Nana'**.
***Chamaecyparis pisifera* 'Plumosa Nana Aurea'**.
***Chamaecyparis pisifera* 'Plumosa Pygmaea'**.
Chamaecyparis pisifera 'Plumosa Pygmaea Aurea'?
Chamaecyparis pisifera 'Pygmaea'. Listed name. Some plants so called may be **C. pisifera 'Squarrosa Minima'** or **C. pisifera 'Plumosa Pygmaea'**.
Chamaecyparis pisifera 'Pygmaea Aurea'?
***Chamaecyparis pisifera* 'Squarrosa'**. MOSS CYPRESS.
***Chamaecyparis pisifera* 'Squarrosa Minima'**.
Chamaecyparis pisifera 'Squarrosa Nana'?
***Chamaecyparis pisifera* 'Squarrosa Sulphurea'**.
Chamaecyparis pisifera 'Squarrosa Veitchii'. See **C. pisifera 'Squarrosa'**.
***Chamaecyparis pisifera* 'Sulphurea'**.
Chamaecyparis pisifera 'Variegata'?
Chamaecyparis 'Plumosa Aurea'. See **C. pisifera 'Plumosa Aurea'**.
Chamaecyparis 'Pomona Dwarf'?
Chamaecyparis 'Silver Queen'. See **C. lawsoniana 'Silver Queen'**.
Chamaecyparis 'Squarrosa Nana'?
Chamaecyparis 'Squarrosa Cyanoviridis'. See **C. pisifera 'Boulevard'**.
Chamaecyparis 'Sulphurea'. Plants so called may be **C. pisifera 'Sulphurea'** or **C. pisifera 'Squarrosa Sulphurea'**.
Chamaecyparis thyoides (L.) BSP. Syn.: *Cupressus thyoides*. WHITE CEDAR.
***Chamaecyparis thyoides* 'Andelyensis'**.
***Chamaecyparis thyoides* 'Ericoides'**.
Chamaedaphne calyculata (L.) Moench. LEATHERLEAF.
Chamaedorea atrovirens Mart.
Chamaedorea cataractarum Mart. Some plants so called are **C. monostachys**.
Chamaedorea costaricana Oerst. COSTA RICAN PARLOR PALM.
Chamaedorea elatior Mart.
Chamaedorea elegans Mart. Syn.: *Collinia elegans*. PARLOR PALM. Plants called *Neanthe bella* belong here.
Chamaedorea ernesti-augusti H. Wendl.
Chamaedorea erumpens H. E. Moore. BAMBOO PALM.
Chamaedorea fragrans (Ruiz & Pav.) Mart.
Chamaedorea geonomiformis H. Wendl.
Chamaedorea glaucifolia H. Wendl.
Chamaedorea graminifolia H. Wendl. May not be cult. Plants so called may be misidentified.
Chamaedorea klotzschiana H. Wendl.
Chamaedorea metallica O. F. Cook ex H. E. Moore.
Chamaedorea microspadix Burret.
Chamaedorea monostachys Burret. Some plants called **C. cataractarum** belong here.
Chamaedorea oblongata Mart.
Chamaedorea radicalis Mart.
Chamaedorea sartorii Liebm.
Chamaedorea schippii Burret.
Chamaedorea seifrizii Burret.
Chamaedorea stolonifera H. Wendl.
Chamaedorea tenella H. Wendl.
Chamaedorea tepejilote Liebm.
Chamaelaucium. See **Chamelaucium**.
Chamaerops excelsa. Listed name for **Trachycarpus fortunei**.
Chamaerops fortunei Hook. See **Trachycarpus fortunei**.
Chamaerops humilis L. HAIR PALM. EUROPEAN FAN PALM. MEDITERRANEAN FAN PALM.

Chamaerops humilis var. ***elatior*** Gussone. Syn.: **C. humilis** var. *macrocarpa*. BIG-FRUIT FAN PALM.
Chamaerops humilis var. ***macrocarpa*** (Tineo ex Gussone) Becc. See **C. humilis** var. **elatior**.
Chamaerops martiana Wallich. See **Trachycarpus martianus**.
Chamelaucium ciliatum Desf.
Chamelaucium uncinatum Schauer. GERALDTON WAXFLOWER.
Chamelaucium uncinatum 'Vista'.
CHAMISE: **Adenostoma fasciculatum**.
CHASTE TREE: **Vitex agnus-castus**.
  BLUE: **Vitex agnus-castus**.
  CHINESE: **Vitex agnus-castus**.
  NEW ZEALAND: **Vitex lucens**.
CHECKERBERRY: **Gaultheria procumbens**.
Cheiranthus scoparius Brouss.
Cheiranthus semperflorens Schousboe.
CHENILLE PLANT: **Acalypha hispida**.
CHERIMOYA: **Annona cherimola**.
CHERRY, AKEBONO: **Prunus × yedoensis 'Akebono'**.
  AMANOGAWA: **Prunus serrulata 'Amanogawa'**.
  AMAYADORI: **Prunus serrulata 'Amayadori'**.
  AUSTRALIAN BUSH: **Syzygium paniculatum**.
  AUTUMN FLOWERING HIGAN: **Prunus subhirtella 'Autumnalis'**.
  BARBADOS: **Eugenia uniflora**.
  BARBADOS: **Malpighia glabra**.
  BASTARD: **Ehretia tinifolia**.
  BENI HIGAN: **Prunus subhirtella 'Rosea'**.
  BENI HOSHI: **Prunus serrulata 'Beni Hoshi'**.
  BIRCH BARK: **Prunus serrula**.
  BIRD: **Prunus avium**.
  BOTAN-ZAKURA: **Prunus serrulata 'Botan-zakura'**.
  BRIGHT RED BUSH: **Syzygium paniculatum 'Red Flame'**.
  CALIFORNIA: **Prunus ilicifolia**.
  CATALINA: **Prunus lyonii**.
  CHEAL'S WEEPING: Misapplied to **Prunus serrulata 'Kiku-shidare-zakura'**.
  CHRISTMAS: **Solanum pseudocapsicum**.
  CORNELIAN: **Cornus mas**.
  DOUBLE-FLOWERED MAZZARD: **Prunus avium 'Plena'**.
  DOUBLE WEEPING FLOWERING: **Prunus serrulata 'Kiku-shidare-zakura'**.
  DOUBLE WEEPING ROSEBUD: **Prunus subhirtella 'Pendula Plena Rosea'**.
  DOWNY: **Prunus tomentosa**.
  DWARF: **Prunus pumila**.
  DWARF BUSH: **Syzygium paniculatum 'Compactum'**.
  FALSE JERUSALEM: **Solanum diflorum**.
  FUGENZO: **Prunus serrulata 'Fugenzo'**.
  GREAT WHITE: **Prunus serrulata 'Tai Haku'**.
  HALLY JOLIVETTE: **Prunus 'Hally Jolivette'**.
  HIGAN: **Prunus subhirtella**.
  HI-ZAKURA: **Prunus serrulata 'Hi-zakura'**.
  HOLLY-LEAF: **Prunus ilicifolia**.
  JACQUEMONT: **Prunus jacquemontii**.
  JAPANESE: **Prunus × yedoensis**.
  JAPANESE FLOWERING: **Prunus serrulata**.
  JERUSALEM: **Solanum pseudocapsicum**.
  KIKU-SHIDARE: **Prunus serrulata 'Kiku-shidare-zakura'**.
  KIKU-SHIDARE-ZAKURA: **Prunus serrulata 'Kiku-shidare-zakura'**.
  KIRIN: **Prunus serrulata 'Kirin'**.
  KWANZAN: **Prunus serrulata 'Kwanzan'**.
  MANCHU: **Prunus tomentosa**.
  MIKURUMA-GAESHI: **Prunus serrulata 'Mikuruma-gaeshi'**.
  MOUNT FUJI: **Prunus serrulata 'Shirotae'**.
  NADEN: **Prunus serrulata 'Takasago'**.
  NANKING: **Prunus tomentosa**.
  NORTH JAPANESE HILL: **Prunus sargentii**.
  OJOCHIN: **Prunus serrulata 'Ojochin'**.
  ORIENTAL WEEPING: **Prunus serrulata 'Kiku-shidare-zakura'**.
  PIE: **Prunus cerasus**.
  PINK HIGAN: **Prunus subhirtella 'Rosea'**.
  PINK SPRING: **Prunus subhirtella 'Rosea'**.
  POTOMAC: **Prunus × yedoensis**.
  PURPLE-LEAF SAND: **Prunus × cistena**.
  ROSEBUD: **Prunus subhirtella**.
  SAND: **Prunus besseyi**.
  SAND: **Prunus depressa**.
  SAND: **Prunus pumila**.
  SARGENT: **Prunus sargentii**.
  SEKIYAMA: **Prunus serrulata 'Kwanzan'**.
  SHIROFUGEN: **Prunus serrulata 'Shirofugen'**.
  SHIRO-HIGAN: **Prunus subhirtella**.
  SHIRO-HIGAN: **Prunus subhirtella 'Rosea'**.
  SHOGETSU: **Prunus serrulata 'Shogetsu'**.
  SIEBOLD: **Prunus serrulata 'Takasago'**.
  SINGLE WEEPING: **Prunus subhirtella 'Pendula'**.
  SINGLE WEEPING ROSEBUD: **Prunus subhirtella 'Pendula'**.
  SOUR: **Prunus cerasus**.
  SPRING: **Prunus subhirtella**.
  SURINAM: **Eugenia uniflora**.
  SWEET: **Prunus avium**.
  TAI HAKU: **Prunus serrulata 'Tai Haku'**.
  TAIWAN: **Prunus campanulata**.
  TAKASAGO: **Prunus serrulata 'Takasago'**.
  TANKO SHINJU: **Prunus serrulata 'Tanko Shinju'**.
  UKON: **Prunus serrulata 'Ukon'**.
  WASHINO-O: **Prunus serrulata 'Washino-o'**.
  WESTERN SAND: **Prunus besseyi**.
  WHITCOMB: **Prunus subhirtella 'Whitcombii'**.
  YOSHINO: **Prunus × yedoensis**.
CHERRY-OF-THE-RIO GRANDE: **Myrciaria edulis**.
CHESTNUT, AMERICAN: **Castanea dentata**.
  CAPE: **Calodendrum capense**.
  CHINESE: **Castanea mollissima**.
  DWARF HORSE: **Aesculus parviflora**.
  EUROPEAN HORSE: **Aesculus hippocastanum**.
  HORSE: **Aesculus hippocastanum**.
  MORETON BAY: **Castanosperum australe**.
  RED HORSE: **Aesculus × carnea**.
  RUBY HORSE: **Aesculus × carnea 'Briotii'**.
  SPANISH: **Castanea sativa**.
CHESTNUT VINE: **Tetrastigma voinieranum**.
Chilianthus arboreus (L. f.) A. DC. See **Buddleia saligna**.
Chilopsis linearis (Cav.) Sweet. FLOWERING WILLOW. DESERT WILLOW.
Chimonanthus fragrans Lindl. See **C. praecox**.
Chimonanthus praecox (L.) Link. Syn.: **C. fragrans**. *Meratia praecox*. WINTERSWEET.
Chimonanthus praecox 'Grandiflorus'. WINTERSWEET.
Chimonobambusa falcata (Nees) Nakai. Syn.: *Arundinaria falcata*. *Bambusa falcata*. BLUE BAMBOO. NINGALA BAMBOO. SICKLE BAMBOO.
Chimonobambusa marmorea (Mitf.) Makino. Syn.: *Arundinaria marmorea*. MARBLED BAMBOO.
Chimonobambusa quadrangularis (Fenzi) Makino. SQUARE-STEM BAMBOO.
CHINABERRY: **Melia azedarach**.
CHINESE EVERGREEN: **Aglaonema modestum**.
CHINESE-LANTERN: **Abutilon hybridum**.
CHINQUAPIN, BUSH: **Chrysolepis sempervirens**.
  GIANT: **Chrysolepis chrysophylla**.
  GOLDEN: **Chrysolepis chrysophylla** var. **minor**.

CHINQUAPIN, continued
   SIERRA: **Chrysolepis sempervirens.**
Chionanthus retusus Lindl. & Paxt. CHINESE FRINGE TREE.
Chionanthus virginicus L. FRINGE TREE. OLD-MAN'S BEARD.
Chiranthodendron pentadactylon Larréategui. Syn.:
   C. platanoides. MONKEY-HAND TREE. MEXICAN HAND PLANT.
Chiranthodendron platanoides (Humb. & Bonpl.) Baill. See
   C. pentadactylon.
Chloranthus spicatus (Thunb.) Makino.
Choisya ternata HBK. MEXICAN ORANGE. MOCK ORANGE.
CHOKEBERRY, RED: **Aronia arbutifolia.**
Chonemorpha fragrans (Moon) Alston. Syn.: *C. macrophylla*.
   Plants called **Trachelospermum fragrans** belong here.
Chonemorpha macrophylla (Roxb.) G. Don. See **C. fragrans.**
Chorisia speciosa St.-Hil. FLOSS-SILK TREE.
Chorisia speciosa 'Majestic Beauty'.
Chorizema cordatum Lindl. HEART-LEAF FLAME PEA. AUSTRALIAN FLAME PEA.
Chorizema ilicifolium Labill. HOLLY FLAME PEA.
Chorizema varium Benth. FLAME MOCK ORANGE. BUSH FLAME PEA.
CHRISTMAS BERRY: **Heteromeles arbutifolia.**
CHRISTMAS-BERRY TREE: **Schinus terebinthifolius.**
CHRISTMAS-CHEER: **Sedum × rubrotinctum.**
CHRISTMAS FLOWER: **Euphorbia pulcherrima.**
CHRISTMAS TREE, CALIFORNIA: **Cedrus deodara.**
   NEW ZEALAND: **Metrosideros excelsus.**
   NEW ZEALAND: **Metrosideros robustus.**
Chrysalidocarpus cabadae H. E. Moore. CABADA PALM.
Chrysalidocarpus lutescens H. Wendl. Syn.: *Areca lutescens*. YELLOW PALM. CANE PALM. BUTTERFLY PALM.
Chrysalidocarpus madagascariensis Becc.
Chrysanthemum frutescens. L. MARGUERITE. PARIS DAISY.
Chrysanthemum frutescens 'Chrysaster'. BOSTON YELLOW DAISY.
Chrysanthemum lacustre Brot. PORTUGUESE DAISY.
Chrysanthemum maximum Ramond. Some plants so called are **C. × superbum.**
Chrysanthemum × superbum Bergmans ex J. Ingram. [?**C. lacustre × C. maximum**]. SHASTA DAISY. Some plants called **C. maximum** belong here.
Chrysolarix amabilis (J. Nels.) H. E. Moore. See **Pseudolarix kaempferi.**
Chrysolepis chrysophylla (Dougl. ex Hook.) Hjelmquist. Syn.: *Castanopsis chrysophylla*. GIANT CHINQUAPIN.
Chrysolepis chrysophylla var. *minor* (Benth.) Munz. Syn.: *Castanopsis chrysophylla* var. *minor*. GOLDEN CHINQUAPIN.
Chrysolepis sempervirens (Kell.) Hjelmquist. Syn.: *Castanopsis sempervirens*. SIERRA CHINQUAPIN. BUSH CHINQUAPIN.
Chrysothamnus nauseosus (Pall.) Britt. RABBITBRUSH.
Chrysothamnus nauseosus subsp. *hololeucus* (Gray) Hall & Clements.
Chrysothamnus nauseosus subsp. *leiospermus* (Gray) Hall & Clements.
CHUPEROSA: **Justicia californica.**
Chusquea coronalis Soderstrom & Calderon. Plants called *C. machrisii* belong here. MACHRIS BAMBOO.
Chusquea machrisii. Listed name for **C. coronalis.**
Cibotium chamissoi Kaulfuss. HAWAIIAN TREE FERN. Some plants so called are **C. glaucum.**
Cibotium glaucum (Sm.) Hook. & Arn. HAWAIIAN TREE FERN. Some plants called **C. chamissoi** belong here.
Cibotium schiedei Schlechtend. & Cham. MEXICAN TREE FERN.
Cienfuegosia hakeifolia (Giordano) Hochr. See **Alyogyne hakeifolia.**
CIGAR PLANT: **Cuphea ignea.**
   BLUE: **Cuphea hyssopifolia.**

CIGAR TREE: **Catalpa speciosa.**
Cineraria candidissima Hort. See **Senecio vira-vira.**
Cineraria maritima L. See **Senecio cineraria.**
Cineraria maritima var. *candidissma* Hort. See **Senecio vira-vira.**
Cinnamomum camphora (L.) J. Presl. Syn.: *Camphora officinalis. C. officinarum*. CAMPHOR TREE.
Cinnamomum daphnoides Siebold & Zucc.
Cinnamomum glanduliferum (Wallich) Nees. NEPAL CAMPHOR TREE.
CINQUEFOIL, SHRUBBY: **Potentilla fruticosa.**
   SHRUBBY: **Potentilla parvifolia.**
CIRIO: **Fouquieria columnaris.**
Cissus antarctica Vent. Syn.: *C. baudiniana. Vitis antarctica. V. baudiniana*. KANGAROO IVY. KANGAROO VINE. KANGAROO TREEBINE.
Cissus baudiniana Brouss. See **Cissus antarctica.**
Cissus capensis (Burm. f.) Willd. See **Rhoicissus capensis.**
Cissus discolor Blume.
Cissus hypoglauca Gray. Syn.: *Vitis hypoglauca*.
Cissus 'Mandaiana'. See: **C. rhombifolia** 'Mandaiana'.
Cissus rhombifolia Vahl. Syn.: *Vitis rhombifolia*. VENEZUELA TREEBINE. VENEZUELA GRAPE IVY. GRAPE IVY.
Cissus rhombifolia 'Mandaiana'.
Cissus striata Ruiz & Pav. Syn.: *Ampelopsis sempervirens*. MINIATURE GRAPE IVY.
Cissus voinierana (Baltet) Viala. See **Tetrastigma voinieranum.**
Cissus vomerensis. Listed name for **Tetrastigma voinieranum.**
Cistus × *aguilari* Pau. [**C. populifolius** × **C. ladanifer**].
Cistus × *aguilari* 'Maculatus'.
Cistus corbariensis Pourr. See **C. × hybridus.**
Cistus corsicus. Listed name for **C. incanus** subsp. **corsicus.**
Cistus crispus L.
Cistus × *cyprius* Lam. [**C. ladanifer** × **C. laurifolius**].
Cistus 'Doris Hibberson'.
Cistus formosus Curtis. See **Halimium lasianthinum** subsp. **formosum.**
Cistus × *hybridus* Pourr. [**C. salviifolius** × **C. populifolius**]. Syn.: *C. corbariensis*. WHITE ROCK ROSE.
Cistus incanus L. Syn.: *C. villosus*.
Cistus incanus subsp. *corsicus* (Loisel.) Heywood.
Cistus ladanifer L. Syn.: *C. ladanifer* forma *maculatus*. LAUDANUM. BROWN-EYED ROCK ROSE. SPOTTED ROCK ROSE.
Cistus ladanifer forma *maculatus* (Dunal) Dansereau. See **C. ladanifer.**
Cistus laurifolius L. LAUREL ROCK ROSE.
Cistus maculatus. Listed name for variants of those species in which petals are sometimes blotched, sometimes unblotched.
Cistus populifolius L.
Cistus × *purpureus* Lam. [**C. ladanifer** × **C. incanus**]. ORCHID-SPOT ROCK ROSE.
Cistus salviifolius L. SAGE-LEAF ROCK ROSE.
Cistus villosus L. See **C. incanus.**
× *Citrofortunella floridana* J. Ingram & H. E. Moore. [**Fortunella japonica** × **Citrus aurantiifolia**]. LIMEQUAT.
× *Citrofortunella floridana* 'Eustis'. EUSTIS LIMEQUAT.
× *Citrofortunella mitis* (Blanco) J. Ingram & H. E. Moore. [**Citrus reticulata** × **Fortunella ?margarita**]. Syn.: *Citrus mitis*. CALAMONDIN. Some plants called *Citrus madurensis* belong here.
CITRON: **Citrus medica.**
Citrus aurantiifolia (Christm.) Swingle. LIME. KEY LIME. MEXICAN LIME. WEST INDIAN LIME.
Citrus aurantiifolia 'Bearss'. BEARSS SEEDLESS LIME.
Citrus aurantiifolia 'Rangpur'. See **C. × limonia.**

Citrus aurantium L. SOUR ORANGE. SEVILLE ORANGE. BITTER ORANGE.
Citrus aurantium subsp. *bergamia* (Risso & Poit.) Wight & Arn. BERGAMOT ORANGE. Perhaps not cult. in Calif. Some plants called bergamot orange are *C. aurantium* 'Bouquet'.
Citrus aurantium 'Bouquet'. BOUQUET ORANGE. Some plants called bergamot orange belong here.
Citrus 'Bearss'. See **C. aurantiifolia 'Bearss'**.
Citrus 'Bouquet'. See **C. aurantium 'Bouquet'**.
Citrus 'Clementine'. See **C. reticulata 'Clementine'**.
Citrus 'Dancy'. See **C. reticulata 'Dancy'**.
Citrus 'Diller'. See **Citrus sinensis 'Diller'**.
Citrus 'Dweet'. See **C. × nobilis 'Dweet'**.
Citrus 'Etrog'. See **C. medica 'Etrog'**.
Citrus 'Eureka'. See **C. limon 'Eureka'**.
Citrus 'Eustis'. See **× Citrofortunella floridana 'Eustis'**.
Citrus grandis (L.) Osbeck. See **C. maxima**.
Citrus grandis 'Chandler'. See **C. maxima 'Chandler'**.
Citrus 'Hamlin'. See **C. sinensis 'Hamlin'**.
Citrus japonica Thunb. See **Fortunella japonica**.
Citrus 'King'. See **C. × nobilis 'King'**.
Citrus limon (L.) Burm. f. LEMON.
Citrus limon 'Eureka'. EUREKA LEMON.
Citrus limon 'Lisbon'. LISBON LEMON.
Citrus limon 'Meyer'. Syn.: *C. meyeri*. MEYER LEMON. DWARF LEMON. CHINESE DWARF LEMON.
Citrus limon 'Ponderosa'. PONDEROSA LEMON. GIANT LEMON.
Citrus limon 'Villafranca'. VILLAFRANCA LEMON.
Citrus × limonia Osbeck. [*C. limon* × *C.* ?*reticulata*]. RANGPUR. RANGPUR LIME. LEMANDARIN.
Citrus 'Lisbon'. See **C. limon 'Lisbon'**.
Citrus madurensis Lour. See **Fortunella japonica**. Some plants called *C. madurensis* are **× Citrofortunella mitis**.
Citrus 'Marsh'. See **C. × paradisi 'Marsh'**.
Citrus maxima (Burm.) Merrill. Syn.: *C. grandis*. POMELO. POMMELO. PUMELO. PUMMELO. SHADDOCK.
Citrus maxima 'Chandler'. CHANDLER'S PUMMELO.
Citrus medica L. CITRON.
Citrus medica 'Etrog'.
Citrus 'Meyer'. See **C. limon 'Meyer'**.
Citrus meyeri Tanaka. See **C. limon 'Meyer'**.
Citrus 'Minneola'. See **C. × tangelo 'Minneola'**.
Citrus mitis Blanco. See **× Citrofortunella mitis**.
Citrus nobilis Andr., not Lour. var. *deliciosa* (Ten.) Swingle. See **C. reticulata**.
Citrus × nobilis Lour. [*C. reticulata* × *C. sinensis*]. TANGOR.
Citrus × nobilis 'Dweet'.
Citrus × nobilis 'King'. KING MANDARIN. KING TANGOR.
Citrus × nobilis 'Temple'. TEMPLE ORANGE. TEMPLE TANGOR.
Citrus 'Otaheite'. See **Citrus otaitensis**.
Citrus otaitensis Risso & Poit. OTAHEITE ORANGE.
Citrus 'Owari'. See **C. reticulata 'Owari'**.
Citrus × paradisi Macfady. [*C. maxima* × *C. sinensis*]. GRAPEFRUIT.
Citrus × paradisi 'Marsh'. MARSH SEEDLESS GRAPEFRUIT.
Citrus × paradisi 'Ruby'.
Citrus 'Ponderosa'. See **C. limon 'Ponderosa'**.
Citrus reticulata Blanco. Syn.: *C. nobilis* var. *deliciosa*. MANDARIN. MANDARIN ORANGE. SATSUMA. SATSUMA ORANGE. TANGERINE.
Citrus reticulata 'China'.
Citrus reticulata 'Clementine'. ALGERIAN MANDARIN.
Citrus reticulata 'Dancy'. DANCY MANDARIN. DANCY TANGERINE.
Citrus reticulata 'King'. See **C. × nobilis 'King'**.
Citrus reticulata 'Kinnow'. KINNOW MANDARIN.
Citrus reticulata 'Owari'. OWARI SATSUMA. OWARI MANDARIN.
Citrus reticulata 'Temple'. See **C. × nobilis 'Temple'**.
Citrus reticulata 'Wilking'.
Citrus 'Robertson'. See **C. sinensis 'Robertson'**.
Citrus 'Ruby'. See **Citrus × paradisi 'Ruby'**.
Citrus 'Sampson'. See **C. × tangelo 'Sampson'**.
Citrus 'Shamouti'. See **C. sinensis 'Shamouti'**.
Citrus sinensis (L.) Osbeck. SWEET ORANGE.
Citrus sinensis 'Diller'. DILLER ORANGE.
Citrus sinensis 'Hamlin'. HAMLIN ORANGE.
Citrus sinensis 'Robertson'. ROBERTSON NAVEL ORANGE.
Citrus sinensis 'Seedless Valencia'. SEEDLESS VALENCIA ORANGE.
Citrus sinensis 'Shamouti'. SHAMOUTI ORANGE.
Citrus sinensis 'Summernavel'. SUMMERNAVEL ORANGE.
Citrus sinensis 'Tarocco'. TAROCCO ORANGE.
Citrus sinensis 'Trovita'. TROVITA ORANGE.
Citrus sinensis 'Valencia'. VALENCIA ORANGE.
Citrus sinensis 'Washington'. WASHINGTON NAVEL ORANGE.
Citrus 'Summernavel'. See **Citrus sinensis 'Summernavel'**.
Citrus × tangelo J. Ingram & H. E. Moore. [*C. × paradisi* × *C. reticulata*]. TANGELO.
Citrus × tangelo 'Minneola'. MINNEOLA TANGELO.
Citrus × tangelo 'Sampson'. SAMPSON TANGELO.
Citrus 'Temple'. See **C. × nobilis 'Temple'**.
Citrus 'Valencia'. See **C. sinensis 'Valencia'**.
Citrus 'Villafranca'. See **C. limon 'Villafranca'**.
Citrus 'Washington'. See **C. sinensis 'Washington'**.
Citrus 'Wilking'. See **C. reticulata 'Wilking'**.
Citrus 'Willow Leaf'. See **C. reticulata 'China'**.
Cladrastis lutea (Michx. f.) K. Koch. AMERICAN YELLOWWOOD.
CLARY, MEADOW: **Salvia pratensis**.
CLEMATIS, DOWNY: **Clematis macropetala**.
  EVERGREEN: **Clematis armandii**.
  GOLDEN: **Clematis tangutica**.
  SCARLET: **Clematis texensis**.
  SWEET AUTUMN: **Clematis dioscoreifolia**.
Clematis afoliata Buchanan.
Clematis alpina (L.) Mill.
Clematis armandii Franch. EVERGREEN CLEMATIS.
Clematis armandii 'Farquhariana'.
Clematis armandii 'Hendersonii Rubra'.
Clematis 'Barbara Dibley'. See **C. (Patens Group) 'Barbara Dibley'**.
Clematis 'Candida'. See **C. (Lanuginosa Group) 'Candida'**.
Clematis chrysocoma Franch. Syn.: *C. spooneri*.
Clematis cirrhosa L.
Clematis columbiana (Nutt.) Torr. & Gray.
Clematis 'Comtesse de Bouchard'. See **C. (Jackmanii Group) 'Comtesse de Bouchaud'**.
Clematis 'Crimson King'. See **C. (Lanuginosa Group) 'Crimson King'**.
Clematis 'Crimson Star'. See **C. (Lanuginosa Group) 'Crimson Star'**.
Clematis davidiana Decne. ex B. Verl. See **C. heracleifolia** var. **davidiana**.
Clematis dioscoreifolia Lev. & Vaniot. SWEET AUTUMN CLEMATIS. Some plants called *C. paniculata* belong here.
Clematis dioscoreifolia var. **robusta** (Carrière) Rehd. Syn: *C. paniculata*.
Clematis 'Duchess of Edinburgh'. See **C. (Florida Group) 'Duchess of Edinburgh'**.
Clematis 'Edith Cavell'. See **C. (Florida Group) 'Edith Cavell'**.
Clematis 'Elsa Spaeth'. See **C. (Lanuginosa Group) 'Elsa Spaeth'**.
Clematis 'Ernest Markham'. See **C. (Viticella Group) 'Ernest Markham'**.
Clematis flammula L.

Clematis florida Thunb.
Clematis (Florida Group) **'Duchess of Edinburgh'**.
Clematis (Florida Group) **'Edith Cavell'**.
Clematis forsteri J. F. Gmelin.
Clematis fusca Turcz.
Clematis **'General MacArthur'**.
Clematis 'Gypsy Queen'. See **C.** (Jackmanii Group) **'Gypsy Queen'**.
Clematis 'Hagley Hybrid'. See **C.** (Jackmanii Group) **'Hagley Hybrid'**.
Clematis henryi D. Oliver. Plants so called may be **C.** × *lawsoniana* **'Henryi'**.
Clematis heracleifolia DC.
Clematis heracleifolia var. *davidiana* (Decne. ex B. Verlot) Hemsl. Syn.: *C. davidiana*.
Clematis 'Huldine'. See **C.** (Viticella Group) **'Huldine'**.
Clematis integrifolia L.
Clematis × *jackmanii* T. Moore. [**C. lanuginosa** × **C. viticella**].
Clematis × *jackmanii* **'Rubra'**.
Clematis (Jackmanii Group) **'Comtesse de Bouchaud'**.
Clematis (Jackmanii Group) **'Gypsy Queen'**.
Clematis (Jackmanii Group) **'Hagley Hybrid'**.
Clematis (Jackmanii Group) **'Mme. Baron-Veillard'**.
Clematis (Jackmanii Group) **'Mme. Edouard André'**.
Clematis (Jackmanii Group) **'Mrs. Cholmondeley'**.
Clematis 'Lady Betty Balfour'. See **C.** (Viticella Group) **'Lady Betty Balfour'**.
Clematis lanuginosa Lindl.
Clematis (Lanuginosa Group) **'Candida'**.
Clematis (Lanuginosa Group) **'Crimson King'**.
Clematis (Lanuginosa Group) **'Crimson Star'**.
Clematis (Lanuginosa Group) **'Elsa Spaeth'**.
Clematis (Lanuginosa Group) **'Lord Neville'**.
Clematis (Lanuginosa Group) **'Nelly Moser'**.
Clematis (Lanuginosa Group) **'Prins Hendrik'**.
Clematis (Lanuginosa Group) **'Ramona'**.
Clematis (Lanuginosa Group) **'Susan P. Emory'**.
Clematis (Lanuginosa Group) **'William Kennett'**.
Clematis lasiantha Nutt.
Clematis 'Lasurstern'. See **C.** (Patens Group) **'Lasurstern'**.
Clematis × *lawsoniana* T. Moore & Jackman. [**C. lanuginosa** × **C. patens**].
Clematis × *lawsoniana* **'Henryi'**.
Clematis 'Lord Neville'. See **C.** (Lanuginosa Group) **'Lord Neville'**.
Clematis macropetala Ledeb. DOWNY CLEMATIS.
Clematis 'Mme. Baron-Veillard'. See **C.** (Jackmanii Group) **'Mme. Baron-Veillard'**.
Clematis 'Mme. Edouard André'. See **C.** (Jackmanii Group) **'Mme. Edouard André'**.
Clematis montana Buch.-Ham. ex DC.
Clematis montana var. *rubens* Wils.
Clematis montana **'Tetrarose'**.
Clematis 'Mrs. Cholmondeley'. See **C.** (Jackmanii Group) **'Mrs. Cholmondeley'**.
Clematis 'Nelly Moser'. See **C.** (Lanuginosa Group) **'Nelly Moser'**.
Clematis orientalis L.
Clematis paniculata Thunb. not J. F. Gmelin. See **C. dioscoreifolia** var. **robusta**. Some plants called *C. paniculata* are **C. dioscoreifolia**.
Clematis patens C. Morr. & Decne.
Clematis (Patens Group) **'Barbara Dibley'**.
Clematis (Patens Group) **'Lasurstern'**.
Clematis (Patens Group) **'President'**.

Clematis petriei Allan.
Clematis 'President'. See **C.** (Patens Group) **'President'**.
Clematis 'Prins Hendrik'. See **C.** (Lanuginosa Group) **'Prins Hendrik'**.
Clematis 'Ramona'. See **C.** (Lanuginosa Group) **'Ramona'**.
Clematis spooneri Rehd. & Wils. See **C. chrysocoma**.
Clematus 'Susan P. Emory'. See **C.** (Lanuginosa Group) **'Susan P. Emory'**.
Clematis tangutica (Maxim.) Korsh. GOLDEN CLEMATIS.
Clematis texensis Buckley. SCARLET CLEMATIS.
Clematis 'Ville de Lyon'. See **C.** (Viticella Group) **'Ville de Lyon'**.
Clematis vitalba L. TRAVELER'S JOY. OLD-MAN'S BEARD.
Clematis viticella L.
Clematis (Viticella Group) **'Ernest Markham'**.
Clematis (Viticella Group) **'Huldine'**.
Clematis (Viticella Group) **'Lady Betty Balfour'**.
Clematis (Viticella Group) **'Ville de Lyon'**.
Clematis 'William Kennett'. See **C.** (Lanuginosa Group) **'William Kennett'**.
Cleome isomeris Greene. See **Isomeris arborea**.
Clerodendron. See **Clerodendrum**.
Clerodendrum balfouri Dombrain. See **C. thomsoniae**.
Clerodendrum bungei Steud. Syn.: *Clerodendrum foetidum*. KASHMIR BOUQUET.
Clerodendrum foetidum Bunge. See **C. bungei**.
Clerodendrum fragrans (Vent.) R. Br. See **C. philippinum**.
Clerodendrum myricoides (Hochst.) R. Br. ex Vatke. Not cult. in Calif. Plants so called are **C. ugandense**.
Clerodendrum philippinum Schauer. Syn.: *C. fragrans*. KASHMIR BOUQUET.
Clerodendrum thomsoniae Balf. Syn.: *C. balfouri*. BLEEDING-HEART GLORY-BOWER.
Clerodendrum trichotomum Thunb. HARLEQUIN GLORY-BOWER.
Clerodendrum ugandense Prain. Some plants called **C. myricoides** belong here.
Clethra alnifolia L. SUMMERSWEET. SWEET PEPPERBUSH.
Clethra alnifolia **'Paniculata'**.
Clethra arborea Ait. LILY-OF-THE-VALLEY TREE.
Cleyera japonica Thunb. Syn.: *Eurya ochnacea*.
Clianthus dampieri A. Cunn. ex Lindl. See **C. formosus**.
Clianthus formosus (G. Don) Ford & Vickery. Syn.: *C. dampieri*. DESERT PEA. STURT DESERT PEA. GLORY PEA.
Clianthus puniceus (G. Don) Soland. ex Lindl. GLORY PEA. PARROT'S-BEAK.
CLIFF-GREEN: **Paxistima canbyi**.
Clinostigma mooreanum (F. J. Muell.) H. Wendl. & Drude. See **Lepidorrhachis mooreana**.
CLOCK VINE, BUSH: **Thunbergia erecta**.
CLOVER, ELK: **Aralia californica**.
Clytostoma binatum (Thunb.) Sandwith. Syn.: *Bignonia purpurea*.
Clytostoma callistegioides (Cham.) Bur. Syn.: *Bignonia callistegioides*. *B. speciosa*. *B. violacea*. LAVENDER TRUMPET VINE. VIOLET TRUMPET VINE.
Cneoridium dumosum (Nutt.) Hook. f. BERRYRUE. BUSHRUE.
Cneorum tricoccon L. SPURGE OLIVE.
Cobaea scandens Cav. CUP-AND-SAUCER VINE.
Coccothrinax argentata (Jacq.) L. H. Bailey. SILVER PALM.
Coccothrinax argentea (Lodd. ex Schult. & Schult. f.) Sarg. ex Becc. Syn.: *Thrinax argentea*. BROOM PALM. SILVER THATCH PALM. SILVER THATCH.
Coccothrinax crinita Becc. SEAM-BERRY PALM. THATCH PALM.
Coccothrinax dussiana L. H. Bailey.
COCCULUS, LAUREL-LEAF: **Cocculus laurifolius**.

Cocculus laurifolius (Roxb.) DC. LAUREL-LEAF SNAILSEED. LAUREL-LEAF COCCULUS.
COCKSPUR: ***Crataegus crus-galli.***
COCONUT, MONKEY: ***Jubaea chilensis.***
Cocos amara Jacq. See **Rhyticocos amara.**
Cocos australis Mart. See **Arecastrum romanzoffianum** var. **australe.** Some plants called *Cocos australis* are **Butia capitata.**
Cocos plumosa Hook. See **Arecastrum romanzoffianum.**
Codiaeum variegatum (L.) Blume. CROTON.
Coffea arabica L. COFFEE. ARABIAN COFFEE.
COFFEE: ***Coffea arabica.***
    ARABIAN: ***Coffea arabica.***
    WILD: ***Polyscias guilfoylei.***
COFFEEBERRY: ***Rhamnus californica.***
    CHAPARRAL: ***Rhamnus californica*** subsp. ***tomentella.***
COFFEE TREE: ***Polyscias guilfoylei.***
    KENTUCKY: ***Gymnocladus dioica.***
Coleonema album (Thunb.) Bartl. & H. L. Wendl. Syn.: *Diosma alba.* WHITE BREATH-OF-HEAVEN. BREATH-OF-HEAVEN.
Coleonema pulchrum Hook. Syn.: *Diosma pulchra.* PINK BREATH-OF-HEAVEN. PINK DIOSMA.
Colletia armata Miers.
Colletia cruciata Gillies & Hook. See **C. paradoxa.**
Colletia paradoxa (Spreng.) Escalante. Syn.: *C. cruciata.* ANCHOR PLANT.
Collinia elegans (Mart.) Liebm. ex Oerst. See **Chamaedorea elegans.**
Colocasia esculenta (L.) Schott. Syn.: *Caladium esculentum.* TARO.
Colutea arborescens L. SHRUBBY BLADDER SENNA.
Comarostaphylis diversifolia (Parry) Greene. Syn.: *Arctostaphylos diversifolia.* SUMMER HOLLY.
Comarostaphylis diversifolia var. ***planifolia*** Jeps.
Combretum grandiflorum G. Don.
COMPTIE: ***Zamia pumila.***
CONFEDERATE VINE: ***Trachelospermum jasminoides.***
    VARIEGATED: ***Trachelospermum jasminoides* 'Variegatum'.**
Convolvulus cneorum L. BUSH MORNING-GLORY. SILVER BUSH.
Convolvulus mauritanicus Boiss. See **C. sabatius.**
Convolvulus sabatius Viviani. Syn.: *C. mauritanicus.* GROUND MORNING-GLORY.
COONTIE: ***Zamia pumila.***
COPA-DE-LECHE: ***Solandra longiflora.***
COPA-DE-ORO: ***Solandra maxima.***
Copernicia alba Morong. Syn.: *C. australis.*
Copernicia australis Becc. See **C. alba.**
Copernicia cerifera (Arruda da Camara) Mart. See **C. prunifera.**
Copernicia prunifera (Mill.) H. E. Moore. Syn.: *C. cerifera.* CARNAUBA PALM.
COPPERLEAF, CALIFORNIA: ***Acalypha californica.***
    PAINTED: ***Acalypha wilkesiana.***
COPROSMA, CREEPING: ***Coprosma × kirkii.***
Coprosma baueri T. Kirk not Endl. See **C. repens.**
***Coprosma* 'Brownsleeves'.**
***Coprosma* 'Coppershine'.**
Coprosma* × *cunninghamii Hook. f. [**C. propinqua** × **C. robusta**].
***Coprosma* 'Greensleeves'.**
Coprosma* × *kirkii Cheeseman. [**C. acerosa** A. Cunn. × **C. repens**]. CREEPING COPROSMA.
***Coprosma* 'Lofty'.**
Coprosma microphylla. Listed name for **C. × kirkii.**
Coprosma petriei Cheeseman.
Coprosma petriei var. ***atropurpurea*** Cockayne & Allan.

Coprosma propinqua A. Cunn.
Coprosma repens A. Rich. Syn.: *C. baueri.* MIRROR PLANT. MIRROR SHRUB. LOOKING-GLASS PLANT.
***Coprosma repens* 'Argentea'.**
***Coprosma repens* 'Aureo-variegata'.**
***Coprosma repens* 'Marginata'.**
Coprosma repens 'Prostrata Variegata'?
***Coprosma repens* 'Variegata'.**
Coprosma robusta Raoul.
***Coprosma* 'Shiner'.**
CORAL PLANT: ***Russelia equisetiformis.***
CORAL TREE, AUSTRALIAN: ***Erythrina vespertilio.***
    BIDWILL: ***Erythrina* × *bidwillii.***
    COCKSPUR: ***Erythrina crista-galli.***
    NAKED: ***Erythrina coralloides.***
    NATAL: ***Erythrina humeana.***
CORAL-BEADS: ***Sedum stahlii.***
CORALBERRY: ***Symphoricarpos orbiculatus.***
    CHENAULT: ***Symphoricarpos* × *chenaultii.***
CORALLITA: ***Antigonon leptopus.***
CORAL VINE: ***Antigonon leptopus.***
Cordia boissieri A. DC.
Cordia superba Cham. Syn.: *C. superba* var. *elliptica.*
Cordia superba var. *elliptica* Cham. See **C. superba.**
Cordyline australis (G. Forst.) Endl. Syn.: *Dracaena australis.* CABBAGE TREE. DRACAENA PALM.
***Cordyline australis* 'Atropurpurea'.** BRONZE DRACAENA.
Cordyline banksii Hook. f.
***Cordyline banksii* 'Purpurea'.**
Cordyline baueri Hook. f.
Cordyline dracaenoides Kunth.
Cordyline haageana K. Koch.
Cordyline indivisa (G. Forst.) Steud. Syn.: *Dracaena indivisa.* BLUE DRACAENA.
***Cordyline indivisa* 'Atropurpurea'.**
Cordyline rubra Huegel ex Kunth.
Cordyline stricta Endl. Syn.: *Dracaena stricta.* AUSTRALIAN DRACAENA. PALM LILY.
Cordyline terminalis (L.) Kunth. TI. HAWAIIAN GOOD-LUCK PLANT.
***Cordyline terminalis* 'Mme. Eugene André'.**
COREOPSIS, GIANT: ***Coreopsis gigantea.***
    TREE: ***Coreopsis gigantea.***
Coreopsis gigantea (Kell.) Hall. GIANT COREOPSIS. TREE COREOPSIS.
Coreopsis maritima (Nutt.) Hook. f.
CORK TREE, AMUR: ***Phellodendron amurense.***
CORKSCREW FLOWER: ***Vigna caracalla.***
CORNEL, DWARF: ***Cornus canadensis.***
Cornus alba L. RED-BARK DOGWOOD. TATARIAN DOGWOOD.
***Cornus alba* 'Sibirica'.** SIBERIAN DOGWOOD.
***Cornus alba* 'Variegata'.**
Cornus* × *californica C. A. Mey. [**C. occidentalis** × **C. stolonifera**]. CREEK DOGWOOD.
Cornus canadensis L. BUNCHBERRY. DWARF CORNEL.
Cornus capitata Wallich. WHITE EVERGREEN DOGWOOD.
Cornus 'Cherokee Chief'. See **C. florida 'Cherokee Chief'.**
Cornus 'Cherokee Princess'. See **C. florida 'Cherokee Princess'.**
Cornus 'Cloud 9'. See **C. florida 'Cloud 9'.**
Cornus controversa Hemsl. ex Prain. GIANT DOGWOOD.
***Cornus* 'Eddie's White Wonder'.** [**C. nuttallii** × **C. florida**].
Cornus florida L. FLOWERING DOGWOOD. EASTERN DOGWOOD.
***Cornus florida* 'Cherokee Chief'.** RED FLOWERING DOGWOOD.
***Cornus florida* 'Cherokee Princess'.**
***Cornus florida* 'Cloud 9'.**
***Cornus florida* 'Fastigiata'.**

Cornus florida 'Mystery'.
Cornus florida 'Pendula'.
Cornus florida 'Pluribracteata'. DOUBLE WHITE DOGWOOD.
Cornus florida 'Prosser Red'.
Cornus florida 'Pygmy'.
Cornus florida 'Rainbow'.
Cornus florida 'Royal Red'.
Cornus florida 'Rubra'. PINK FLOWERING DOGWOOD. RED FLOWERING DOGWOOD.
Cornus florida 'Tricolor'?
Cornus florida 'Variegata'.
Cornus florida 'Welchii'. TRICOLOR DOGWOOD.
Cornus florida 'White Cloud'.
Cornus kousa Hance. KOUSA DOGWOOD.
Cornus kousa var. *chinensis* Osborn. CHINESE KOUSA DOGWOOD.
Cornus kousa 'Milky Way'.
Cornus mas L. CORNELIAN CHERRY.
Cornus nuttallii Aud. PACIFIC DOGWOOD. WESTERN DOGWOOD.
Cornus nuttallii 'Eddiei'.
Cornus nuttallii 'Eddie's Wonder'. May be C. 'Eddie's White Wonder'.
Cornus nuttallii 'Gold Spot'. See **C. nuttallii** 'Eddiei'.
Cornus nuttallii 'Pilgrim'.
Cornus 'Rubra'. See **C. florida** 'Rubra'.
Cornus sanguinea L. BLOOD-TWIG DOGWOOD.
Cornus sericea L. See **C. stolonifera**.
Cornus 'Sibirica'. See **C. alba** 'Sibirica'.
Cornus stolonifera Michx. Syn.: *C. sericea*. RED-OSIER DOGWOOD. AMERICAN DOGWOOD.
Cornus stolonifera 'Flaviramea'. YELLOW-TWIG DOGWOOD.
Corokia × *cheesemanii* Carse. [**C. buddleoides** A. Cunn. × **C. cotoneaster**].
Corokia cotoneaster Raoul.
Corokia macrocarpa T. Kirk.
Corokia 'Red Wonder'.
Corokia 'Yellow Wonder'.
Coronilla emerus L. SCORPION SENNA.
CORREA, PINK AUSTRALIAN: **Correa pulchella**.
    YELLOW AUSTRALIAN: **Correa backhousiana**.
Correa alba Andr.
Correa backhousiana Hook. YELLOW AUSTRALIAN CORREA. Some plants called *C. ferruginea* and **C.** × **magnifica** belong here.
Correa ferruginea. Listed name for **C. backhousiana**.
Correa × *harrisii* Paxt. [**C. pulchella** × **C. reflexa**].
Correa × *magnifica* Paxt. Hybrid of unknown origin. May not be cult. Plants so called are usually **C. backhousiana**.
Correa neglecta Ashby. See **C. pulchella**.
Correa pulchella Mackay ex Sweet. Syn.: *C. neglecta*. AUSTRALIAN FUCHSIA. PINK AUSTRALIAN CORREA.
Correa reflexa (Labill.) Vent. Syn.: *C. speciosa*.
Correa 'Silver Bells'.
Correa speciosa J. Donn ex Andr. See **C. reflexa**.
CORSAGE VINE: **Stephanotis floribunda**.
Cortaderia argentea (Nees) Stapf. See **C. selloana**.
Cortaderia jubata (Lem.) Stapf.
Cortaderia selloana (Schult & Schult f.) Aschers & Graebn. Syn.: *C. argentea*. PAMPAS GRASS.
Cortaderia selloana 'Rubra'?
Corylopsis pauciflora Siebold & Zucc. BUTTERCUP WINTER HAZEL. WINTER HAZEL.
Corylopsis spicata Siebold & Zucc. SPIKE WINTER HAZEL. WINTER HAZEL.
Corylus avellana L. EUROPEAN FILBERT. EUROPEAN HAZELNUT.
Corylus avellana 'Atropurpurea'. See **C. avellana** 'Fusco-rubra'.
Corylus avellana 'Atropurpurea Superba'?
Corylus avellana 'Contorta'. HARRY LAUDER'S WALKING-STICK.
Corylus avellana 'Fusco-rubra'.
Corylus californica (A. DC.) Rose. See **C. cornuta** var. *californica*.
Corylus colurna L. TURKISH FILBERT. TURKISH HAZELNUT. TURKISH HAZEL.
Corylus 'Contorta'. See **C. avellana** 'Contorta'.
Corylus cornuta Marsh. BEAKED HAZELNUT.
Corylus cornuta var. *californica* (A. DC.) Sharp. Syn.: *C. californica*. WESTERN HAZELNUT.
Corylus maxima Mill. FILBERT.
Corylus maxima 'Purpurea'. PURPLE GIANT FILBERT.
Corynabutilon vitifolium (Cav.) Kearney. Syn.: *Abutilon vitifolium*.
Corynocarpus laevigata J. R. Forst. & G. Forst. NEW ZEALAND LAUREL.
COSMETIC-BARK TREE: **Murraya paniculata**.
Cotinus coggygria Scop. Syn.: *Rhus cotinus*. SMOKE TREE.
Cotinus coggygria 'Atropurpureus'. See **C. coggygria** 'Purpureus'.
Cotinus coggygria 'Purpureus'.
Cotinus coggygria 'Royal Purple'.
COTONEASTER, BEARBERRY: **Cotoneaster dammeri**.
    CHERRY BUSH: **Cotoneaster zabelii**.
    CRANBERRY: **Cotoneaster apiculatus**.
    CREEPING: **Cotoneaster adpressus**.
    DWARF SILVER-LEAF: **Cotoneaster buxifolius**.
    EARLY: **Cotoneaster adpressus** var. *praecox*.
    FRANCHET: **Cotoneaster franchetii**.
    GROUND: **Cotoneaster horizontalis** var. *perpusillus*.
    NECKLACE: **Cotoneaster conspicuus**.
    PARNEY: **Cotoneaster lacteus**.
    RED CLUSTERBERRY: **Cotoneaster lacteus**.
    ROCK: **Cotoneaster horizontalis**.
    ROCKSPRAY: **Cotoneaster microphyllus**.
    SILVER-LEAF: **Cotoneaster pannosus**.
    SPREADING: **Cotoneaster divaricatus**.
    THYME ROCKSPRAY: **Cotoneaster microphyllus** var. *thymifolius*.
    WILLOW-LEAF: **Cotoneaster salicifolius**.
Cotoneaster adpressus Bois. CREEPING COTONEASTER.
Cotoneaster adpressus 'Little Gem'.
Cotoneaster adpressus var. *praecox* Bois & Berthault. Syn.: *C. praecox*. EARLY COTONEASTER.
Cotoneaster apiculatus Rehd. & Wils. CRANBERRY COTONEASTER.
Cotoneaster apiculatus 'Nanus'?
Cotoneaster arborescens Zabel. See **C. lindleyi**.
Cotoneaster bullatus Bois.
Cotoneaster bullatus var. *macrophyllus* Rehd. & Wils.
Cotoneaster buxifolius Wallich ex Lindl. DWARF SILVER-LEAF COTONEASTER. Plants called *C. pannosus* 'Nanus', and in Calif., **C. glaucophyllus**, belong here.
Cotoneaster buxifolius forma *vellaeus*. Rehd. & Wils.
Cotoneaster cochleatus (Franch.) Klotz. See **C. microphyllus** 'Cochleatus'.
Cotoneaster congestus Baker. Syn.: *C. microphyllus* var. *glacialis*.
Cotoneaster congestus 'Likiang'.
Cotoneaster conspicuus Marquand. Syn.: *C. conspicuus* var. *decorus*. NECKLACE COTONEASTER.
Cotoneaster conspicuus var. *decorus* Russell. See **C. conspicuus**.
Cotoneaster 'Coolidgei'?
Cotoneaster 'Cornubia'. [*C. frigidus* × *C. ?salicifolius*].
Cotoneaster dammeri Schneid. Syn.: *C. humifusus*. BEARBERRY COTONEASTER.
Cotoneaster dammeri 'Lowfast'.
Cotoneaster dammeri var. *radicans* Schneid.
Cotoneaster dammeri 'Skogholmen'.

Cotoneaster davidianus Nichols. See **C. horizontalis.**
Cotoneaster decorus. Listed name for **C. conspicuus.**
Cotoneaster divaricatus Rehd. & Wils. SPREADING COTONEASTER.
Cotoneaster franchetii Bois. FRANCHET COTONEASTER.
Cotoneaster franchetii var. **sternianus** Turill. Some plants called **C. wardii** may belong here.
Cotoneaster frigidus Wallich ex Lindl.
Cotoneaster frigidus 'Pendulus'. Some plants called C. 'Pendulus' may belong here.
Cotoneaster glaucophyllus Franch. Probably not cult. in Calif. Plants so called are **C. buxifolius.**
Cotoneaster glaucophyllus forma **serotinus** (Hutchinson) Stapf. Syn.: *C. serotinus.*
Cotoneaster harrovianus Wils.
Cotoneaster henryanus (Schneid.) Rehd. & Wils.
Cotoneaster horizontalis Decne. Syn.: *C. davidianus.* ROCK COTONEASTER.
Cotoneaster horizontalis 'Little Gem'?
Cotoneaster horizontalis var. **perpusillus** Schneid. GROUND COTONEASTER.
Cotoneaster horizontalis 'Robustus'.
Cotoneaster horizontalis 'Variegatus'.
Cotoneaster horizontalis 'Wilsonii'.
Cotoneaster humifusus Duthie ex Veitch. See **C. dammeri.**
Cotoneaster 'Hybridus Pendulus'. Some plants called C. 'Pendulus' may belong here.
Cotoneaster lacteus W. W. Sm. Syn.: *C. parneyi.* RED CLUSTERBERRY COTONEASTER. PARNEY COTONEASTER.
Cotoneaster 'Likiang'. See **C. congestus 'Likiang'.**
Cotoneaster lindleyi Steud. Syn.: *C. arborescens.*
Cotoneaster 'Lowfast'. See **C. dammeri 'Lowfast'.**
Cotoneaster melanocarpus Lodd.
Cotoneaster melanotrichus cochleatus. Listed name. Plants so called may be **C. microphyllus 'Cochleatus'.**
Cotoneaster microphyllus Wallich ex Lindl. ROCKSPRAY COTONEASTER.
Cotoneaster microphyllus 'Cochleatus'. Syn.: *C. cochleatus.*
Cotoneaster microphyllus congestus. Listed name. Plants so called may be **C. congestus** or **C. microphyllus.**
Cotoneaster microphyllus 'Emerald Spray'.
Cotoneaster microphyllus var. *glacialis* Hook. f. See **C. congestus.**
Cotoneaster microphyllus 'Minor'?
Cotoneaster microphyllus var. **thymifolius** (Lindl.) Koehne. Syn.: *C. thymifolius.* THYME ROCKSPRAY COTONEASTER.
Cotoneaster multiflorus Bunge.
Cotoneaster pannosus Franch. SILVER-LEAF COTONEASTER.
Cotoneaster pannosus 'Nanus'. Listed name for **C. buxifolius.**
Cotoneaster parneyi Hoyt. See **C. lacteus.**
Cotoneaster 'Pendulus'. Plants so called may be **C. frigidus 'Pendulus', C. 'Hybridus Pendulus',** or **C. × watereri 'Pendulus'.**
Cotoneaster praecox Meunissier. See **C. adpressus** var. **praecox.**
Cotoneaster praecox 'Boer'?
Cotoneaster prostratus Baker. See **C. rotundifolius.**
Cotoneaster 'Repens'. See **C. salicifolius 'Repens'.**
Cotoneaster 'Rothschildianus'. [*C. frigidus × C. salicifolius*].
Cotoneaster rotundifolius Wallich ex Lindl. Syn.: *C. prostratus.*
Cotoneaster rugosus E. Pritzel. See **C. salicifolius** var. **rugosus.**
Cotoneaster 'Saldam'. [*C. salicifolius × C.* sp.].
Cotoneaster salicifolius Franch. WILLOW-LEAF COTONEASTER.
Cotoneaster salicifolius 'Autumn Fire'. See **C. salicifolius 'Herbstfeuer'.**
Cotoneaster salicifolius var. **floccosus** Rehd. & Wils.
Cotoneaster salicifolius 'Gnom'.

Cotoneaster salicifolius henryanus. Listed name. Plants so called may be **C. henryanus.**
Cotoneaster salicifolius 'Herbstfeuer'.
Cotoneaster salicifolius 'Repens'.
Cotoneaster salicifolius var. **rugosus** (E. Pritzel) Rehd. & Wils. Syn.: *C. rugosus.*
Cotoneaster salicifolius 'Saldam'. See **C. 'Saldam'.**
Cotoneaster serotinus Hutchinson. See **C. glaucophyllus** forma **serotinus.**
Cotoneaster simonsii Baker.
Cotoneaster 'Skogholm'. See **C. dammeri 'Skogsholmen'.**
Cotoneaster thymifolius Lindl. See **C. microphyllus** var. **thymifolius.**
Cotoneaster wardii W. W. Sm. May not be in cult. Plants so called are usually **C. franchetii** var. **sternianus.**
Cotoneaster × watereri Exell. [*C. frigidus × C. henryanus*].
Cotoneaster × watereri 'Cornubia'. See **C. 'Cornubia'.**
Cotoneaster × watereri 'Herbstfeuer'. See **C. salicifolius 'Herbstfeuer'.**
Cotoneaster × watereri 'Pendulus'. Some plants called C. 'Pendulus' may belong here.
Cotoneaster wilsonii Nakai. May not be cult. Plants so called are usually **C. horizontalis 'Wilsonii'.**
Cotoneaster zabelii Schneid. CHERRY BUSH COTONEASTER.
COTTON, CYPRESS LAVENDER: **Santolina chamaecyparissus.**
  GREEN LAVENDER: **Santolina virens.**
  LAVENDER: **Santolina chamaecyparissus.**
COTTONLEAF: **Helichrysum petiolatum.**
COTTONWOOD: **Populus deltoides.**
  BLACK: **Populus trichocarpa.**
  EASTERN: **Populus deltoides.**
  FREMONT: **Populus fremontii.**
  MOUNTAIN: **Populus × acuminata.**
  WESTERN: **Populus fremontii.**
Cotyledon orbiculata L.
Cotyledon undulata Haw.
Cotyledon undulata 'Superba'.
Cowania mexicana D. Don var. **stansburiana** (Torr.) Jeps. Syn.: *C. stansburiana.* CLIFF ROSE.
Cowania stansburiana Torr. See **C. mexicana** var. **stansburiana.**
COWBERRY: **Vaccinium vitis-idaea.**
COW-ITCH: **Lagunaria patersonii.**
COYOTE BRUSH: **Baccharis pilularis** subsp. **consanguinea.**
  DWARF: **Baccharis pilularis.**
CRAB: See APPLE.
CRABAPPLE BUSH: **Crossosoma californicum.**
CRANBERRY: **Vaccinium vitis-idaea.**
  EUROPEAN: **Vaccinium oxycoccus.**
  HIGHBUSH: **Viburnum trilobum.**
  MOUNTAIN: **Vaccinium vitis-idaea** var. **minus.**
  ROCK: **Vaccinium vitis-idaea** var. **minus.**
  SMALL: **Vaccinium oxycoccus.**
CRANBERRY BUSH: **Viburnum trilobum.**
  EUROPEAN: **Viburnum opulus.**
CRANBERRY TREE: **Viburnum trilobum.**
CRANE FLOWER: **Strelitzia reginae.**
Crassula arborescens (Mill.) Willd. JADE PLANT.
Crassula argentea Thunb. See **C. ovata.**
Crassula falcata H. Wendl. See **C. perfoliata** var. **falcata.**
Crassula lactea Soland.
Crassula multicava Lem.
Crassula obliqua Soland. See **C. ovata.**
Crassula ovata (Mill.) Druce. Syn.: *C. argentea. C. obliqua. C. portulacea.* CAULIFLOWER EARS. JADE TREE.
Crassula ovata 'Hummel's Sunset'.
Crassula ovata 'Sunset'.

Crassula ovata 'Sunset Gold'.
Crassula perfoliata L. var. *falcata* (Wendl.) Toelken. Syn.: *C. falcata*.
Crassula portulacea Lam. See **C. ovata**.
Crassula tetragona L.
Crataegus 'Autumn Glory'. [?**C. laevigata** × **C. pubescens**].
Crataegus carrierei Vauvel. See **C.** × **lavallei**.
Crataegus cordata Ait. See **C. phaenopyrum**.
Crataegus crus-galli L. COCKSPUR. COCKSPUR THORN.
Crataegus douglasii Lindl. BLACK HAWTHORN.
Crataegus laevigata (Poir.) DC. Syn.: *C. oxyacantha*. ENGLISH HAWTHORN.
Crataegus laevigata 'Coccinea Flore Pleno'. PAUL'S SCARLET HAWTHORN.
Crataegus laevigata 'Paulii'. See **C. laevigata** 'Coccinea Flore Pleno'.
Crataegus laevigata 'Paul's Scarlet'. See **C. laevigata** 'Coccinea Flore Pleno'.
Crataegus laevigata 'Plena'. DOUBLE WHITE ENGLISH HAWTHORN.
Crataegus laevigata 'Punicea Flore Pleno'. DOUBLE PINK ENGLISH HAWTHORN.
Crataegus laevigata 'Rosea'. SINGLE PINK ENGLISH HAWTHORN.
Crataegus laevigata 'Rosea Plena'. See **C. laevigata** 'Punicea Flore Pleno'.
Crataegus × *lavallei* Herincq ex Lav. [**C. crus-galli** × **C. pubescens**]. Syn.: *C. carrierei*. LAVALLE THORN. CARRIÈRE HAWTHORN.
Crataegus monogyna Jacq. SINGLE-SEED HAWTHORN. ENGLISH HAWTHORN.
Crataegus monogyna 'Biflora'. GLASTONBURY THORN.
Crataegus × *mordenensis* Boom. [**C. laevigata** × **C. succulenta**].
Crataegus × *mordenensis* 'Toba'.
Crataegus oxyacantha of authors, not L. See **C. laevigata**.
Crataegus oxyacantha 'Coccinea Flore Pleno'. See **C. laevigata** 'Coccinea Flore Pleno'.
Crataegus oxyacantha 'Paulii'. See **C. laevigata** 'Coccinea Flore Pleno'.
Crataegus oxyacantha 'Paul's Scarlet'. See **C. laevigata** 'Coccinea Flore Pleno'.
Crataegus oxyacantha 'Plena'. See **C. laevigata** 'Plena'.
Crataegus oxyacantha pubescens. Listed name, probably for **C. pubescens**.
Crataegus oxyacantha 'Rosea'. See **C. laevigata** 'Rosea'.
Crataegus oxyacantha 'Rosea Plena'. See **C. laevigata** 'Punicea Flore Pleno'.
Crataegus phaenopyrum (L. f.) Medic. Syn.: *C. cordata*. WASHINGTON THORN.
Crataegus pinnatifida Bunge. CHINESE HAWTHORN.
Crataegus pinnatifida var. *major* N. E. Br. LARGE CHINESE HAWTHORN.
Crataegus pubescens (HBK) Steud. MEXICAN HAWTHORN.
Crataegus succulenta Schrad.
Crataegus 'Toba'. See **C.** × **mordenensis** 'Toba'.
CREAMBUSH: **Holodiscus discolor**.
CREOSOTE BUSH: **Larrea divaricata**.
Crinodendron dependens (Ruiz & Pav.) Kuntze. See **C. patagua**.
Crinodendron patagua Molina. Syn.: *C. dependens*. *Tricuspidaria dependens*. LILY-OF-THE-VALLEY TREE.
Crossandra infundibuliformis (L.) Nees.
Crossosoma californicum Nutt. CRABAPPLE BUSH.
CROSS VINE: **Bignonia capreolata**.
Crotalaria agatiflora Schweinfurth ex Engler. CANARY-BIRD BUSH.
CROTON: **Codiaeum variegatum**.

CROWBERRY, BLACK: **Empetrum nigrum**.
CROWN-OF-GOLD TREE: **Cassia spectabilis**.
CROWN-OF-THORNS: **Euphorbia milii**.
Cryptocarya miersii. Listed name for **Beilschmiedia miersii**.
Cryptocarya rubra (Molina) Skeels.
CRYPTOMERIA, DWARF: **Cryptomeria japonica** 'Nana'.
  JAPANESE: **Cryptomeria japonica**.
  LOBB: **Cryptomeria japonica** 'Lobbii'.
  PLUME: **Cryptomeria japonica** 'Elegans'.
Cryptomeria japonica (L. f.) D. Don. JAPANESE CEDAR. JAPANESE REDWOOD. JAPANESE CRYPTOMERIA.
Cryptomeria japonica 'Compacta'.
Cryptomeria japonica 'Elegans'. PLUME CRYPTOMERIA. PLUME CEDAR.
Cryptomeria japonica 'Elegans Compacta'.
Cryptomeria japonica 'Globosa Nana'.
Cryptomeria japonica 'Lobbii'. LOBB CRYPTOMERIA.
Cryptomeria japonica 'Lobbii Nana'.
Cryptomeria japonica 'Monstrosa'.
Cryptomeria japonica 'Nana'. DWARF CRYPTOMERIA.
Cryptomeria japonica 'Pygmaea'. See **C. japonica** 'Nana'.
Cryptomeria japonica 'Spiraliter Falcata'.
Cryptomeria japonica 'Vilmoriniana'.
Ctenanthe pilosa (Schauer) Eichl. Some plants called *Bamburanta arnoldiana* may belong here.
CUCUMBER TREE: **Magnolia acuminata**.
  LARGE-LEAVED: **Magnolia macrophylla**.
  YELLOW: **Magnolia acuminata** var. *cordata*.
Cunninghamia lanceolata (Lamb.) Hook. CHINA FIR.
Cunninghamia lanceolata 'Glauca'.
Cunonia capensis L. AFRICAN RED ALDER.
CUP-AND-SAUCER VINE: **Cobaea scandens**.
Cupania anacardioides A. Rich. See **Cupaniopsis anacardioides**.
Cupania sapida. Listed name for **Cupaniopsis anacardioides**.
Cupaniopsis anacardioides (A. Rich.) Radlk. Syn.: *Cupania anacardioides*. CARROTWOOD. TUCKEROO. Some plants called **Blighia sapida** belong here.
CUPFLOWER: **Nierembergia scoparia**.
Cuphea aequipetala Cav.
Cuphea 'Fire Fly'.
Cuphea hyssopifolia HBK. BLUE CIGAR PLANT. FALSE HEATHER.
Cuphea hyssopifolia 'Alba'.
Cuphea hyssopifolia 'Rosea'.
Cuphea ignea A. DC. Syn.: *C. platycentra*. CIGAR PLANT. FIRECRACKER PLANT.
Cuphea micropetala HBK.
Cuphea platycentra Lem. See **C. ignea**.
CUP-OF-GOLD: **Solandra maxima**.
× *Cupressocyparis leylandii* (A. B. Jackson & Dallim.) Dallim. & A. B. Jackson [**Chamaecyparis nootkatensis** × **Cupressus macrocarpa**]. Syn.: *Cupressus leylandii*. LEYLAND CYPRESS.
× *Cupressocyparis leylandii* 'Green Spire'.
× *Cupressocyparis leylandii* 'Haggerston Grey'.
× *Cupressocyparis leylandii* 'Leighton Green'.
× *Cupressocyparis leylandii* 'Naylor's Blue'.
Cupressus abramsiana C. B. Wolf. See **C. goveniana** var. *abramsiana*.
Cupressus arizonica Greene. ARIZONA CYPRESS. ROUGH-BARKED ARIZONA CYPRESS. Some cvs. listed under **C. arizonica** may belong to **C. arizonica** var. *glabra*.
Cupressus arizonica 'Compacta'.
Cupressus arizonica var. *glabra* (Sudw.) Little. Syn.: *C. glabra*. SMOOTH ARIZONA CYPRESS. SMOOTH-BARKED ARIZONA CYPRESS. Some cvs. listed under **C. arizonica** may belong here.

Cupressus arizonica 'Glauca'.
Cupressus arizonica 'Pyramidalis'.
Cupressus arizonica var. *stephensonii* (C. B. Wolf) Little. Syn.: *C. stephensonii*. CUYAMACA CYPRESS.
Cupressus cashmeriana Royle ex Carrière. KASHMIR CYPRESS.
Cupressus 'Crippsii'. See **C. macrocarpa** 'Crippsii'.
Cupressus forbesii Jeps. See **C. guadalupensis** var. *forbesii*.
Cupressus funebris Endl. See **Chamaecyparis funebris**.
Cupressus glabra Sudw. See **C. arizonica** var. *glabra*.
Cupressus goveniana Gord. GOWEN CYPRESS.
Cupressus goveniana var. *abramsiana* (C. B. Wolf) Little. Syn.: *C. abramsiana*. SANTA CRUZ CYPRESS.
Cupressus guadalupensis S. Wats. GUADALUPE CYPRESS.
Cupressus guadalupensis var. *forbesii* (Jeps.) Little. Syn.: *C. forbesii*. FORBES CYPRESS. TECATE CYPRESS.
Cupressus leylandii A. B. Jackson & Dallim. See × **Cupressocyparis leylandii**.
Cupressus lusitanica Mill. MEXICAN CYPRESS. PORTUGEUSE CYPRESS.
Cupressus macrocarpa Hartw. MONTEREY CYPRESS.
Cupressus macrocarpa 'Aurea'.
Cupressus macrocarpa 'Contorta'?
Cupressus macrocarpa 'Crippsii'.
Cupressus macrocarpa 'Donard Gold'. DONARD GOLD CYPRESS.
Cupressus macrocarpa 'Lambertiana Aurea'.
Cupressus obtusa K. Koch. See **Chamaecyparis obtusa**.
Cupressus 'Roylei'. See **C. sempervirens** 'Indica'.
Cupressus sempervirens L. ITALIAN CYPRESS.
Cupressus sempervirens 'Fastigiata'. See **C. sempervirens** 'Stricta'.
Cupressus sempervirens 'Glauca'.
Cupressus sempervirens var. *horizontalis* (Mill.) Gord. SPREADING ITALIAN CYPRESS.
Cupressus sempervirens 'Indica'.
Cupressus sempervirens 'Pyramidalis'. See **C. sempervirens** 'Stricta'.
Cupressus sempervirens 'Stricta'. COLUMNAR ITALIAN CYPRESS. A collective name for all fastigiate or columnar cvs. not otherwise named.
Cupressus stephensonii C. B. Wolf. See **C. arizonica** var. *stephensonii*.
Cupressus thyoides L. See **Chamaecyparis thyoides**.
CURIOSITY PLANT: **Cereus peruvianus** 'Monstrosus'.
CURRANT, BUFFALO: **Ribes aureum**.
  CATALINA: **Ribes viburnifolium**.
  CHAPARRAL: **Ribes malvaceum**.
  EVERGREEN: **Ribes viburnifolium**.
  FLOWERING: **Ribes sanguineum** var. *glutinosum*.
  FUCHSIA-FLOWERED: **Ribes speciosum**.
  GOLDEN: **Ribes aureum**.
  GOLDEN: **Ribes aureum** var. *gracillimum*.
  INDIAN: **Symphoricarpos orbiculatus**.
  PINK WINTER: **Ribes sanguineum**.
  RED-FLOWERING: **Ribes sanguineum**.
  WHITE-FLOWERED: **Ribes indecorum**.
CUSHIONBUSH: **Calocephalus brownii**.
Cussonia spicata Thunb. SPIKED CABBAGE TREE. CABBAGE TREE.
Cyathea medullaris (J. R. Forst.) Swartz. See **Sphaeropteris medullaris**.
CYCAD, CROZIER: **Cycas circinalis**.
  SAGO: **Zamia pumila**.
Cycas cairnsiana F. J. Muell.
Cycas circinalis L. SAGO PALM. FERN PALM. QUEEN SAGO. CROZIER CYCAD.
Cycas revoluta Thunb. SAGO PALM.
Cydonia japonica (Thunb.) Pers. and most listed cvs. See **Chaenomeles japonica** and cvs.
Cydonia japonica 'Rubra'. See **Chaenomeles speciosa** 'Rubra'.
Cydonia japonica 'Toyo-Nishiki'. See **Chaenomeles speciosa** 'Toyo-Nishiki'.
Cydonia oblonga Mill. Syn.: *C. vulgaris*. QUINCE.
Cydonia sinensis (Dum.-Cours.) Thouin. Syn.: *Chaenomeles sinensis*. CHINESE QUINCE.
Cydonia vulgaris Pers. See **C. oblonga**.
Cyperus papyrus L. PAPYRUS. PAPER PLANT.
Cyphomandra betacea (Cav.) Sendtn. See **C. crassifolia**.
Cyphomandra crassifolia (Ort.) Macb. Syn.: *C. betacea*. TREE TOMATO. TOMATO TREE.
CYPRESS, ALASKA: **Chamaecyparis nootkatensis**.
  ARIZONA: **Cupressus arizonica**.
  AZURE: **Chamaecyparis lawsoniana** 'Azurea'.
  BALD: **Taxodium distichum**.
  BLUE LAWSON: **Chamaecyparis lawsoniana** 'Allumii'.
  BOULEVARD: **Chamaecyparis pisifera** 'Boulevard'.
  CLUB-MOSS: **Chamaecyparis lawsoniana** 'Lycopodioides'.
  COLUMNAR ITALIAN: **Cupressus sempervirens** 'Stricta'.
  CRIPPS GOLDEN: **Chamaecyparis obtusa** 'Crippsii'.
  CUYAMACA: **Cupressus arizonica** var. *stephensonii*.
  DONARD GOLD: **Cupressus macrocarpa** 'Donard Gold'.
  DWARF ALASKA: **Chamaecyparis nootkatensis** 'Compacta'.
  DWARF GOLDEN HINOKI: **Chamaecyparis obtusa** 'Nana Aurea'.
  DWARF GOLDEN THREAD-BRANCH: **Chamaecyparis pisifera** 'Filifera Aurea Nana'.
  DWARF HINOKI: **Chamaecyparis obtusa** 'Nana'.
  DWARF LAWSON: **Chamaecyparis lawsoniana** 'Minima'.
  DWARF NOOTKA: **Chamaecyparis nootkatensis** 'Compacta'.
  ELLWOOD: **Chamaecyparis lawsoniana** 'Ellwoodii'.
  FALSE: **Chamaecyparis** spp.
  FERNSPRAY: **Chamaecyparis obtusa** 'Filicoides'.
  FLETCHER: **Chamaecyparis lawsoniana** 'Fletcheri'.
  FORBES: **Cupressus guadalupensis** var. *forbesii*.
  FORSTECK: **Chamaecyparis lawsoniana** 'Forsteckensis'.
  GOLDEN HINOKI: **Chamaecyparis obtusa** 'Aurea'.
  GOLDEN LAWSON: **Chamaecyparis lawsoniana** 'Lutea'.
  GOLDEN PLUME: **Chamaecyparis pisifera** 'Plumosa Aurea'.
  GOLDEN THREAD-BRANCH: **Chamaecyparis pisifera** 'Filifera Aurea'.
  GOWEN: **Cupressus goveniana**.
  GUADALUPE: **Cupressus guadalupensis**.
  HINOKI: **Chamaecyparis obtusa**.
  ITALIAN: **Cupressus sempervirens**.
  KASHMIR: **Cupressus cashmeriana**.
  KNOWEFIELD: **Chamaecyparis lawsoniana** 'Knowefieldensis'.
  LAWSON: **Chamaecyparis lawsoniana**.
  LEYLAND: × **Cupressocyparis leylandii**.
  LITTLE BLUE: **Chamaecyparis lawsoniana** 'Minima Glauca'.
  MEXICAN: **Cupressus lusitanica**.
  MONTEREY: **Cupressus macrocarpa**.
  MONTEZUMA: **Taxodium mucronatum**.
  MOSS: **Chamaecyparis pisifera** 'Squarrosa'.
  MOURNING: **Chamaecyparis funebris**.
  NEST: **Chamaecyparis lawsoniana** 'Nestoides'.
  NEST: **Chamaecyparis lawsoniana** 'Nidiformis'.
  NOOTKA: **Chamaecyparis nootkatensis**.
  PLUME: **Chamaecyparis pisifera** 'Plumosa'.
  PORTUGUESE: **Cupressus lusitanica**.
  PYRAMIDAL BLUE: **Chamaecyparis lawsoniana** 'Allumii'.
  ROUGH-BARKED ARIZONA: **Cupressus arizonica**.
  SANTA CRUZ: **Cupressus goveniana** var. *abramsiana*.
  SAWARA: **Chamaecyparis pisifera**.
  SCARAB: **Chamaecyparis lawsoniana** 'Allumii'.
  SILVER QUEEN: **Chamaecyparis lawsoniana** 'Silver Queen'.
  SLENDER HINOKI: **Chamaecyparis obtusa** 'Gracilis'.

CYPRESS, continued
   SMOOTH ARIZONA: **Cupressus arizonica** var. **glabra**.
   SMOOTH-BARKED ARIZONA: **Cupressus arizonica** var. **glabra**.
   SPREADING ITALIAN: **Cupressus sempervirens** var. **horizontalis**.
   STEWART GOLDEN: **Chamaecyparis lawsoniana 'Stewartii'**.
   SWAMP: **Taxodium distichum**.
   TECATE: **Cupressus guadalupensis** var. **forbesii**.
   THREAD-BRANCH: **Chamaecyparis pisifera 'Filifera'**.
   WEEPING: **Chamaecyparis lawsoniana 'Pendula'**.
   WESTERMANN: **Chamaecyparis lawsoniana 'Westermannii'**.
   WISSEL: **Chamaecyparis lawsoniana 'Wisselii'**.
CYPRUS TURPENTINE: **Pistacia terebinthus**.
Cytisus 'Andreanus'. See **C. scoparius 'Andreanus'**.
Cytisus ardoini Fournier.
Cytisus battandieri Maire. ATLAS BROOM.
Cytisus canariensis (L.) Kuntze. Syn.: *Genista canariensis*. CANARY ISLAND BROOM. GENISTA (of florists).
Cytisus 'Carla'. See **C. (Dallimore Hybrid) 'Carla'**.
Cytisus (Dallimore Hybrids). A group designation for hybrids derived from **C. × dallimorei** and other spp.
Cytisus (Dallimore Hybrid) 'Burkwoodii'. BURKWOOD BROOM.
Cytisus (Dallimore Hybrid) 'Carla'.
Cytisus (Dallimore Hybrid) 'Dorothy Walpole'.
Cytisus (Dallimore Hybrid) 'Geoffrey Skipwith'.
Cytisus (Dallimore Hybrid) 'Killiney Red'.
Cytisus (Dallimore Hybrid) 'Lilac Time'.
Cytisus (Dallimore Hybrid) 'Lord Lambourne'.
Cytisus (Dallimore Hybrid) 'Marie Burkwood'.
Cytisus (Dallimore Hybrid) 'Peter Pan'.
Cytisus (Dallimore Hybrid) 'Pomona'.
Cytisus (Dallimore Hybrid) 'Red Wings'.
Cytisus (Dallimore Hybrid) 'San Francisco'.
Cytisus (Dallimore Hybrid) 'Stanford'.
Cytisus (Dallimore Hybrid) 'St. Mary's.
Cytisus × dallimorei Rolfe. [**C. multiflorus × C. scoparius 'Andreanus'**].
Cytisus decumbens (Durande) Spach. PROSTRATE BROOM.
Cytisus demissus Boiss. See **C. hirsutus** var. **demissus**.
Cytisus fragrans Lam. See **C. supranubius**.
Cytisus 'Geoffrey Skipwith'. See **C. (Dallimore Hybrid) 'Geoffrey Skipwith'**.
Cytisus hirsutus L. var. **demissus** (Boiss.) Halácsy. Syn.: *C. demissus*.
Cytisus 'Hollandia'. [**C. × praecox × C. (Dallimore Hybrid) 'Burkwoodii'**].
Cytisus × kewensis Bean. [**C. ardoini × C. multiflorus**]. KEW BROOM.
Cytisus lydia. Listed name for **Genista lydia**.
Cytisus monspessulanus L. MONTPELIER BROOM. FRENCH BROOM.
Cytisus 'Moonlight'. See **C. × praecox 'Warminster'** or **C. scoparius 'Moonlight'**.
Cytisus multiflorus (L'Her. ex Ait.) Sweet. WHITE SPANISH BROOM.
Cytisus × praecox Bean. [**C. purgans × C. multiflorus**].
Cytisus × praecox 'Albus'.
Cytisus × praecox 'Hollandia'. See **C. 'Hollandia'**.
Cytisus × praecox 'Luteus'.
Cytisus × praecox 'Moonlight'. See **C. × praecox 'Warminster'**.
Cytisus × praecox 'Warminster'. MOONLIGHT BROOM. WARMINSTER BROOM.
Cytisus × praecox 'Zeelandia'. See **C. 'Zeelandia'**.
Cytisus procumbens (Waldstein & Kitaibel ex Willd.) K. Spreng. Syn.: *Genista procumbens*.
Cytisus purgans (L.) Spach. PROVENCE BROOM.

Cytisus purpureus Scop. PURPLE BROOM.
Cytisus racemosus Nichols., not Marnock. See **C. × spachianus**.
Cytisus scoparius (L.) Link. Syn.: *Genista scoparia*. SCOTCH BROOM.
Cytisus scoparius hybrids. See **C. (Dallimore Hybrids)**.
Cytisus scoparius 'Andreanus'. Syn.: *Genista andreana*. NORMANDY BROOM.
Cytisus scoparius 'Burkwoodii'. See **C. (Dallimore Hybrid) 'Burkwoodii'**.
Cytisus scoparius 'Lord Lambourne'. See **C. (Dallimore Hybrid) 'Lord Lambourne'**.
Cytisus scoparius 'Moonlight'. MOONLIGHT BROOM.
Cytisus scoparius 'Pomona'. See **C. (Dallimore Hybrid) 'Pomona'**.
Cytisus scoparius 'Red Wings'. See **C. (Dallimore Hybrid) 'Red Wings'**.
Cytisus scoparius 'San Francisco'. See **C. (Dallimore Hybrid) 'San Francisco'**.
Cytisus scoparius 'Stanford'. See **C. (Dallimore Hybrid) 'Stanford'**.
Cytisus scoparius 'St. Mary's'. See **C. (Dallimore Hybrid) 'St. Mary's'**.
Cytisus × spachianus Webb [?*C. canariensis × C. stenopetalus*]. Syn.: *C. racemosus*. EASTER BROOM.
Cytisus stenopetalus (Webb) Christ.
Cytisus supranubius (L. f.) Kuntze. Syn.: *C. fragrans*. *Genista fragrans*. SWEET BROOM.
Cytisus 'Warminster'. See **C. × praecox 'Warminster'**.
Cytisus 'Zeelandia'. [**C. × praecox × C. × dallimorei**].

D

Daboecia 'Alba'. See **D. cantabrica 'Alba'**.
Daboecia azorica Tutin & E. F. Warb.
Daboecia cantabrica (Huds.) K. Koch. Syn.: *D. poliifolia*. IRISH BELL HEATH. IRISH HEATH.
Daboecia cantabrica 'Alba'.
Daboecia cantabrica 'Atropurpurea'.
Daboecia cantabrica 'Pallida'.
Daboecia cantabrica 'Praegerae'.
Daboecia cantabrica 'Rosea'. See **D. cantabrica 'Pallida'**.
Daboecia poliifolia (Juss.) D. Don. See **D. cantabrica**.
Daboecia poliifolia 'Alba'. See **D. cantabrica 'Alba'**.
Daboecia poliifolia 'Rosea'. See **D. cantabrica 'Pallida'**.
Daboecia 'Praegerae'. See **D. cantabrica 'Praegerae'**.
DAGGER PLANT: **Yucca aloifolia**.
DAHLIA, TREE: **Dahlia imperialis**.
Dahlia imperialis Roezl ex Ortgies. TREE DAHLIA.
Dais cotinifolia L. POMPON TREE.
DAISY, AFRICAN: **Arctotis stoechadifolia**.
   AFRICAN SHRUB: **Felicia fruticosa**.
   BLUE: **Felicia amelloides**.
   BOSTON YELLOW: **Chrysanthemum frutescens 'Chrysaster'**.
   CLANWILLIAM: **Euryops speciosissimus**.
   GIANT TREE: **Montanoa arborescens**.
   GIANT TREE: **Montanoa grandiflora**.
   GOLDEN SHRUB: **Euryops pectinatus**.
   PARIS: **Chrysanthemum frutescens**.
   PORTUGUESE: **Chrysanthemum lacustre**.
   SHASTA: **Chrysanthemum × superbum**.
   SOUTH AFRICAN TREE: **Euryops speciosissimus**.
   TRAILING AFRICAN: **Osteospermum fruticosum**.
DAISYBUSH: **Olearia × haastii**.

Dalea spinosa Gray. SMOKE TREE.
Dammaropsis kingiana Warb.
Danae racemosa (L.) Moench. Syn.: *Ruscus racemosus*. ALEXANDRIAN LAUREL.
DAPHNE, FEBRUARY: **Daphne mezereum**.
   GARLAND: **Daphne cneorum**.
   LILAC: **Daphne genkwa**.
   MANTEN: **Daphne × mantensiana**.
   MANTEN: **Daphne × mantensiana 'Manten'**.
   PINK WINTER: **Daphne odora 'Aureo-marginata'**.
   SOMERSET: **Daphne × burkwoodii 'Somerset'**.
   WHITE WINTER: **Daphne odora 'Alba'**.
   WINTER: **Daphne odora**.
Daphne 'Alba'. See **D. odora 'Alba'**.
Daphne blagayana Freyer.
Daphne × burkwoodii Turrill. [*D. caucasica* × *D. cneorum*].
Daphne × burkwoodii 'Somerset'. SOMERSET DAPHNE.
Daphne caucasica Pall.
Daphne cneorum L. GARLAND DAPHNE. GARLAND FLOWER.
Daphne cneorum 'Alba'.
Daphne cneorum 'Eximia'.
Daphne cneorum 'Ruby Glow'.
Daphne collina Sm.
Daphne collina var. *neapolitana* (Lodd.) Lindl. See **D. × neapolitana**.
Daphne genkwa Siebold & Zucc. LILAC DAPHNE.
Daphne giraldii Nitsche.
Daphne laureola L. SPURGE LAUREL.
Daphne × mantensiana Manten ex T. Taylor & Vrugtman. [**D. × burkwoodii** × **D. retusa**]. MANTEN DAPHNE.
Daphne × mantensiana 'Manten'. MANTEN DAPHNE.
Daphne mezereum L. FEBRUARY DAPHNE.
Daphne mezereum forma *alba* (Weston) Schelle.
Daphne × neapolitana Lodd. [?**D. oleoides** Schrab. × **D. cneorum**]. Syn.: *D. collina* var. *neapolitana*.
Daphne odora Thunb. WINTER DAPHNE.
Daphne odora 'Alba'. WHITE WINTER DAPHNE.
Daphne odora 'Aureo-marginata'. PINK WINTER DAPHNE.
Daphne odora 'Marginata'. See **D. odora 'Aureo-marginata'**.
Daphne odora 'Rose Queen'.
Daphne retusa Hemsl.
Daphne 'Somerset'. See **D. × burkwoodii 'Somerset'**.
Dasylirion wheeleri S. Wats.
DATE, CHINESE: **Ziziphus jujuba**.
   EDIBLE: **Phoenix dactylifera**.
DATE PLUM: **Diospyros lotus**.
DATE PLUM: **Diospyros virginiana**.
DATIL: **Yucca baccata**.
Datura arborea L. See **Brugmansia arborea**.
Datura × candida (Pers.) Safford. See **Brugmansia × candida**.
Datura sanguinea Ruiz & Pav. See **Brugmansia sanguinea**.
Datura suaveolens Humb. & Bonpl. ex Willd. See **Brugmansia suaveolens**.
Daubentonia grandiflora. Listed name for **Sesbania grandiflora**.
Daubentonia punicea (Cav.) DC. See **Sesbania punicea**.
Daubentonia tripetiana Poit. See **Sesbania punicea**.
Daubentonia tripetii Poit. See **Sesbania punicea**.
Davidia involucrata Baill. DOVE TREE.
Decumaria barbara L. CLIMBING HYDRANGEA.
DEERBRUSH: **Ceanothus integerrimus**.
DEERWEED: **Lotus scoparius**.
Delonix regia (Bojer ex Hook.) Raf. Syn.: *Poinciana regia*. ROYAL POINCIANA. FLAMBOYANT.
Delostoma lobbii Seem.
Delostoma roseum (Karst. & Triana) K. Schum.
Dendriopoterium menendezii Sventenius.

Dendrocalamus asper (Schult. f.) Backer ex K. Heyne.
Dendrocalamus giganteus Munro.
Dendrocalamus latiflorus Munro. May not be cult. Some plants so called may be **Bambusa oldhamii**.
Dendrocalamus membranaceus Munro. WHITE BAMBOO.
Dendrocalamus strictus (Roxb.) Nees. MALE BAMBOO. CALCUTTA BAMBOO.
Dendromecon harfordii Kell. See **D. rigida** subsp. **harfordii**.
Dendromecon rigida Benth. TREE POPPY. BUSH POPPY.
Dendromecon rigida subsp. **harfordii** (Kell.) Raven. Syn.: *D. harfordii*. ISLAND TREE POPPY.
Dendropanax chevalieri (R. Vig.) Merrill.
Dendropanax trifidus (Thunb.) Makino ex Hara. Syn.: *Gilibertia japonica*.
Desfontainea spinosa Ruiz & Pav.
DEUTZIA, DOUBLE ROSE: **Deutzia scabra 'Plena'**.
   FUZZY: **Deutzia scabra**.
   SLENDER: **Deutzia gracilis**.
Deutzia crenata Siebold & Zucc.
Deutzia × elegantissima (Lemoine) Rehd. [**D. purpurascens** × **D. scabra**].
Deutzia gracilis Siebold & Zucc. SLENDER DEUTZIA.
Deutzia gracilis var. *rosea* Lemoine. See **D. × rosea**.
Deutzia × hybrida Lemoine. [**D. longifolia** Franch. × **D. discolor** Hemsl.].
Deutzia × hybrida 'Magicien'.
Deutzia × hybrida 'Mont Rose'.
Deutzia × lemoinei Bois. [**D. parviflora** Bunge × **D. gracilis**].
Deutzia 'Magicien'. See **D. × hybrida 'Magicien'**.
Deutzia monbeigii W. W. Sm.
Deutzia 'Mont Rose'. See **D. × hybrida 'Mont Rose'**.
Deutzia 'Pride of Rochester'. See **D. scabra 'Pride of Rochester'**.
Deutzia pulchra S. Vidal.
Deutzia purpurascens (L. Henry) Rehd.
Deutzia × rosea (Lemoine) Rehd. [**D. gracilis** × **D. purpurascens**]. Syn.: *D. gracilis* var. *rosea*.
Deutzia scabra Thunb. FUZZY DEUTZIA.
Deutzia scabra 'Candidissima'.
Deutzia scabra 'Plena'. DOUBLE ROSE DEUTZIA.
Deutzia scabra 'Pride of Rochester'.
DEVIL'S-WALKING-STICK: **Aralia spinosa**.
DEVILWOOD: **Osmanthus americanus**.
DEWDROP, GOLDEN: **Duranta repens**.
Dichorisandra thyrsiflora Mikan. BLUE GINGER.
DICKSONIA, NEW ZEALAND: **Dicksonia squarrosa**.
   TASMANIAN: **Dicksonia antarctica**.
Dicksonia antarctica Labill. TASMANIAN TREE FERN. TASMANIAN DICKSONIA.
Dicksonia squarrosa (G. Forst.) Swartz. NEW ZEALAND DICKSONIA. NEW ZEALAND TREE FERN.
Dicliptera suberecta (André) Bremekamp. Syn.: *Jacobinia suberecta*.
Dieffenbachia amoena Bull. Not cult. Plants so called have not been identified.
Dieffenbachia exotica. Listed name. Plants so called have not been identified.
Dieffenbachia maculata (Lodd.) G. Don. Syn.: *D. picta*. SPOTTED DUMB CANE.
Dieffenbachia maculata 'Rudolph Roehrs'.
Dieffenbachia maculata 'Superba'.
Dieffenbachia picta Schott, and listed cvs. See **D. maculata** and listed cvs.
Dieffenbachia seguine (Jacq.) Schott. DUMB CANE. MOTHER-IN-LAW PLANT.
Dietes bicolor Sweet. Syn.: *Moraea bicolor*.

Dietes iridioides (L.) Sweet ex Klatt. Syn.: *Moraea iridioides.* AFRICAN IRIS.
Digitalis canariensis L. See **Isoplexis canariensis.**
DIOON, CHESTNUT: **Dioon edule.**
   GIANT: **Dioon spinulosum.**
Dioon dohenii. Listed name. Plants so called have not been identified.
Dioon edule Lindl. CHESTNUT DIOON.
Dioon spinulosum Dyer. GIANT DIOON.
Dioscorea elephantipes (L'Hér.) Engler. Syn.: *Testudinaria elephantipes.* ELEPHANT'S FOOT. HOTTENTOT-BREAD.
Dioscorea macrostachya Benth.
DIOSMA, PINK: **Coleonema pulchrum.**
Diosma alba Thunb. See **Coleonema album.**
Diosma ericoides L. BREATH-OF-HEAVEN. BUCHU.
Diosma pulchra. Listed name for **Coleonema pulchrum.**
Diosma reevesii?
Diospyros 'Fuyu'. See **D. kaki 'Fuyu'.**
Diospyros 'Hachiya'. See **D. kaki 'Hachiya'.**
Diospyros kaki L. f. KAKI. JAPANESE PERSIMMON.
Diospyros kaki 'Chocolate Brown'.
Diospyros kaki 'Fuyu'. FUYU PERSIMMON.
Diospyros kaki 'Hachiya'. HACHIYA PERSIMMON.
Diospyros kaki 'Tamopan'.
Diospyros lotus L. DATE PLUM. DATE PLUM PERSIMMON.
Diospyros virginiana L. VIRGINIA PERSIMMON. POSSUMWOOD. POSSUM APPLE. DATE PLUM.
DIPLACUS, AZALEA-FLOWERED: **Diplacus grandiflorus.**
   ISLAND: **Diplacus parviflorus.**
   MISSION RED: **Diplacus puniceus.**
   SOUTHERN: **Diplacus longiflorus.**
   WOOLLY: **Diplacus longiflorus** var. **calycinus.**
Diplacus aurantiacus (Curtis) Jeps. Syn.: *Mimulus aurantiacus.* BUSH MONKEY FLOWER.
Diplacus calycinus Eastw. See **D. longiflorus** var. **calycinus.**
Diplacus grandiflorus (Lindl. & Paxt.) Groenl. Syn.: *Mimulus bifidus.* PLUMAS MONKEY FLOWER. AZALEA-FLOWERED DIPLACUS.
Diplacus longiflorus Nutt. Syn.: *Mimulus longiflorus.* MONKEY FLOWER. SOUTHERN MONKEY FLOWER. SOUTHERN DIPLACUS.
Diplacus longiflorus var. ***calycinus*** (Eastw.) Jeps. Syn.: *D. calycinus. Mimulus longiflorus* subsp. *calycinus.* WOOLLY DIPLACUS. WOOLLY MONKEY FLOWER.
Diplacus longiflorus var. ***rutilus*** (A. L. Grant) McMinn. Syn.: *D. rutilus. Mimulus longiflorus* var. *rutilus.* RED MONKEY FLOWER.
Diplacus parviflorus Greene. Syn.: *Mimulus flemingii.* ISLAND MONKEY FLOWER. ISLAND DIPLACUS.
Diplacus puniceus Nutt. Syn.: *Mimulus puniceus.* MISSION RED DIPLACUS.
Diplacus rutilus (A. L. Grant) McMinn. See **D. longiflorus** var. **rutilus.**
Dipladenia × *amoena* T. Moore. See **Mandevilla** × **amabilis.**
Dipladenia rosea. Listed name, perhaps for **Mandevilla** × **amabilis.**
Dipladenia splendens (Hook. f.) A. DC. See **Mandevilla splendens.**
Diplazium esculentum (Retzius) Swartz. VEGETABLE FERN.
Diplopappus filifolius (Vent.) DC. See **Felicia fruticosa.**
Diplopappus fruticulosus Less. See **Felicia fruticosa.**
Diplothemium caudescens Mart. See **Polyandrococos caudescens.**
Dirca occidentalis Gray. WESTERN LEATHERWOOD.
Dissotis rotundifolia (Sm.) Triana.

Distictis buccinatoria (DC.) A. Gentry. Syn.: *Bignonia cherere. Phaedranthus buccinatorius.* BLOOD-RED TRUMPET VINE.
Distictis cinerea Greenm. See **D. laxiflora.**
Distictis lactiflora (Vahl.) DC.
Distictis laxiflora (DC.) Greenm. Syn.: *D. cinerea.* VANILLA TRUMPET VINE.
Distictis 'Rivers'. ROYAL TRUMPET VINE.
Distictis riversii. Listed name for **D. 'Rivers'.**
Distylium racemosum Siebold & Zucc. ISU TREE.
Distylium racemosum 'Variegatum'.
DITTANY, CRETE: **Origanum dictamnus.**
Dizygotheca elegantissima (Veitch) R. Vig. & Guillaum. Syn.: *Aralia elegantissima.* FINGER ARALIA. THREAD-LEAF FALSE ARALIA.
Dizygotheca kerchoveana (Veitch) N. Taylor.
DODONAEA, PURPLE-LEAVED: **Dodonaea viscosa 'Purpurea'.**
Dodonaea microzyga F. J. Muell.
Dodonaea viscosa (L.) Jacq. CLAMMY HOPSEED. RED HOPSEED. HOPBUSH. HOPSEED BUSH.
Dodonaea viscosa 'Atropurpurea'. See **D. viscosa 'Purpurea'.**
Dodonaea viscosa 'Purpurea'. PURPLE-LEAVED DODONAEA. HOPSEED TREE.
Dodonaea viscosa 'Saratoga'.
DOGWOOD, AMERICAN: **Cornus stolonifera.**
   BLOOD-TWIG: **Cornus sanguinea.**
   CHINESE KOUSA: **Cornus kousa** var. **chinensis.**
   CREEK: **Cornus** × **californica.**
   DOUBLE WHITE: **Cornus florida 'Pluribracteata'.**
   EASTERN: **Cornus florida.**
   FLOWERING: **Cornus florida.**
   GIANT: **Cornus controversa.**
   KOUSA: **Cornus kousa.**
   PACIFIC: **Cornus nuttallii.**
   PINK FLOWERING: **Cornus florida 'Rubra'.**
   RED-BARK: **Cornus alba.**
   RED FLOWERING: **Cornus florida 'Cherokee Chief'.**
   RED FLOWERING: **Cornus florida 'Rubra'.**
   RED-OSIER: **Cornus stolonifera.**
   SIBERIAN: **Cornus alba 'Sibirica'.**
   TATARIAN: **Cornus alba.**
   TRICOLOR: **Cornus florida 'Welchii'.**
   WESTERN: **Cornus nuttallii.**
   WHITE EVERGREEN: **Cornus capitata.**
   YELLOW-TWIG: **Cornus stolonifera 'Flaviramea'.**
Dolichandra cynanchoides Cham. Syn.: *Macfadyena cynanchoides.*
Dolichos lignosus L. AUSTRALIAN PEA VINE.
DOMBEYA, PINK-BALL: **Dombeya** × **cayeuxii.**
   PINK-BALL: **Dombeya wallichii.**
Dombeya burgessiae Gerrard ex Harv. Syn.: *D. mastersii. D. nairobensis.*
Dombeya calantha K. Schum.
Dombeya × ***cayeuxii*** André. [*D. burgessiae* × *D. wallichii*]. PINK-BALL DOMBEYA.
Dombeya mastersii Hook. f. See **D. burgessiae.**
Dombeya nairobensis Engler. See **D. burgessiae.**
Dombeya wallichii (Lindl.) Benth. & Hook. ex B. D. Jackson. PINK-BALL DOMBEYA.
DONKEY-TAIL: **Sedum morganianum.**
Doryanthes excelsa Correa. GLOBE SPEAR LILY.
Doryanthes palmeri W. Hill. PALMER SPEAR LILY. SPEAR FLOWER.
DOVE TREE: **Davidia involucrata.**
Dovyalis caffra (Hook. f. & Harv.) Warb. Syn.: *Aberia caffra.* KEI APPLE.
Doxantha capreolata (L.) Miers. See **Bignonia capreolata.**

Doxantha unguis-cati (L.) Rehd. See **Macfadyena unguis-cati**.
DRACAENA, AUSTRALIAN: **Cordyline stricta**.
  BLUE: **Cordyline indivisa**.
  BRONZE: **Cordyline australis 'Atropurpurea'**.
  STRIPED: **Dracaena deremensis 'Warneckii'**.
Dracaena australis G. Forst. See **Cordyline australis**.
Dracaena cincta Baker. Some plants listed as **D. marginata** may belong here.
Dracaena concinna Kunth. Some plants listed as **D. marginata** may belong here.
Dracaena deremensis Engler.
Dracaena deremensis **'Craigii'**.
Dracaena deremensis **'Gold King'**.
Dracaena deremensis **'Janet Craig'**.
Dracaena deremensis **'Janet Craig Compacta'**.
Dracaena deremensis **'Longii'**.
Dracaena deremensis **'Warneckii'**. STRIPED DRACAENA.
Dracaena deremensis **'Warneckii Compacta'**.
Dracaena draco L. DRAGON TREE.
Dracaena fragrans (L.) Ker-Gawl.
Dracaena fragrans **'Lindenii'**.
Dracaena fragrans **'Massangeana'**.
Dracaena fragrans **'Rothiana'**.
Dracaena fragrans **'Victoria'**.
Dracaena godseffiana Baker. See **D. surculosa**.
Dracaena goldieana Baker.
Dracaena hookerana K. Koch.
Dracaena hookerana 'Rothiana'. See **D. fragrans 'Rothiana'**.
Dracaena indivisa G. Forst. See **Cordyline indivisa**.
Dracaena indivisa 'Atropurpurea'. See **Cordyline indivisa 'Atropurpurea'**.
Dracaena 'Janet Craig'. See **D. deremensis 'Janet Craig'**.
Dracaena marginata Lam. Probably not cult. Plants so called may be either **D. cincta** or **D. concinna**.
Dracaena 'Massangeana'. See **D. fragrans 'Massangeana'**.
Dracaena **'Masseffana'**. [**D. fragrans 'Massangeana'** × **D. surculosa**].
Dracaena 'Mme. André'. See **Cordyline terminalis 'Mme. Eugene André'**.
Dracaena reflexa Lam. Syn.: *Pleomele reflexa*.
Dracaena reflexa **'Song of India'**.
Dracaena 'Rothiana'. See **D. fragrans 'Rothiana'**.
Dracaena sanderana Sander ex M. T. Mast.
Dracaena stricta Hort. See **Cordyline stricta**.
Dracaena surculosa Lindl. Syn.: *D. godseffiana*.
Dracaena surculosa **'Florida Beauty'**.
Dracaena surculosa **'Kelleri'**.
Dracaena thalioides E. Morr. Syn.: *Pleomele thalioides*.
Dracaena 'Warneckii'. See **D. deremensis 'Warneckii'**.
DRACENA. See DRACAENA.
DRAGON TREE: **Dracaena draco**.
Drimys lanceolata (Poir.) Baill. PEPPER TREE.
Drimys winteri J. R. Forst. & G. Forst. WINTER'S BARK.
Dryandra formosa R. Br.
Duranta ellisia Jacq. See **D. repens**.
Duranta erecta L. See **D. repens**.
Duranta erecta 'Grandiflora'. See **D. repens 'Grandiflora'**.
Duranta lorentzii Griseb.
Duranta plumieri L. See **D. repens**.
Duranta repens L. Syn.: *D. ellisia*. *D. erecta*. *D. plumieri*. SKY FLOWER. GOLDEN DEWDROP. PIGEON BERRY.
Duranta repens **'Grandiflora'**.
Duranta stenostachya Tod. BRAZILIAN SKY FLOWER.
DUSTY-MILLER: **Artemisia stellerana**.
DUSTY-MILLER: **Centaurea cineraria**.
DUSTY-MILLER: **Senecio cineraria**.
DUSTY-MILLER: **Senecio vira-vira**.
  CATALINA: **Eriophyllum nevinii**.
DUTCHMAN'S PIPE: **Aristolochia durior**.
DUTCHMAN'S PIPE: **Aristolochia tomentosa**.
  CALIFORNIA: **Aristolochia californica**.
Dyschoriste thunbergiiflora (S. Moore) Lindau. HELMET SHRUB.

E

EASTER-LILY VINE: **Beaumontia grandiflora**.
EBONY, GREEN: **Jacaranda acutifolia**.
  MOUNTAIN: **Bauhinia variegata**.
Echeveria agavoides Lem. Syn.: *Urbinia agavoides*.
Echium candicans L. f.
Echium fastuosum Jacq. PRIDE-OF-MADEIRA.
Echium fastuosum **'Blue Improved'**.
Echium pininana Webb & Berth. PRIDE-OF-TENERIFFE.
Echium wildpretii H. Pearson ex Hook. f. TOWER-OF-JEWELS.
EGLANTINE: **Rosa rubiginosa**.
Ehretia anacua (Terán & Berl.) I. M. Johnst. Syn.: *E. elliptica*. SUGARBERRY. ANAQUA.
Ehretia dicksonii Hance.
Ehretia elliptica A. DC. See **E. anacua**.
Ehretia hottentotica Burchell. See **E. rigida**.
Ehretia rigida (Thunb.) Druce. Syn.: *E. hottentotica*.
Ehretia thyrsiflora (Siebold & Zucc.) Nakai.
Ehretia tinifolia L. BASTARD CHERRY.
ELAEAGNUS, GOLDEN: **Elaeagnus pungens 'Maculata'**.
  SILVER-EDGE: **Elaeagnus pungens 'Marginata'**.
  THORNY: **Elaeagnus pungens**.
  VARIEGATED THORNY: **Elaeagnus pungens 'Variegata'**.
Elaeagnus angustifolia L. OLEASTER. RUSSIAN OLIVE. SILVER BERRY.
Elaeagnus **'Clemsonii'**.
Elaeagnus **'Coral Silver'**.
Elaeagnus × *ebbingei* Boom. [**E. macrophylla** × **E. pungens**]. EBBINGE'S SILVERBERRY.
Elaeagnus 'Fruitlandii'. See **E. pungens 'Fruitlandii'**.
Elaeagnus macrophylla Thunb.
Elaeagnus macrophylla 'Ebbingei'. Listed name for **E. × ebbingei**.
Elaeagnus 'Maculata'. See **E. pungens 'Maculata'**.
Elaeagnus multiflora Thunb. GUMI.
Elaeagnus philippensis Perrottet.
Elaeagnus pungens Thunb. SILVERBERRY. THORNY ELAEAGNUS.
Elaeagnus pungens **'Aurea'**.
Elaeagnus pungens **'Fruitlandii'**. FRUITLAND SILVERBERRY.
Elaeagnus pungens **'Maculata'**. GOLDEN ELAEAGNUS.
Elaeagnus pungens **'Marginata'**. SILVER-EDGE ELAEAGNUS.
Elaeagnus pungens **'Variegata'**. VARIEGATED THORNY ELAEAGNUS.
Elaeagnus umbellata Thunb.
Elaeagnus umbellata **'Cardinal'**.
Elaeodendron orientale Jacq. See **Cassine orientalis**.
ELDER: See ELDERBERRY.
ELDER, BOX: **Acer negundo**.
  SILVER VARIEGATED BOX: **Acer negundo 'Variegatum'**.
  YELLOW: **Tecoma stans**.
ELDERBERRY, BLACK: **Sambucus melanocarpa**.
  BLUE: **Sambucus caerulea**.
  COAST RED: **Sambucus callicarpa**.
  EUROPEAN RED: **Sambucus racemosa**.
  MEXICAN: **Sambucus mexicana**.

ELDERBERRY, continued
    MOUNTAIN RED: **Sambucus microbotrys.**
    RED: **Sambucus racemosa.**
ELEPHANT BUSH: **Portulacaria afra.**
ELEPHANT-EAR: **Philodendron hastatum.**
    BLUE: **Xanthosma sagittifolium.**
    PURPLE: **Xanthosma violaceum.**
ELEPHANT-FOOT TREE: **Beaucarnea recurvata.**
ELEPHANT'S-EAR: **Alocasia macrorrhiza.**
ELEPHANT'S FOOD: **Portulacaria afra.**
ELEPHANT'S FOOT: **Dioscorea elephantipes.**
ELK GRASS: **Xerophyllum tenax.**
ELM, AMERICAN: **Ulmus americana.**
    BELGIAN: **Ulmus × hollandica 'Belgica'.**
    BREA: **Ulmus parvifolia 'Drake'.**
    BUISMAN: **Ulmus 'Christine Buisman'.**
    CAMPERDOWN: **Ulmus glabra 'Camperdownii'.**
    CEDAR: **Ulmus crassifolia.**
    CHICHESTER: **Ulmus × vegeta.**
    CHINESE: **Ulmus parvifolia.**
    DRAKE EVERGREEN: **Ulmus parvifolia 'Drake'.**
    DUTCH: **Ulmus × hollandica.**
    ENGLISH: **Ulmus procera.**
    EVERGREEN: **Ulmus parvifolia.**
    EVERGREEN CHINESE: **Ulmus parvifolia.**
    HOLLAND: **Ulmus × hollandica.**
    HUNTINGDON: **Ulmus × vegeta.**
    SCOTCH: **Ulmus glabra.**
    SIBERIAN: **Ulmus pumila.**
    SMOOTH-LEAF: **Ulmus carpinifolia.**
    WYCH: **Ulmus glabra.**
Embothrium coccineum J. R. Forst. & G. Forst. Syn.: *E. lanceolatum.* CHILEAN FIRE TREE. CHILEAN FIREBUSH.
Embothrium lanceolatum Ruiz & Pav. See **E. coccineum.**
Empetrum nigrum L. BLACK CROWBERRY.
EMPRESS TREE: **Paulownia tomentosa.**
ENCELIA, CALIFORNIA: **Encelia californica.**
Encelia californica Nutt. BUSH SUNFLOWER. CALIFORNIA ENCELIA.
Encelia farinosa Gray. INCIENSO. BRITTLEBUSH.
Encelia virginensis A. Nels. subsp. **actonii** (Elmer) Keck. ACTON SUNFLOWER.
Encephalartos ferox G. Bertoloni.
Encephalartos gratus Prain.
Encephalartos lebomboensis Verdoorn.
Encephalartos transvenosus Stapf & Davy.
ENGELMANN CREEPER: **Parthenocissus quinquefolia 'Engelmannii'.**
ENKIANTHUS, RED-VEINED: **Enkianthus campanulatus.**
Enkianthus campanulatus (Miq.) Nichols. RED-VEINED ENKIANTHUS.
Enkianthus campanulatus var. **palibinii** (Craib) Bean. Syn.: *E. palibinii.*
Enkianthus cernuus (Siebold & Zucc.) Makino.
Enkianthus cernuus var. **rubens** (Siebold & Zucc.) Makino.
Enkianthus palibinii Craib. See **E. campanulatus** var. **palibinii.**
Enkianthus perulatus (Miq.) Schneid.
Ensete maurelii. Listed name for an unidentified species.
Ensete ventricosum (Welw.) E. E. Cheesman. Syn.: *Musa ensete.* ABYSSINIAN BANANA.
EPAULETTE TREE: **Pterostyrax hispida.**
Epigaea asiatica Maxim. ASIATIC GROUND LAUREL.
Epigaea repens L. TRAILING ARBUTUS.
Epipremnum aureum (Linden & André) Bunting. Syn.: *Pothos aureus. Raphidophora aurea. Scindapsus aureus.* GOLDEN POTHOS.
Eranthemum nervosum (Vahl) R. Br. See **E. pulchellum.**
Eranthemum pulchellum Andr. Syn.: *E. nervosum.* BLUE SAGE.

Eremophila maculata (Ker-Gawl.) F. J. Muell.
Erica annectens Guthrie & Bolus.
Erica arborea L. TREE HEATH.
Erica arborea 'Alpina'.
Erica australis L. SOUTHERN HEATH.
Erica australis 'Mr. Robert'.
Erica australis 'Riverslea'.
Erica baccans L.
Erica bauera Andr.
Erica blanda Andr. See **E. doliiformis.** Some plants called **E. blanda** are **E. verticillata.**
Erica blenna Salisb.
Erica caffra L.
Erica canaliculata Andr. BLACK-EYED HEATHER. Plants called **E. melanthera** belong here.
Erica canaliculata 'Boscaweniana'. BOSCAWEN'S BLACK-EYED HEATHER.
Erica canaliculata 'Rosea'. CHRISTMAS HEATHER.
Erica canaliculata 'Rubra'. May be the same as **E. canaliculata 'Rosea'.**
Erica carnea L. and most listed cvs. See **E. herbacea.**
Erica carnea 'Arthur Johnson'. See **E. × darleyensis 'Arthur Johnson'.**
Erica carnea 'George Rendall'. See **E. × darleyensis 'George Rendall'.**
Erica chamissonis Klotzsch ex Benth. ORCHID BELL.
Erica chloroloma Lindl.
Erica ciliaris L. DORSET HEATH.
Erica ciliaris 'Mrs. C. H. Gill'.
Erica ciliaris 'Stoborough'.
Erica ciliaris 'Wych'.
Erica cinerea L. BELL HEATHER.
Erica cinerea 'Alba'.
Erica cinerea 'Alba Minor'.
Erica cinerea 'Atrorubens'.
Erica cinerea 'Atrosanguinea'.
Erica cinerea 'C. D. Eason'.
Erica cinerea 'C. D. Ford'.
Erica cinerea 'Colligan Bridge'.
Erica cinerea 'Cuprea'. Listed name, perhaps for **Calluna vulgaris 'Cuprea'.**
Erica cinerea 'Domino'.
Erica cinerea 'Eden Valley'.
Erica cinerea 'Foxhollow Mahogany'.
Erica cinerea 'Golden Drop'.
Erica cinerea 'Golden Hue'.
Erica cinerea 'P. S. Patrick'.
Erica cinerea 'Rose Queen'.
Erica cinerea 'Rosea'.
Erica cinerea 'Violacea'.
Erica codonodes Lindl. See **E. lusitanica.**
Erica cruenta Soland.
Erica 'Darley Dale'. See **E. × darleyensis 'Darley Dale'.**
Erica × darleyensis Bean. [**E. herbacea × E. erigena**]. MEDITERRANEAN HYBRID HEATHER.
Erica × darleyensis 'Arthur Johnson'.
Erica × darleyensis 'Darley Dale'.
Erica × darleyensis 'George Rendall'.
Erica × darleyensis 'Silberschmelze'.
Erica 'Dawn'. See **E. × watsonii 'Dawn'.**
Erica diaphana K. Spreng.
Erica doliiformis Salisb. Syn.: *E. blanda.* RED EVERBLOOMING HEATHER. EVERBLOOMING FRENCH HEATHER. Some plants called **E. doliiformis** may be **E. verticillata.**
Erica erigena R. Ross. Syn.: *E. mediterranea* of most authors, not L. MEDITERRANEAN HEATH.

Erica erigena 'Brightness'.
Erica erigena 'Maxima'. DWARF HEATHER.
Erica erigena 'Superba'.
Erica erigena 'W. T. Rackliff'.
Erica 'Felix Faure'. PINK FRENCH HEATHER.
Erica formosa Thunb.
Erica 'George Rendall'. See *E.* × *darleyensis* 'George Rendall'.
Erica glandulosa Thunb.
Erica globosa Andr.
Erica 'Gwen'. See *E.* × *watsonii* 'Gwen'.
Erica herbacea L. Syn.: *E. carnea. E. mediterranea.* WINTER HEATH. MEDITERRANEAN HEATH.
Erica herbacea 'Alba'.
Erica herbacea 'Aurea'.
Erica herbacea 'Cecilia M. Beale'.
Erica herbacea 'Eileen Porter'.
Erica herbacea 'James Backhouse'.
Erica herbacea 'King George'. KING GEORGE SPRING HEATHER.
Erica herbacea 'Praecox Rubra'.
Erica herbacea 'Ruby Glow'. RUBY GLOW HEATHER.
Erica herbacea 'Sherwood Creeping'. See *E. herbacea* 'Sherwoodii'.
Erica herbacea 'Sherwood Early Red'.
Erica herbacea 'Sherwoodii'.
Erica herbacea 'Snow Queen'.
Erica herbacea 'Springwood'. See *E. herbacea* 'Springwood White'.
Erica herbacea 'Springwood Pink'.
Erica herbacea 'Springwood White'. SPRINGWOOD WHITE HEATHER.
Erica herbacea 'Startler'.
Erica herbacea 'Vivellii'.
Erica herbacea 'Winter Beauty'.
Erica hirtiflora Curtis. Some plants called *E. regerminans* belong here.
Erica holosericea Salisb.
Erica hyalina. Listed name for *E. hyemalis*.
Erica hybrida 'Dawn'. Listed name for *E.* × *watsonii* 'Dawn'.
Erica hyemalis Nichols. FRENCH HEATHER. WHITE-TIPPED WINTER HEATHER.
Erica 'John McLaren'. See *E. mammosa* 'John McLaren'.
Erica 'Kevernensis'. See *E. vagans* 'St. Keverne'.
Erica lusitanica K. Rudolphi. Syn.: *E. codonodes.* SPANISH HEATH.
Erica mackaiana Babington.
Erica mackaiana 'Plena'.
Erica mammosa L.
Erica mammosa 'John McLaren'.
Erica mammosa 'Jubilee'.
Erica mauritanica L. Syn.: *E. viridipurpurea.*
Erica 'Maxima'. See *E. erigena* 'Maxima'.
Erica 'Maxwell'. See *E. vagans* 'Mrs. D. F. Maxwell'.
Erica mediterranea L. See *E. herbacea.*
Erica mediterranea of most authors, not L., see *E. erigena.*
Erica mediterranea 'Brightness'. See *E. erigena* 'Brightness'.
Erica mediterranea 'Maxima'. See *E. erigena* 'Maxima'.
Erica mediterranea 'W. T. Rackliff'. See *E. erigena* 'W. T. Rackliff'.
Erica melanthera L. Not in cult. Plants so called are *E. canaliculata.*
Erica mollis Andr.
Erica oatsii Rolfe.
Erica patersonia Andr.
Erica persoluta L.
Erica perspicua Wendl.
Erica peziza Lodd.

Erica pinea Thunb.
Erica plukenetii L.
Erica × *praegeri* Ostenfeld. [*E. mackaiana* × *E. tetralix*].
Erica quadrangularis Salisb.
Erica regerminans L. Not cult. in Calif. Plants so called are *E. hirtiflora* or other small flowered species.
Erica regia Bartl.
Erica regia 'Variegata'.
Erica 'Riverslea'. See *E. australis* 'Riverslea'.
Erica sessiliflora L. f.
Erica 'Silberschmelze'. See *E.* × *darleyensis* 'Silberschmelze'.
Erica speciosa Andr.
Erica 'Springwood Pink'. See *E. herbacea* 'Springwood Pink'.
Erica 'Springwood White'. See *E. herbacea* 'Springwood White'.
Erica stricta Andr. See *E. terminalis.*
Erica 'Superba'. See *E. erigena* 'Superba'.
Erica taxifolia Ait.
Erica terminalis Salisb. Syn.: *E. stricta.* CORSICAN HEATHER.
Erica tetralix L. CROSS-LEAVED HEATH.
Erica tetralix 'Alba'.
Erica tetralix 'Alba Mollis'.
Erica tetralix 'Con Underwood'.
Erica tetralix 'Connie Underwood'. See *E. tetralix* 'Con Underwood'.
Erica tetralix 'Constance Underwood'. See *E. tetralix* 'Con Underwood'.
Erica tetralix 'Darleyensis'. Not to be confused with *E.* × *darleyensis.*
Erica tetralix 'Mackaiana Plena'. Listed name, probably for *E. mackaiana* 'Plena'.
Erica tetralix 'Mollis'.
Erica tetralix 'Praegeri'. See *E.* × *praegeri.*
Erica triflora L.
Erica vagans L. CORNISH HEATH.
Erica vagans 'Alba'.
Erica vagans 'Birch Glow'.
Erica vagans 'Diana Hornibrook'.
Erica vagans 'George Underwood'.
Erica vagans 'Lyonesse'.
Erica vagans 'Mrs. D. F. Maxwell'.
Erica vagans 'Nana'.
Erica vagans 'St. Keverne'.
Erica × *veitchii* Bean. [*E. arborea* × *E. lusitanica*].
Erica ventricosa Thunb.
Erica versicolor Wendl. Some plants called *E. blanda* or *E. doliiformis* belong here.
Erica verticillata Bergius. Some plants called *E. blanda* or *E. doliiformis* belong here.
Erica vestita Thunb.
Erica viridipurpurea L. See *E. mauritanica.*
Erica vulgaris L. See *Calluna vulgaris.*
Erica × *watsonii* Benth. [*E. ciliaris* × *E. tetralix*].
Erica × *watsonii* 'Dawn'.
Erica × *watsonii* 'Gwen'.
Erica × *williamsii* Druce. [*E. tetralix* × *E. vagans*].
Erica 'W. T. Rackliff'. See *E. erigena* 'W. T. Rackliff'.
Erinacea anthyllis Link.
Eriobotrya deflexa (Hemsl.) Nakai. Syn.: *Photinia deflexa.* BRONZE LOQUAT.
Eriobotrya 'Golden Nugget'. See *E. japonica* 'Golden Nugget'.
Eriobotrya japonica (Thunb.) Lindl. LOQUAT.
Eriobotrya japonica 'Advance'.
Eriobotrya japonica 'Champagne'.
Eriobotrya japonica 'Gold Nugget'.
Eriobotrya japonica 'MacBeth'.

Eriodictyon crassifolium Benth. YERBA SANTA.
Eriodictyon trichocalyx Heller.
Eriogonum arborescens Greene. SANTA CRUZ ISLAND BUCKWHEAT.
Eriogonum × blissianum Mason. [**E. giganteum × E. arborescens**]. BLISS'S HYBRID BUCKWHEAT.
Eriogonum cinereum Benth. ASHYLEAF BUCKWHEAT.
Eriogonum crocatum A. Davids. SAFFRON BUCKWHEAT. YELLOW BUCKWHEAT.
Eriogonum fasciculatum Benth. CALIFORNIA BUCKWHEAT. WILD BUCKWHEAT.
Eriogonum giganteum S. Wats. SAINT CATHERINE'S LACE.
Eriogonum grande Greene. Syn.: *E. latifolium* subsp. *grande*. PINK BUCKWHEAT.
Eriogonum grande var. **rubescens** (Greene) Munz. Syn.: *E. latifolium* subsp. *rubescens*. *E. rubescens*. RED BUCKWHEAT.
Eriogonum latifolium Sm. subsp. *grande* (Greene) Stokes. See **E. grande**.
Eriogonum latifolium subsp. *rubescens* (Greene) Stokes. See **E. grande** var. **rubescens**.
Eriogonum lobbii Torr. & Gray. GRANITE BUCKWHEAT.
Eriogonum parvifolium Sm. SEA CLIFF BUCKWHEAT.
Eriogonum rubescens Greene. See **E. grande** var. **rubescens**.
Eriogonum umbellatum Torr. SULPHUR FLOWER.
Eriogonum wrightii Torr. ex Benth. subsp. **subscaposum** (S. Wats.) Stokes. WRIGHT BUCKWHEAT.
Eriophyllum confertiflorum (DC.) Gray.
Eriophyllum nevinii Gray. CATALINA DUSTY-MILLER.
Eriophyllum staechadifolium Lag.
Erythea armata (S. Wats.) S. Wats. See **Brahea armata**.
Erythea brandegeei C. Purpus. See **Brahea brandegeei**.
Erythea edulis (H. Wendl. ex S. Wats.) S. Wats. See **Brahea edulis**.
Erythea elegans Franceschi ex Becc. See **Brahea elegans**.
Erythrina abyssinica Lam. ex DC. Syn.: *E. tomentosa*.
Erythrina acanthocarpa E. Mey.
Erythrina americana Mill.
Erythrina arborescens Roxb.
Erythrina atitlanensis Krukoff & Barneby.
Erythrina barqueroana Krukoff & Barneby.
Erythrina berteroana Urban.
Erythrina × bidwillii Lindl. [**E. herbacea × E. crista-galli**]. BIDWILL CORAL TREE.
Erythrina caffra Thunb. Syn.: *E. constantiana*. KAFFIR BOOM.
Erythrina chiapasana Krukoff.
Erythrina chiriquensis Krukoff.
Erythrina constantiana Micheli. See **E. caffra**.
Erythrina corallodendrum L. Probably not cult. in Calif. Plants so called may be misidentified.
Erythrina coralloides DC. NAKED CORAL TREE.
Erythrina costaricensis Micheli.
Erythrina crista-galli L. COCKSPUR CORAL TREE.
Erythrina embryana. Listed name for **E. caffra**.
Erythrina falcata Benth.
Erythrina flabelliformis Kearney.
Erythrina folkersii Krukoff & Moldenke.
Erythrina fusca Lour. Syn.: *E. glauca*. *E. ovalifolia*.
Erythrina gibbsae E. G. Baker. See **E. latissima**.
Erythrina glauca Willd. See **E. fusca**.
Erythrina guatemalensis Krukoff.
Erythrina herbacea L. CARDINAL-SPEAR. CHEROKEE BEAN.
Erythrina hondurensis Standl.
Erythrina huehuetenangensis Krukoff & Barneby.
Erythrina humeana K. Spreng. Syn.: *E. princeps*. NATAL CORAL TREE.
Erythrina humeana var. **raja** (Meissn.) Harv.

Erythrina indica Lam. See **E. variegata**.
Erythrina lanceolata Standl.
Erythrina latissima E. Mey. Syn.: *E. gibbsae*.
Erythrina lysistemon Hutchinson. Some plants called *E. princeps* belong here.
Erythrina macrophylla A. DC.
Erythrina mitis Jacq. Syn.: *E. umbrosa*.
Erythrina monosperma Gaud.-Beaup. See **E. tahitensis**.
Erythrina ovalifolia Roxb. See **E. fusca**.
Erythrina phlebocarpa F. M. Bailey. See **E. variegata**.
Erythrina poeppigiana (Walpers) O. F. Cook. May not be cult. in Calif. Plants so called may be misidentified.
Erythrina poianthes Brot. See **E. speciosa**.
Erythrina princeps A. Dietrich. See **E. humeana**. Some plants called *E. princeps* are **E. lysistemon**.
Erythrina reticulata Presl. See **E. speciosa**.
Erythrina sandwicensis Degener. See **E. tahitensis**.
Erythrina speciosa Andr. Syn.: *E. reticulata*. *E. poianthes*.
Erythrina steyermarkii Krukoff & Barneby.
Erythrina tahitensis Nadeaud. Syn.: *E. monosperma*. *E. sandwicensis*.
Erythrina tomentosa R. Br. ex A. Rich. See **E. abyssinica**.
Erythrina umbrosa HBK. See **E. mitis**.
Erythrina variegata L. Syn.: *E. indica*. *E. phlebocarpa*.
Erythrina vespertilio Benth. AUSTRALIAN CORAL TREE.
Erythrina zeyheri Harv.
ESCALLONIA, PINK: **Escallonia laevis**.
    RED: **Escallonia rubra**.
    WHITE: **Escallonia bifida**.
Escallonia 'Alice'.
Escallonia 'Apple Blossom'.
Escallonia 'Balfourii'. See **E. × exoniensis 'Balfourii'**.
Escallonia berteriana DC. See **E. pulverulenta**.
Escallonia bifida Link & Otto. Syn.: *E. montevidensis*. *E. floribunda*. WHITE ESCALLONIA.
Escallonia 'C. F. Ball'. See **E. rubra 'C. F. Ball'**.
Escallonia 'Eric Walther'. See **E. × franciscana 'Eric Walther'**.
Escallonia × exoniensis Veitch. [**E. rosea × E. rubra**].
Escallonia × exoniensis 'Balfourii'.
Escallonia × exoniensis 'Fradesii'.
Escallonia floribunda Hook. f. See **E. bifida**.
Escallonia 'Fradesii'. See **E. × exoniensis 'Fradesii'**.
Escallonia × franciscana Eastw. [**E. illinita × E. rubra**]. Some plants called **E. rosea** belong here.
Escallonia × franciscana 'Eric Walther'.
Escallonia glasneviensis?
Escallonia grahamiana Hook. & Arn. See **E. illinita**.
Escallonia 'Gwendolyn Anley'. See **E. virgata 'Gwendolyn Anley'**.
Escallonia illinita Presl. Syn.: *E. grahamiana*. *E. viscosa*.
Escallonia 'Ingramii'. See **E. rubra 'Ingramii'**.
Escallonia 'Iveyi'.
Escallonia 'Jubilee'. [?**E. laevis × E. rubra**].
Escallonia laevis (Vell.) Sleumer. Syn.: *E. organensis*. PINK ESCALLONIA.
Escallonia 'Langleyensis'.
Escallonia leucantha Rémy.
Escallonia 'Lou Allen'.
Escallonia macrantha Hook. & Arn. See **E. rubra** var. **macrantha**.
Escallonia montevidensis (Cham. & Schlechtend.) DC. See **E. bifida**.
Escallonia organensis G. Gardner. See **E. laevis**.
Escallonia philippiana (Engler) M. T. Mast. See **E. virgata**.
Escallonia 'Pink Princess'?
Escallonia pulverulenta (Ruiz & Pav.) Pers. Syn.: *E. berteriana*.

Escallonia punctata DC. See **E. rubra**.
Escallonia 'Rockii'.
Escallonia rosea Griseb. Some plants so called may be **E. × franciscana**.
Escallonia rubra (Ruiz & Pav.) Pers. Syn.: *E. punctata*. RED ESCALLONIA.
Escallonia rubra 'C. F. Ball'.
Escallonia rubra 'Ingramii'.
Escallonia rubra var. **macrantha** (Hook. & Arn.) Reiche. Syn.: *E. macrantha*.
Escallonia rubra 'William Watson'.
Escallonia virgata (Ruiz & Pav.) Pers. Syn.: *E. philippiana*.
Escallonia virgata 'Gwendolyn Anley'.
Escallonia viscosa Forbes. See **E. illinita**.
Escallonia 'William Watson'. See **E. rubra 'William Watson'**.
EUCALYPTUS, DOLLAR-LEAF: **Eucalyptus cinerea**.
    DOLLAR-LEAF: **Eucalyptus polyanthemos**.
    DOLLAR-LEAF: **Eucalyptus pulverulenta**.
    FLAMING: **Eucalyptus ficifolia**.
    MOITCH: **Eucalyptus rudis**.
    ROUND-LEAVED: **Eucalyptus polyanthemos**.
    SCARLET: **Eucalyptus ficifolia**.
Eucalyptus acmenioides Schauer. WHITE MAHOGANY. Some plants called **E. triantha** belong here.
Eucalyptus affinis Deane & Maiden.
Eucalyptus albens Benth. WHITE BOX.
Eucalyptus × *algeriensis* Trabut.
Eucalyptus amygdalina Labill. BLACK PEPPERMINT.
Eucalyptus angulosa Schauer. See **E. incrassata**.
Eucalyptus annulata Benth.
Eucalyptus archeri Maiden & Blakely. ARCHER'S ALPINE GUM.
Eucalyptus argillacea W. Fitzg. Syn.: *E. leucophylla*. KIMBERLY GRAY BOX.
Eucalyptus bauerana Schauer. BLUE BOX.
Eucalyptus blakelyi Maiden. BLAKELY'S RED GUM.
Eucalyptus bloxsomei Maiden. YELLOW-JACKET.
Eucalyptus botryoides Sm.
Eucalyptus burdettiana Blakely & Steedman.
Eucalyptus burracoppinensis Maiden & Blakely. BURRACOPPIN MALLEE.
Eucalyptus caesia Benth. GUNGURRU.
Eucayptus calophylla R. Br. ex Lindl. MARRI. RED GUM.
Eucalyptus calycogona Turcz. GOOSEBERRY MALLEE.
Eucalyptus camaldulensis Dehnhardt. Syn.: *E. rostrata*. RED GUM. RIVER RED GUM.
Eucalyptus cambageana Maiden.
Eucalyptus cinerea F. J. Muell. ex Benth. DOLLAR-LEAF EUCALYPTUS. MEALY STRINGYBARK. SILVER-DOLLAR TREE.
Eucalyptus citriodora Hook. Syn.: *E. maculata* var. *citriodora*. LEMON GUM.
Eucalyptus cladocalyx F. J. Muell. Syn.: *E. corynocalyx*. SUGAR GUM.
Eucalyptus cladocalyx 'Nana'. DWARF SUGAR GUM.
Eucalyptus clavigera A. Cunn. ex Schauer. APPLE GUM.
Eucalyptus cloeziana F. J. Muell. QUEENSLAND MESSMATE. YELLOW MESSMATE.
Eucalyptus coccifera Hook. f. TASMANIAN SNOW GUM.
Eucalyptus 'Compacta'. See **E. globulus 'Compacta'**.
Eucalyptus cornuta Labill. YATE TREE.
Eucalyptus coronata C. A. Gardner. Syn.: *E. mitrata*.
Eucalyptus corynocalyx F. J. Muell. See **E. cladocalyx**.
Eucalyptus cosmophylla F. J. Muell. CUP GUM.
Eucalyptus crebra F. J. Muell. NARROW-LEAVED IRONBARK. Some plants called **E. racemosa** belong here.
Eucalyptus crucis Maiden. BLUE-LEAVED GUM. SOUTHERN CROSS SILVER MALLEE.

Eucalyptus dalrympleana Maiden. MOUNTAIN GUM.
Eucalyptus deanei Maiden. DEANE'S GUM.
Eucalyptus deglupta Blume. MINDANAO GUM.
Eucalyptus delegatensis R. T. Baker. ALPINE ASH.
Eucalyptus desmondensis Maiden & Blakely. DESMOND'S MALLEE.
Eucalyptus diptera C. Andr. TWO-WINGED GIMLET. BASTARD GIMLET.
Eucalyptus diversicolor F. J. Muell. KARRI. KARRI GUM.
Eucalyptus dwyeri Maiden & Blakely. DWYER'S MALLEE GUM.
Eucalyptus elata Dehnhardt. Syn.: *E. lindleyana*. *E. longifolia*. RIVER PEPPERMINT.
Eucalyptus eremophila (Diels) Maiden. TALL SAND MALLEE. HORNED MALLEE.
Eucalyptus erythrocorys F. J. Muell. ILLYARIE. RED CAP GUM.
Eucalyptus erythronema Turcz. RED-FLOWERED MALLEE.
Eucalyptus eugenioides of Australian authors, not Sieber ex A. Spreng. See **E. globoidea**.
Eucalyptus eximia Schauer. YELLOW BLOODWOOD.
Eucalyptus ficifolia F. J. Muell. RED-FLOWERING GUM. FLAMING EUCALYPTUS. SCARLET-FLOWERING GUM. SCARLET EUCALYPTUS.
Eucalyptus ficifolia 'Coral'.
Eucalyptus forrestiana Diels. FUCHSIA GUM. FORREST'S MARLOCK.
Eucalyptus gardneri Maiden. GARDNER'S MALLET.
Eucalyptus glaucescens Maiden & Blakely. TINGIRINGI GUM.
Eucalyptus globoidea Blakely. Syn.: *E. eugenioides*. WHITE STRINGYBARK.
Eucalyptus globulus Labill. BLUE GUM. TASMANIAN BLUE GUM.
Eucalyptus globulus 'Compacta'. BUSHY BLUE GUM. DWARF TASMANIAN BLUE GUM.
Eucalyptus grandis W. Hill ex Maiden. ROSE GUM.
Eucalyptus grossa F. J. Muell. COARSE-LEAVED MALLEE. COARSE-FLOWERED MALLEE.
Eucalyptus gunnii Hook. f. CIDER GUM.
Eucalyptus incrassata Labill. Syn.: *E. angulosa*. LERP MALLEE. RIDGE-FRUITED MALLEE.
Eucalyptus johnstonii Maiden. JOHNSTON'S GUM.
Eucalyptus kruseana F. J. Muell. KRUSE'S MALLEE.
Eucalyptus lansdowneana F. J. Muell. & J. E. Br. CRIMSON MALLEE BOX.
Eucalyptus lehmannii (Preiss ex Schauer) Benth. BUSHY YATE.
Eucalyptus leucophylla Domin. See **E. argillacea**.
Eucalyptus leucoxylon F. J. Muell. WHITE IRONBARK.
Eucalyptus leucoxylon var. *macrocarpa* J. E. Br. LARGE-FRUITED WHITE IRONBARK.
Eucalyptus leucoxylon 'Rosea'. See **E. leucoxylon**.
Eucalyptus lindleyana Blakely, not DC. See **E. elata**.
Eucalyptus linearis Dehnhardt. See **E. pulchella**.
Eucalyptus longifolia Lindl., not Link & Otto. See **E. elata**.
Eucalyptus loxophleba Benth. YORK GUM.
Eucalyptus macarthurii Deane & Maiden. CAMDEN WOOLLYBUTT.
Eucalyptus macrandra F. J. Muell. ex Benth. LONG-FLOWERED MARLOCK.
Eucalyptus macrocarpa Hook. BLUEBUSH.
Eucalyptus macrorhyncha F. J. Muell. RED STRINGYBARK.
Eucalyptus maculata Hook. SPOTTED GUM.
Eucalyptus maculata var. *citriodora* (Hook.) L. H. Bailey. See **E. citriodora**.
Eucalyptus maculosa R. T. Baker. See **E. mannifera** subsp. **maculosa**.
Eucalyptus maidenii F. J. Muell. MAIDEN'S GUM.
Eucalyptus mannifera Mudie subsp. **maculosa** (R. T. Baker) L. Johnson. Syn.: *E. maculosa*. RED-SPOTTED GUM.
Eucalyptus megacornuta C. A. Gardner. WARTED YATE.
Eucalyptus melliodora A. Cunn. ex Schauer. YELLOW BOX.
Eucalyptus micrantha DC. See **E. racemosa**.
Eucalyptus microtheca F. J. Muell. FLOODED BOX.

Eucalyptus mitrata C. A. Gardner. See **E. coronata**.
Eucalyptus moorei Maiden & Cambage. NARROW-LEAVED SALLY.
Eucalyptus myrtifolia Link?
Eucalyptus neglecta Maiden. OMEO ROUND-LEAVED GUM.
Eucalyptus nicholii Maiden & Blakely. NICHOL'S WILLOW-LEAVED PEPPERMINT.
Eucalyptus niphophila Maiden & Blakely. SNOW GUM.
Eucalyptus nutans F. J. Muell. RED-FLOWERED MOORT.
Eucalyptus occidentalis Endl. FLAT-TOPPED YATE.
Eucalyptus oldfieldii F. J. Muell. OLDFIELD'S MALLEE.
Eucalyptus orbifolia F. J. Muell. ROUND-LEAVED MALLEE.
Eucalyptus orpetii. Listed name for a plant said to have originated in Calif. as a hybrid between **E. caesia** and **E. macrocarpa.**
Eucalyptus ovata Labill. SWAMP GUM.
Eucalyptus oxymitra Blakely. SHARP-CAPPED MALLEE.
Eucalyptus parvifolia Cambage. SMALL-LEAVED GUM.
Eucalyptus pauciflora Sieber ex A. Spreng. GHOST GUM. CABBAGE GUM. SNOW GUM.
Eucalyptus peltata Benth.
Eucalyptus perriniana F. J. Muell. ex Rodway. ROUND-LEAVED SNOW GUM. SPINNING GUM.
Eucalyptus pilularis Sm. BLACKBUTT.
Eucalyptus planchoniana F. J. Muell. BASTARD TALLOW WOOD.
Eucalyptus platypus Hook. ROUND-LEAVED MOORT.
Eucalyptus polyanthemos Schauer. AUSTRALIAN BEECH. DOLLAR-LEAF EUCALYPTUS. RED BOX. ROUND-LEAVED EUCALYPTUS. SILVER-DOLLAR GUM. SILVER-DOLLAR TREE.
Eucalyptus populifolia Hook. See **E. populnea.**
Eucalyptus populnea F. J. Muell. Syn.: *E. populifolia.* BRIMBLE BOX.
Eucalyptus preissiana Schauer. BELL-FRUITED MALLEE.
Eucalyptus pruinosa Schauer. SILVER-LEAVED BOX.
Eucalyptus pryoriana L. Johnson.
Eucalyptus ptychocarpa F. J. Muell. SWAMP BLOODWOOD.
Eucalyptus pulchella Desf. Syn.: *E. linearis.* WHITE PEPPERMINT.
Eucalyptus pulverulenta Sims. SILVER-DOLLAR TREE. DOLLAR-LEAF EUCALYPTUS. SILVER MOUNTAIN GUM.
Eucalyptus punctata DC. GRAY GUM.
Eucalyptus pyriformis Turcz. PEAR-FRUITED MALLEE.
Eucalyptus racemosa Cav. Syn.: *E. micrantha.* SNAPPY GUM. Some plants called **E. racemosa** are **E. crebra.**
Eucalyptus regnans F. J. Muell. GIANT GUM.
Eucalyptus resinifera Sm. RED MAHOGANY.
Eucalyptus rhodantha Blakely & Steedman. ROSE MALLEE.
Eucalyptus robusta Sm. SWAMP MAHOGANY.
Eucalyptus rostrata Schlechtend. See **E. camaldulensis.**
Eucalyptus rubida Deane & Maiden. CANDLE-BARK GUM.
Eucalyptus rudis Endl. MOITCH. DESERT GUM. SWAMP GUM.
Eucalyptus saligna Sm. SYDNEY BLUE GUM.
Eucalyptus salmonophloia F. J. Muell. SALMON GUM.
Eucalyptus sepulcralis F. J. Muell. BLUE WEEPING GUM.
Eucalyptus sideroxylon A. Cunn. ex Woolls. Syn.: *E. sideroxylon* var. *rosea.* MULGA IRONBARK. PINK IRONBARK. RED IRONBARK.
Eucalyptus sideroxylon var. *rosea* Ingham. See **E. sideroxylon.**
Eucalyptus simmondsii Maiden.
Eucalyptus spathulata Hook. NARROW-LEAVED GIMLET. SWAMP MALLEE.
Eucalyptus staigerana F. J. Muell. ex F. M. Bailey. LEMON-SCENTED IRONBARK.
Eucalyptus steedmanii C. A. Gardner. STEEDMAN'S GUM.
Eucalyptus stellulata Sieber ex DC. BLACK SALLY.
Eucalyptus stoatei C. A. Gardner. SCARLET PEAR GUM.
Eucalyptus stricklandii Maiden. STRICKLAND'S GUM.
Eucalyptus tasmanica Blakely. See **E. tenuiramis.**

Eucalyptus tenuiramis Miq. Syn.: *E. tasmanica.* SILVER PEPPERMINT.
Eucalyptus tereticornis Sm. Syn.: *E. umbellata.* FOREST RED GUM.
Eucalyptus tetragona (R. Br.) F. J. Muell. WHITE-LEAVED MARLOCK.
Eucalyptus tetraptera Turcz. SQUARE-FRUITED MALLEE.
Eucalyptus torquata Luehm. CORAL GUM.
Eucalyptus 'Torwood'. [**E. torquata** × **E. woodwardii**].
Eucalyptus triantha Link. May not be cult. Some plants so called are **E. acmenioides.**
Eucalyptus umbellata (Gaertner) Domin. See **E. tereticornis.**
Eucalyptus urnigera Hook. f. URN-FRUITED GUM.
Eucalyptus vernicosa Hook. f. VARNISH-LEAVED GUM.
Eucalyptus viminalis Labill. MANNA GUM. WHITE GUM. RIBBON GUM.
Eucalyptus viridis R. T. Baker. GREEN MALLEE.
Eucalyptus watsoniana F. J. Muell.
Eucalyptus woodwardii Maiden. WOODWARD'S BLACKBUTT.
Eucryphia glutinosa (Poepp. & Endl.) Baill.
Eucryphia × intermedia Bausch. [**E. glutinosa** × **E. lucida**].
Eucryphia lucida (Labill.) Baill.
Eucryphia 'Nymansay'. See **E. × nymansensis.**
Eucryphia × nymansensis Bausch. [**E. cordifolia** Cav. × **E. glutinosa**]. Plants called *E.* 'Nymansay' belong here.
Eugenia aggregata (Vell.) Kaiersk. CHERRY-OF-THE-RIO GRANDE. Some plants called **Myrciaria edulis** belong here.
Eugenia apiculata DC. See **Luma apiculata.**
Eugenia australis Wendl. ex Link. See **Syzygium paniculatum.**
Eugenia australis 'Newport'. See **Syzygium paniculatum 'Newport'.**
Eugenia buxifolia (Swartz) Willd. See **E. foetida.**
Eugenia 'Compacta'. See **Syzygium paniculatum 'Compactum'.**
Eugenia edulis Vell. See **Myrciaria edulis.**
Eugenia foetida Pers. Syn.: *E. buxifolia.* SPANISH-STOPPER.
Eugenia 'Globulus'. See **Syzygium paniculatum 'Globulus'.**
Eugenia jambos L. See **Syzygium jambos.**
Eugenia myrtifolia Sims. and all listed cvs. See **Syzygium paniculatum** and cvs.
Eugenia paniculata (Gaertner) Britten. See **Syzygium paniculatum.**
Eugenia paniculata var. *australis* (Wendl. ex Link) L. H. Bailey. See **Syzygium paniculatum.**
Eugenia 'Red Flame'. See **Syzygium paniculatum 'Red Flame'.**
Eugenia smithii Poir. See **Acmena smithii.**
Eugenia uniflora L. BARBADOS CHERRY. SURINAM CHERRY. PITANGA.
EULALIA: **Miscanthus sinensis.**
EUODIA, HUPEH: **Euodia hupehensis.**
Euodia daniellii (J. Benn.) Hemsl.
Euodia hupehensis Dode. HUPEH EUODIA.
EUONYMUS, BOXLEAF: **Euonymus japonica 'Microphylla'.**
  DUC D'ANJOU: **Euonymus japonica 'Duc d'Anjou'.**
  DWARF WINGED: **Euonymus alata 'Compacta'.**
  EVERGREEN: **Euonymus japonica.**
  GOLDEN: **Euonymus japonica 'Aureo-marginata'.**
  GOLD-SPOT: **Euonymus japonica 'Aureo-variegata'.**
  JAPANESE: **Euonymus japonica.**
  KEW: **Euonymus fortunei 'Kewensis'.**
  PURPLE-LEAF: **Euonymus fortunei 'Colorata'.**
  SILVER QUEEN: **Euonymus fortunei 'Silver Queen'.**
  TRAILING: **Euonymus fortunei** var. **radicans.**
  WHITE-MARGIN: **Euonymus japonica 'Albo-marginata'.**
  WINGED: **Euonymus alata.**
  WINTER CREEPER: **Euonymus fortunei.**
Euonymus acuta. Listed name for **E. fortunei.**
Euonymus alata (Thunb.) Siebold. WINGED BURNING BUSH.

WINGED EUONYMUS. WINGED SPINDLE TREE.
Euonymus alata 'Compacta'. DWARF WINGED EUONYMUS.
Euonymus alata 'Monstrosa'.
Euonymus 'Albo-marginata'. See **E. japonica 'Albo-marginata'.**
Euonymus 'Argenteo-variegata'. See **E. japonica 'Argenteo-variegata'** or **E. fortunei 'Gracilis'.**
Euonymus 'Aureo-marginata'. See **E. japonica 'Aureo-marginata.'**
Euonymus 'Aureo-variegata'. See **E. japonica 'Aureo-variegata'.**
Euonymus 'Colorata'. See **E. fortunei 'Colorata'.**
Euonymus 'Compacta'. See **E. alata 'Compacta'.**
Euonymus 'Duc d'Anjou'. See **E. japonica 'Duc d'Anjou'.**
Euonymus 'Dupont'. See **E. kiautschovica 'Dupont'.**
Euonymus europaea L. EUROPEAN SPINDLE TREE.
Euonymus europaea 'Aldenhamensis'.
Euonymus fortunei (Turcz.) Handel-Mazz. Syn.: *E. radicans* var. *acuta*. WINTER CREEPER EUONYMUS.
Euonymus fortunei 'Azusa'.
Euonymus fortunei 'Carrierei'.
Euonymus fortunei 'Colorata'. PURPLE-LEAF EUONYMUS.
Euonymus fortunei 'Emerald Cushion'.
Euonymus fortunei 'Emerald Gaiety'.
Euonymus fortunei 'Emerald 'n Gold'.
Euonymus fortunei 'Emerald Pride'.
Euonymus fortunei 'Erecta'. See **E. fortunei 'Carrierei'.**
Euonymus fortunei 'Golden Prince'.
Euonymus fortunei 'Gracilis'.
Euonymus fortunei 'Kewensis'. KEW EUONYMUS.
Euonymus fortunei 'Longwood'.
Euonymus fortunei 'Minima'.
Euonymus fortunei 'Pyramidalis'. See **E. fortunei 'Sarcoxie'.**
Euonymus fortunei var. **radicans** (Siebold ex Miq.) Rehd. Syn.: *E. radicans*. TRAILING EUONYMUS. COMMON WINTER CREEPER.
Euonymus fortunei radicans 'Argentea'. See **E. fortunei 'Gracilis'.**
Euonymus fortunei radicans 'Argenteo-variegata'. See **E. fortunei 'Gracilis'.**
Euonymus fortunei radicans 'Carrierei'. See **E. fortunei 'Carrierei'.**
Euonymus fortunei radicans 'Variegata'. See **E. fortunei 'Gracilis.'**
Euonymus fortunei 'Sarcoxie'.
Euonymus fortunei 'Silver Queen'. SILVER QUEEN EUONYMUS. SILVER QUEEN WINTER CREEPER.
Euonymus fortunei 'Variegata'. See **E. fortunei 'Gracilis'.**
Euonymus fortunei var. **vegeta** (Rehd.) Rehd. Syn.: *E. radicans* var. *vegeta*. BIG-LEAF WINTER CREEPER.
Euonymus fortunei 'Winter Gem'.
Euonymus grandiflora Wallich.
Euonymus hamiltoniana Wallich. Syn.: *E. sieboldiana*.
Euonymus japonica Thunb. JAPANESE EUONYMUS. EVERGREEN EUONYMUS.
Euonymus japonica 'Albo-marginata'. WHITE-MARGIN EUONYMUS.
Euonymus japonica 'Argenteo-variegata'.
Euonymus japonica 'Aurea'.
Euonymus japonica 'Aureo-marginata'. GOLDEN EUONYMUS.
Euonymus japonica 'Aureo-variegata'. GOLD-SPOT EUONYMUS. GOLDEN-CENTER EUONYMUS.
Euonymus japonica 'Duc d'Anjou'. DUC D'ANJOU EUONYMUS.
Euonymus japonica 'Giltedge'?
Euonymus japonica 'Gold Center'. Listed name. May be **E. japonica 'Aureo-variegata'.**
Euonymus japonica 'Gold Spot'. Listed name. May be **E. japonica 'Aureo-variegata'.**
Euonymus japonica 'Grandifolia'.
Euonymus japonica 'Mediopicta'.
Euonymus japonica 'Microphylla'. Syn.: *E. microphylla*. BOXLEAF EUONYMUS.
Euonymus japonica 'Microphylla Variegata'. May be **E. japonica 'Variegata'.**
Euonymus japonica 'President Gauthier'.
Euonymus japonica 'Pulchella'. See **E. japonica 'Microphylla'.**
Euonymus japonica 'Silver King'?
Euonymus japonica 'Silver Queen'. See **E. fortunei 'Silver Queen'.**
Euonymus japonica 'Variegata'.
Euonymus japonica 'Viridi-variegata'. See **E. japonica 'Duc d'Anjou'.**
Euonymus 'Kewensis'. See **E. fortunei 'Kewensis'.**
Euonymus kiautschovica Loes. Syn.: *E. patens*. Some plants called *E. sieboldiana* belong here.
Euonymus kiautschovica 'Desert Green'.
Euonymus kiautschovica 'DuPont'.
Euonymus kiautschovica 'Manhatten'.
Euonymus kiautschovica 'Newport'.
Euonymus microphylla Carrière. See **E. japonica 'Microphylla'.**
Euonymus patens Rehd. and all listed cvs. See **E. kiautschovica** and cvs.
Euonymus 'President Gauthier'. See **E. japonica 'President Gauthier'.**
Euonymus radicans Siebold ex Miq. See **E. fortunei** var. *radicans*.
Euonymus radicans var. *acuta* Rehd. See **E. fortunei.**
Euonymus radicans 'Argenteo-marginata'. See **E. fortunei 'Gracilis'.**
Euonymus radicans 'Colorata'. See **E. fortunei 'Colorata'.**
Euonymus radicans 'Kewensis'. See **E. fortunei 'Kewensis'.**
Euonymus radicans 'Pyramidalis'. See **E. fortunei 'Sarcoxie'.**
Euonymus radicans 'Variegata'. See **E. fortunei 'Gracilis'.**
Euonymus radicans var. *vegeta* Rehd. See **E. fortunei** var. *vegeta*.
Euonymus radicans vegeta 'Argenteo-marginata'. Listed name, probably for **E. fortunei 'Gracilis'.**
Euonymus sieboldiana Blume. See **E. hamiltoniana.** Some plants called *E. sieboldiana* are **E. kiautschovica.**
Euonymus 'Silver Queen'. See **E. fortunei 'Silver Queen'.**
Euonymus 'Viridis'. Listed name, probably for **E. fortunei** var. *radicans*.
Eupatorium ianthinum (E. Morr.) Hemsl. See **E. sordidum.**
Eupatorium sordidum Less. Syn.: *E. ianthinum*.
EUPHORBIA, ABYSSINIAN: **Euphorbia trigona.**
Euphorbia abyssinica J. F. Gmelin. May not be cult. Plants so called are usually **E. acrurensis** or **E. neglecta.**
Euphorbia acrurensis N. E. Br. Some plants called *E. abyssinica* belong here.
Euphorbia biglandulosa Desf. See **E. rigida.**
Euphorbia bojeri Hook. See **E. milii.**
Euphorbia bupleurifolia Jacq.
Euphorbia cotinifolia L.
Euphorbia mauritanica L.
Euphorbia milii Desmoul. Syn.: *E. bojeri*. CROWN-OF-THORNS.
Euphorbia milii var. **splendens** (Bojer ex Hook.) Ursch & Leandri. Syn.: *E. splendens*.
Euphorbia neglecta N. E. Br. Some plants called *E. abyssinica* belong here.
Euphorbia pulcherrima Willd. ex Klotsch. Syn.: *Poinsettia pulcherrima*. POINSETTIA. CHRISTMAS FLOWER. MEXICAN FLAMELEAF.
Euphorbia pulcherrima 'Henrietta Ecke'.
Euphorbia pulcherrima 'Pixie'.
Euphorbia rigida Bieb. Syn.: *E. biglandulosa*.

Euphorbia splendens Bojer ex Hook. See **E. milii** var. **splendens**.
Euphorbia trigona Haw. AFRICAN MILK TREE. ABYSSINIAN EUPHORBIA.
Euphorbia veneta Willd. See **E. wulfenii**.
Euphorbia wulfenii Hoppe. Syn.: *E. veneta*.
Eurya emarginata (Thunb.) Makino.
Eurya japonica Thunb.
Eurya ochnacea (DC.) Szysz. See **Cleyera japonica**.
EURYOPS, GRAY-LEAVED: **Euryops pectinatus**.
Euryops acraeus M. D. Henders. Some plants called **E. evansii** belong here.
Euryops athanasiae (L. f.) Harv. See **E. speciosissimus**.
Euryops evansii Schlechter. May not be cult. Plants so called may be **E. acraeus**.
Euryops pectinatus (L.) Cass. GOLDEN SHRUB DAISY. GRAY-LEAVED EURYOPS.
Euryops speciosissimus DC. Syn.: *E. athanasiae*. SOUTH AFRICAN TREE DAISY. CLANWILLIAM DAISY.
Euterpe edulis Mart. ASSAI PALM.
Evodia. See **Euodia**.
Exbucklandia populnea (R. Br.) R. W. Br. Syn.: *Bucklandia populnea. Symingtonia populnea*.
Exochorda grandiflora Hook. See **E. racemosa**.
Exochorda korolkowii (Lavall.) Schneid.
Exochorda × *macrantha* (Lemoine) Schneid. [**E. korolkowii** × **E. racemosa**].
Exochorda × *macrantha* 'The Bride'.
Exochorda racemosa (Lindl.) Rehd. Syn.: *E. grandiflora*. PEARLBUSH.

Fabiana imbricata Ruiz & Pav.
Fagus americana Sweet. See **F. grandifolia**.
Fagus grandifolia Ehrh. Syn.: *F. americana*. AMERICAN BEECH.
Fagus 'Pendula'. See **F. sylvatica** 'Pendula'.
Fagus 'Riversii'. See **F. sylvatica** 'Riversii'.
Fagus sylvatica L. EUROPEAN BEECH.
Fagus sylvatica 'Asplenifolia'. FERN-LEAF BEECH.
Fagus sylvatica 'Atropunicea'. BRONZE BEECH. PURPLE BEECH.
Fagus sylvatica 'Borneyensis'.
Fagus sylvatica 'Cuprea'. COPPER BEECH.
Fagus sylvatica 'Dawyckii'.
Fagus sylvatica 'Laciniata'. CUT-LEAF BEECH.
Fagus sylvatica 'Pendula'. WEEPING BEECH.
Fagus sylvatica 'Purpurea'. See **F. sylvatica** 'Atropunicea'.
Fagus sylvatica 'Purpurea Pendula'. WEEPING COPPER BEECH.
Fagus sylvatica 'Riversii'. RIVERS PURPLE BEECH.
Fagus sylvatica 'Rohanii'.
Fagus sylvatica 'Rotundifolia'.
Fagus sylvatica 'Spaethiana'.
Fagus sylvatica 'Tricolor'. TRICOLOR BEECH.
Fagus sylvatica 'Zlatia'. GOLDEN BEECH.
FAIRY DUSTER: **Calliandra eriophylla**.
Fallugia paradoxa (D. Don) Endl. APACHE-PLUME.
Fascicularia pitcairniifolia (Verlot) Mez.
× *Fatshedera lizei* (Cochet) Guillaum. [**Fatsia japonica** × **Hedera helix**]. BOTANICAL-WONDER.
× *Fatshedera lizei* 'Variegata'.
Fatsia japonica (Thunb.) Decne. & Planch. Syn.: *Aralia japonica. A. sieboldii*. JAPANESE ARALIA.

Fatsia japonica 'Moseri'.
FEATHER BUSH: **Lysiloma thornberi**.
Feijoa sellowiana (Berg) Berg. PINEAPPLE GUAVA.
Feijoa sellowiana 'Choiceana'.
Feijoa sellowiana 'Coolidgei'. COOLIDGE PINEAPPLE GUAVA.
Feijoa sellowiana 'Pineapple Gem'.
Felicia aethiopica (Burman) Bolus & Wolley-Dod.
Felicia amelloides (L.) Voss. BLUE MARGUERITE. BLUE DAISY.
Felicia amelloides 'Blue Gabriel'.
Felicia amelloides 'Martha Chandler'.
Felicia amelloides 'San Gabriel'.
Felicia amelloides 'Santa Anita'.
Felicia fruticosa (L.) Nichols. Syn.: *Aster fruticosus. Diplopappus filifolius. D. fruticulosus*. AFRICAN SHRUB DAISY.
Felicia 'San Gabriel'. See **F. amelloides** 'San Gabriel'.
Felicia 'Santa Anita'. See **F. amelloides** 'Santa Anita'.
FELT-BUSH: **Kalanchoe beharensis**.
FELT-PLANT: **Kalanchoe beharensis**.
FERN, ASPARAGUS: **Asparagus setaceus**.
  AUSTRALIAN TREE: **Sphaeropteris cooperi**.
  BLACK-STEM TREE: **Sphaeropteris medullaris**.
  DWARF HAWAIIAN TREE: **Sadleria cyatheoides**.
  HAWAIIAN TREE: **Cibotium chamissoi**.
  HAWAIIAN TREE: **Cibotium glaucum**.
  HAWAIIAN TREE: **Sadleria cyatheoides**.
  MEXICAN TREE: **Cibotium schiedei**.
  NEW ZEALAND TREE: **Dicksonia squarrosa**.
  SAGO TREE: **Sphaeropteris medullaris**.
  TASMANIAN TREE: **Dicksonia antarctica**.
  VEGETABLE: **Diplazium esculentum**.
FERN TREE: **Filicium decipiens**.
FETTERBUSH: **Pieris floribunda**.
Ficus aspera G. Forst. Syn.: *F. parcellii*. CLOWN FIG.
Ficus auriculata Lour. Syn.: *F. roxburghii*. ROXBURGH FIG.
Ficus australis Willd. See **F. rubiginosa**.
Ficus australis 'Variegata'. See **F. rubiginosa** 'Variegata'.
Ficus benjamina L. Syn.: *F. nitida. F. retusa* var. *nitida*. WEEPING FIG. WEEPING CHINESE BANYAN. BENJAMINA.
Ficus benjamina 'Exotica'.
Ficus callosa Willd.
Ficus carica L. EDIBLE FIG.
Ficus 'Decora'. See **F. elastica** 'Decora'.
Ficus deltoidea Jack. Syn.: *F. diversifolia*. MISTLETOE FIG.
Ficus diversifolia Blume. See **F. deltoidea**.
Ficus doescheri. Listed name for **F. elastica**.
Ficus drupacea Thunb. Syn.: *F. drupacea* var. *pubescens. F. mysorensis. F. mysorensis* var. *pubescens. F. mysorensis* var. *subrepanda*. MYSORE FIG. MYSORE RUBBER TREE.
Ficus drupacea var. *pubescens* (Roth) Corner. See **F. drupacea**.
Ficus elastica Roxb. ex Hornem. Syn.: *F. elastica* var. *rubra*. RUBBER PLANT. INDIA RUBBER TREE.
Ficus elastica 'Decora'. BROAD-LEAVED INDIAN RUBBER PLANT.
Ficus elastica var. *rubra* L. H. & E. Z. Bailey. See **F. elastica**.
Ficus elastica 'Variegata'.
Ficus 'Exotica'. See **F. benjamina** 'Exotica'.
Ficus hillii F. M. Bailey. See **F. microcarpa**.
Ficus lyrata Warb. Syn.: *F. pandurata*. BANJO FIG. FIDDLE-LEAF FIG.
Ficus macrophylla Desf. ex Pers. MORETON BAY FIG. AUSTRALIAN BANYAN.
Ficus microcarpa L. f. Syn.: *F. hillii*. LAUREL FIG. INDIAN LAUREL. Some plants called **F. retusa**, *F. nitida* and *F. retusa* var. *nitida* belong here.
Ficus microphylla. Listed name for **F. rubiginosa**.
Ficus 'Minima'. See **F. pumila**.

Ficus mysorensis B. Heyne ex Roth. See **F. drupacea**.
Ficus mysorensis var. *pubescens* (Roth) Corner. See
   F. drupacea.
Ficus mysorensis var. *subrepanda* Wallich ex King. See
   F. drupacea.
Ficus nekbudu Warb. Syn.: *F. utilis*. ZULU FIG. KAFFIR FIG.
Ficus nitida Thunb. See **F. benjamina**. Some plants called
   F. nitida are **F. microcarpa**.
Ficus pandurata Sander, not Hance. See **F. lyrata**.
Ficus parcellii Veitch ex Cogn. & Marchal. See **F. aspera**.
Ficus petiolaris HBK.
Ficus pumila L. Syn.: *F. repens*. CLIMBING FIG. CREEPING FIG.
   Plants which have small juvenile leaves and are called
   F. pumila 'Minima', *F. repens* 'Minima', and *F.* 'Minima'
   belong here.
Ficus pumila 'Minima'. See **F. pumila**. Plants called *F. pumila*
   'Minima' are **F. pumila** with small juvenile leaves.
Ficus radicans Desf. See **F. sagitatta**.
Ficus religiosa L. SACRED FIG. PEEPUL. BO TREE.
Ficus repens Hort., not Willd. See **F. pumila**.
Ficus repens 'Minima'. Listed name for **F. pumila**.
Ficus retusa L. May not be cult. Plants so called are
   F. microcarpa.
Ficus retusa var. *nitida* (Thunb.) Miq. See **F. benjamina**.
   Some plants called *F. retusa* var. *nitida* are **F. microcarpa**.
Ficus roxburghii Wallich ex Miq. See **F. auriculata**.
Ficus rubiginosa Desf. ex Vent. Syn.: *F. australis*. PORT JACKSON
   FIG. RUSTY FIG. LITTLE-LEAF FIG.
Ficus rubiginosa pubescens. Listed name, probably for
   F. rubiginosa.
Ficus rubiginosa 'Variegata'.
Ficus 'Rubra'. See **F. elastica**.
Ficus sagitatta Vahl. Syn.: *F. radicans*.
Ficus socotrana Balf. f.
Ficus sycomorus L. SYCOMORE FIG. MULBERRY FIG.
Ficus thonningii Blume.
Ficus tikoua Bur. WAIPAHU FIG.
Ficus utilis Sim. See **F. nekbudu**.
Ficus watkinsiana F. M. Bailey. WATKINS FIG.
FIG, BANJO: **Ficus lyrata**.
   CLIMBING: **Ficus pumila**.
   CLOWN: **Ficus aspera**.
   CREEPING: **Ficus pumila**.
   DEVIL'S: **Solanum hispidum**.
   EDIBLE: **Ficus carica**.
   FIDDLE-LEAF: **Ficus lyrata**.
   INDIAN: **Opuntia ficus-indica**.
   KAFFIR: **Ficus nekbudu**.
   LAUREL: **Ficus microcarpa**.
   LITTLE-LEAF: **Ficus rubiginosa**.
   MISTLETOE: **Ficus deltoidea**.
   MORETON BAY: **Ficus macrophylla**.
   MULBERRY: **Ficus sycomorus**.
   MYSORE: **Ficus drupacea**.
   PORT JACKSON: **Ficus rubiginosa**.
   ROXBURGH: **Ficus auriculata**.
   RUSTY: **Ficus rubiginosa**.
   SACRED: **Ficus religiosa**.
   SYCOMORE: **Ficus sycomorus**.
   WAIPAHU: **Ficus tikoua**.
   WATKINS: **Ficus watkinsiana**.
   WEEPING: **Ficus benjamina**.
   ZULU: **Ficus nekbudu**.
   FILBERT: **Corylus maxima**.
      EUROPEAN: **Corylus avellana**.
      PURPLE GIANT: **Corylus maxima 'Purpurea'**.
      TURKISH: **Corylus colurna**.
Filicium decipiens (Wight & Arn.) Thwaites. FERN TREE.
FIR, ALGERIAN: **Abies numidica**.
   ALPINE: **Abies lasiocarpa**.
   BALSAM: **Abies balsamea**.
   BRISTLE-CONE: **Abies bracteata**.
   CASCADE: **Abies amabilis**.
   CHINA: **Cunninghamia lanceolata**.
   CORKBARK: **Abies lasiocarpa** var. **arizonica**.
   DOUGLAS: **Pseudotsuga menziesii**.
   EUROPEAN SILVER: **Abies alba**.
   FRASER: **Abies fraseri**.
   GIANT: **Abies grandis**.
   GREEK: **Abies cephalonica**.
   JAPANESE: **Abies firma**.
   KOREAN: **Abies koreana**.
   LOWLAND: **Abies grandis**.
   MOMI: **Abies firma**.
   NOBLE: **Abies procera**.
   NORDMANN: **Abies nordmanniana**.
   PACIFIC SILVER: **Abies amabilis**.
   PLUM: **Podocarpus andinus**.
   RED: **Abies magnifica**.
   ROCKY MOUNTAIN: **Abies lasiocarpa**.
   SANTA LUCIA: **Abies bracteata**.
   SHASTA: **Abies magnifica** var. **shastensis**.
   SILVER: **Abies amabilis**.
   SPANISH: **Abies pinsapo**.
   VEITCH'S SILVER: **Abies veitchii**.
   WHITE: **Abies concolor**.
FIREBUSH: **Streptosolen jamesonii**.
   CHILEAN: **Embothrium coccineum**.
FIRE-CRACKER, BRAZILIAN: **Manettia inflata**.
FIRECRACKER PLANT: **Cuphea ignea**.
FIRE-CRACKER VINE: **Manettia inflata**.
FIRETHORN: See PYRACANTHA.
FIRE TREE, CHILEAN: **Embothrium coccineum**.
FIREWHEEL TREE: **Stenocarpus sinuatus**.
Firmiana platanifolia (L. f.) Marsili. See **F. simplex**.
Firmiana simplex (L.) W. F. Wight. Syn.: *F. platanifolia*.
   Sterculia platanifolia. CHINESE PARASOL TREE. JAPANESE
   VARNISH TREE.
FLAMBOYANT: **Delonix regia**.
FLAMEBUSH, BRAZILIAN: **Calliandra tweedii**.
   MEXICAN: **Calliandra tweedii**.
   TRINIDAD: **Calliandra tweedii**.
FLAME FLOWER: **Pyrostegia venusta**.
FLAME GOLD: **Koelreuteria elegans**.
FLAMELEAF, MEXICAN: **Euphorbia pulcherrima**.
FLAME-OF-THE-FOREST: **Spathodea campanulata**.
FLAME TREE: **Brachychiton acerifolium**.
   CHINESE: **Koelreuteria bipinnata**.
   CHINESE: **Koelreuteria elegans**.
   PINK: **Brachychiton discolor**.
FLAME VINE: **Pyrostegia venusta**.
   MEXICAN: **Senecio confusus**.
FLAMINGO PLANT: **Justicia carnea**.
FLAMING TRUMPET: **Pyrostegia venusta**.
FLANNEL BUSH: **Fremontodendron californicum**.
   MEXICAN: **Fremontodendron mexicanum**.
   NAPA: **Fremontodendron californicum** subsp. **napense**.
FLANNEL FLOWER: **Actinotus helianthi**.
FLATPOD, CANARY ISLAND: **Adenocarpus foliolosus**.
FLAX, GOLDEN: **Linum flavum**.
   NEW ZEALAND: **Phormium tenax**.
   RED NEW ZEALAND: **Phormium tenax 'Rubrum'**.

FLAX, continued
   VARIEGATED NEW ZEALAND: **Phormium tenax 'Variegatum'**.
FLAX BUSH, YELLOW: **Reinwardtia indica**.
FLEECE FLOWER, BUKHARA: **Polygonum baldschuanicum**.
FLEECE VINE, CHINESE: **Polygonum aubertii**.
FLOSS-SILK TREE: **Chorisia speciosa**.
Fockea tugelensis N. E. Br.
Forestiera neomexicana Gray. DESERT OLIVE.
FORSYTHIA, BORDER: **Forsythia × intermedia**.
   FORTUNE: **Forsythia suspensa** var. **fortunei**.
   GREEN-STEM: **Forsythia viridissima**.
   WEEPING: **Forsythia suspensa**.
   WEEPING: **Forsythia suspensa** var. **fortunei**.
   WHITE: **Abeliophyllum distichum**.
Forsythia 'Arnold Dwarf'. See **F. × intermedia 'Arnold Dwarf'**.
Forsythia 'Beatrix Farrand'. See **F. × intermedia 'Beatrix Farrand'**.
Forsythia europaea Degen & Bald.
Forsythia fortunei Lindl. See **F. suspensa** var. **fortunei**.
Forsythia giraldiana Lingelsh.
Forsythia × intermedia Zabel. [*F. suspensa × F. viridissima*]. BORDER FORSYTHIA.
Forsythia × intermedia 'Arnold Dwarf'.
Forsythia × intermedia 'Beatrix Farrand'.
Forsythia × intermedia 'Karl Sax'.
Forsythia × intermedia 'Lynwood'.
Forsythia × intermedia 'Lynwood Gold'. See **F. × intermedia 'Lynwood'**.
Forsythia × intermedia 'Spectabilis'.
Forsythia × intermedia 'Spring Glory'.
Forsythia 'Karl Sax'. See **F. × intermedia 'Karl Sax'**.
Forsythia 'Lynwood Gold'. See **F. × intermedia 'Lynwood'**.
Forsythia ovata Nakai.
Forsythia 'Spectabilis'. See **F. × intermedia 'Spectabilis'**.
Forsythia 'Spring Glory'. See **F. × intermedia 'Spring Glory'**.
Forsythia suspensa (Thunb.) Vahl. WEEPING FORSYTHIA.
Forsythia suspensa var. *fortunei* (Lindl.) Rehd. Syn.: *F. fortunei*. FORTUNE FORSYTHIA. FORTUNE GOLDEN-BELLS. WEEPING FORSYTHIA.
Forsythia suspensa 'Fortunei'. See **F. suspensa** var. **fortunei**.
Forsythia viridissima Lindl. GREEN-STEM FORSYTHIA.
Forsythia viridissima 'Bronxensis'.
Fortunella japonica (Thunb.) Swingle. Syn.: *Citrus japonica. C. madurensis*. ROUND KUMQUAT. MARUMI KUMQUAT.
Fortunella margarita (Lour.) Swingle. OVAL KUMQUAT. NAGAMI KUMQUAT.
FOUNTAIN BUSH: **Russelia equisetiformis**.
FOUNTAIN PLANT: **Russelia equisetiformis**.
Fouquieria columnaris (Kell.) Kell. ex Curran. Syn.: *Idria columnaris*. IDRIA. CIRIO. BOOJUM TREE.
Fouquieria purpusii Brandegee.
Fouquieria splendens Engelm. OCOTILLO.
FOXBERRY: **Vaccinium vitis-idaea**.
FRAMBOISE: **Rubus idaeus**.
FRANGIPANI: **Plumeria rubra**.
FRANKLIN TREE: **Franklinia alatamaha**.
Franklinia alatamaha Marsh. Syn.: *Gordonia alatamaha*. FRANKLIN TREE. GORDONIA.
Fraxinus americana L. WHITE ASH.
Fraxinus americana 'Rose Hill'.
Fraxinus anomala Torr. ex S. Wats. SINGLE-LEAF ASH.
Fraxinus arizonica Hort. See **F. velutina** var. **glabra**.
Fraxinus dipetala Hook. & Arn. TWO-PETAL ASH. FLOWERING ASH. FOOTHILL ASH.
Fraxinus excelsior L. EUROPEAN ASH.
Fraxinus excelsior 'Kimberly'.
Fraxinus excelsior 'Pendula'. WEEPING EUROPEAN ASH.
Fraxinus glabra Thornb. ex Rehd. See **F. velutina** var. **glabra**.
Fraxinus holotricha Koehne.
Fraxinus holotricha 'Moraine'. See **Fraxinus 'Moraine'**.
Fraxinus lanceolata Borkh. See **F. pennsylvanica** var. **lanceolata**.
Fraxinus lanceolata 'Marshall'. See **Fraxinus pennsylvanica** var. **lanceolata** 'Marshall'.
Fraxinus latifolia Benth. Syn.: *F. oregona*. OREGON ASH.
Fraxinus malacophylla Hemsl.
Fraxinus malacophylla 'Queen Mary'.
Fraxinus 'Moraine'. MORAINE ASH.
Fraxinus oregona Nutt. See **F. latifolia**.
Fraxinus ornus L. FLOWERING ASH. MANNA ASH.
Fraxinus oxycarpa Willd.
Fraxinus oxycarpa 'Raywood'. CLARET ASH. RAYWOOD ASH.
Fraxinus pennsylvanica Marsh. Syn.: *F. pennsylvanica* var. *subintegerrima*. RED ASH.
Fraxinus pennsylvanica var. *lanceolata* (Borkh.) Sarg. Syn.: *F. lanceolata*. GREEN ASH.
Fraxinus pennsylvanica var. *lanceolata* 'Marshall'. MARSHALL SEEDLESS GREEN ASH.
Fraxinus pennsylvanica var. *subintegerrima* (Vahl) Fern. See **F. pennsylvanica**.
Fraxinus quadrangulata Michx. BLUE ASH.
Fraxinus 'Raywood'. See **F. oxycarpa 'Raywood'**.
Fraxinus uhdei (Wenz.) Lingelsh. SHAMEL ASH. EVERGREEN ASH.
Fraxinus uhdei 'Majestic Beauty'.
Fraxinus uhdei 'Sexton'.
Fraxinus uhdei 'Tomlinson'. TOMLINSON'S ASH.
Fraxinus velutina Torr. VELVET ASH. ARIZONA ASH.
Fraxinus velutina var. *coriacea* (S. Wats.) Rehd. MONTEBELLO ASH.
Fraxinus velutina var. *glabra* Rehd. Syn.: *F. glabra*. ARIZONA ASH. SMOOTH ARIZONA ASH.
Fraxinus velutina var. *glabra* 'Modesto'. MODESTO ASH.
Fraxinus velutina 'Rio Grande'.
FRECKLE-FACE: **Hypoestes phyllostachys**.
FREMONTIA, CALIFORNIA: **Fremontodendron californicum**.
   MEXICAN: **Fremontodendron mexicanum**.
   NAPA: **Fremontodendron californicum** subsp. **napense**.
   PINE HILL: **Fremontodendron californicum** subsp. **decumbens**.
   SOUTHERN: **Fremontodendron mexicanum**.
Fremontia californica Torr. See **Fremontodendron californicum**.
Fremontia californica var. *napensis* (Eastw.) McMinn. See **Fremontodendron californicum** subsp. **napense**.
Fremontia mexicana (A. Davids.) Macb. See **Fremontodendron mexicanum**.
Fremontia napensis Eastw. See **Fremontodendron californicum** subsp. **napense**.
Fremontodendron 'California Glory'.
Fremontodendron 'California Gold'.
Fremontodendron californicum (Torr.) Cov. Syn.: *Fremontia californica*. FLANNEL BUSH. CALIFORNIA FREMONTIA.
Fremontodendron californicum subsp. *decumbens* (Lloyd) Munz. Syn.: *F. decumbens*. PINE HILL FREMONTIA.
Fremontodendron californicum subsp. *napense* (Eastw.) Munz. Syn.: *Fremontia californica* var. *napensis. F. napensis*. NAPA FLANNEL BUSH. NAPA FREMONTIA.
Fremontodendron decumbens Lloyd. See **F. californicum** subsp. **decumbens**.
Fremontodendron mexicanum A. Davids. Syn.: *Fremontia mexicana*. MEXICAN FLANNEL BUSH. MEXICAN FREMONTIA. SOUTHERN FREMONTIA.
FRIENDSHIP PLANT: **Bilbergia nutans**.

FRINGE TREE: **Chionanthus virginicus.**
  CHINESE: **Chionanthus retusus.**
FUCHSIA, AUSTRALIAN: **Correa pulchella.**
  CALIFORNIA: **Zauschneria californica.**
  CAPE: **Phygelius capensis.**
  TRAILING: **Fuchsia procumbens.**
  TREE: **Fuchsia arborescens.**
  TREE: **Fuchsia paniculata.**
Fuchsia arborescens Sims. Some plants so called are
  F. paniculata. TREE FUCHSIA.
Fuchsia × **bacillaris** Lindl.
Fuchsia boliviana Carrière.
Fuchsia fulgens DC.
Fuchsia 'Gartenmeister Bonsted'.
Fuchsia × **hybrida** Hort. ex Vilm. [?**F. fulgens** ×
  F. magellanica].
Fuchsia isis. Listed name for a plant with bright scarlet, small
  flowers.
Fuchsia magellanica Lam.
Fuchsia microphylla HBK.
Fuchsia paniculata Lindl. Some plants called **F. arborescens**
  belong here. TREE FUCHSIA.
Fuchsia procumbens R. Cunn. ex A. Cunn. TRAILING FUCHSIA.
Fuchsia reflexa. Listed name for **F.** × **bacillaris.**
Fuchsia splendens Zucc.
Fuchsia thymifolia HBK.
Fuchsia triphylla L.
Furcraea selloa K. Koch.
FURZE: **Ulex europaeus.**

Galphimia glauca Cav. Syn.: *Thryallis glauca.*
Galvezia speciosa (Nutt.) Gray. BUSH SNAPDRAGON.
Gamolepis chrysanthemoides DC.
GARDENIA: **Gardenia augusta.**
  DWARF: **Gardenia augusta 'Stricta Nana'.**
  EVERBLOOMING: **Gardenia augusta 'Veitchii'.**
  MINIATURE: **Gardenia augusta 'Radicans'.**
  TRAILING: **Gardenia augusta 'Radicans'.**
  VEITCH: **Gardenia augusta 'Veitchii'.**
Gardenia augusta (L.) Merrill. Syn.: *G. florida.*
  G. jasminoides. CAPE JASMINE. GARDENIA.
Gardenia augusta 'August Beauty'.
Gardenia augusta 'Belmont'.
Gardenia augusta 'Mystery'. MYSTERY CAPE JASMINE.
Gardenia augusta 'Radicans'. MINIATURE GARDENIA. TRAILING
  GARDENIA.
Gardenia augusta 'Stricta Nana'. DWARF GARDENIA.
Gardenia augusta 'Veitchii'. VEITCH GARDENIA. EVERBLOOMING
  GARDENIA.
Gardenia augusta 'Veitchii Improved'.
Gardenia 'August Beauty'. See *G. augusta* 'August Beauty'.
Gardenia 'Belmont'. See **Gardenia augusta 'Belmont'.**
Gardenia citriodora Hook. See **Mitriostigma axillare.**
Gardenia florida L. See **G. augusta.**
Gardenia jasminoides Ellis and listed cvs. See **G. augusta** and
  cvs.
Gardenia 'Mystery'. See *G. augusta* 'Mystery'.
Gardenia radicans Thunb. See *G. augusta* 'Radicans'.
Gardenia 'Stricta Nana'. See *G. augusta* 'Stricta Nana'.
Gardenia thunbergia L. f.

Gardenia 'Veitchii'. See *G. augusta* 'Veitchii'.
GARLAND FLOWER: **Daphne cneorum.**
GARLAND FLOWER: **Hedychium coronarium.**
Garrya elliptica Dougl. ex Lindl. COAST SILK-TASSEL.
Garrya elliptica 'James Roof'.
Garrya flavescens S. Wats. var. **pallida** (Eastw.) Bacig. ex
  Ewan. PALE SILK-TASSEL.
Garrya fremontii Torr. FREMONT SILK-TASSEL.
× **Gaulnettya wisleyensis** Marchant. [**Gaultheria shallon** ×
  Pernettya mucronata]. Syn.: *Gaulthettya wisleyensis.*
× **Gaulnettya wisleyensis 'Wisley Pearl'.**
Gaultheria cuneata (Rehd. & Wils.) Bean.
Gaultheria humifusa (R. Graham) Rydb. ALPINE WINTERGREEN.
Gaultheria itoana Hayata.
Gaultheria nummularioides D. Don.
Gaultheria ovatifolia Gray. OREGON WINTERGREEN.
Gaultheria procumbens L. CHECKERBERRY. WINTERGREEN.
  MOUNTAIN TEA.
Gaultheria shallon Pursh. SALAL. LEMONLEAF.
× *Gaulthettya wisleyensis* (Marchant) Rehd. See
  × **Gaulnettya wisleyensis.**
Geijera parviflora Lindl. AUSTRALIAN WILLOW. WILGA.
Gelsemium sempervirens (L.) Ait. CAROLINA JESSAMINE.
GENISTA (of florists): **Cytisus canariensis.**
Genista aethnensis (Bivona) DC. MT. ETNA BROOM.
Genista andreana Puissant. See **Cytisus scoparius 'Andreanus'.**
Genista anglica L.
Genista canariensis L. See **Cytisus canariensis.**
Genista cinerea (Vill.) DC.
Genista dalmatica Bartl. See **G. sylvestris.**
Genista delphinensis J. Verl. See **G. sagittalis.**
Genista fragrans Hort. See **Cytisus supranubius.**
Genista germanica L.
Genista germanica 'Prostrata'. Listed name for **G. germanica.**
Genista hirsuta decumbens?
Genista hispanica L. SPANISH GORSE. SPANISH BROOM.
Genista horrida (Vahl) DC.
Genista humifusa?
Genista lydia Boiss.
Genista monosperma (L.) Lam. BRIDAL-VEIL BROOM.
Genista pilosa L.
Genista procumbens Waldstein & Kitaibel ex Willd. See
  Cytisus procumbens.
Genista racemosa. Listed name for **Cytisus** × **spachianus.**
Genista sagittalis L. Syn.: *G. delphinensis.*
Genista scoparia (L.) Lam. See **Cytisus scoparius.**
Genista sylvestris Scop. Syn.: *G. dalmatica.*
Genista tinctoria L. DYER'S GREENWEED.
Genista villarsii Clementi.
GERANIUM: **Pelargonium** × **hortorum.**
  APPLE: **Pelargonium odoratissimum.**
  APPLE-SCENTED: **Pelargonium odoratissimum.**
  BEDDING: **Pelargonium** × **hortorum.**
  CALIFORNIA: **Senecio petasitis.**
  FANCY: **Pelargonium** × **domesticum.**
  GARDEN: **Pelargonium** × **hortorum.**
  GOOSEBERRY: **Pelargonium grossularioides.**
  GRAPE-LEAVED: **Pelargonium vitifolium.**
  HERB-SCENTED: **Pelargonium tomentosum.**
  HORSESHOE: **Pelargonium zonale.**
  IVY: **Pelargonium peltatum.**
  LADY WASHINGTON: **Pelargonium** × **domesticum.**
  LEMON: **Pelargonium crispum.**
  LIME: **Pelargonium** × **nervosum.**
  MARTHA WASHINGTON: **Pelargonium** × **domesticum.**
  NUTMEG: **Pelargonium** × **fragrans.**

GERANIUM, continued
   NUTMEG-SCENTED: **Pelargonium × fragrans**.
   OAK-LEAVED: **Pelargonium quercifolium**.
   PANSY-FLOWERED: **Pelargonium × domesticum**.
   PEPPERMINT: **Pelargonium tomentosum**.
   REGAL: **Pelargonium × domesticum**.
   ROSE: **Pelargonium graveolens**.
   ROSE-SCENTED: **Pelargonium capitatum**.
   SHOW: **Pelargonium × domesticum**.
   SWEET-SCENTED: **Pelargonium graveolens**.
   VILLAGE-OAK: **Pelargonium quercifolium**.
GERMANDER: **Teucrium chamaedrys**.
   BUSH: **Teucrium fruticans**.
GIANT-CLUB: **Cereus peruvianus 'Monstrosus'**.
Gigantochloa apus (Blume ex Schult. f.) Kurz. APOOS. TALI.
Gilibertia japonica (Junghuhn) Harms. See **Dendropanax trifidus**.
GIMLET, BASTARD: **Eucalyptus diptera**.
   NARROW-LEAVED: **Eucalyptus spathulata**.
   TWO-WINGED: **Eucalyptus diptera**.
GINGER, BLUE: **Dichorisandra thyrsiflora**.
   BUTTERFLY: **Hedychium coronarium**.
   KAHILI: **Hedychium gardneranum**.
   SHELL: **Alpinia zerumbet**.
Ginkgo biloba L. MAIDENHAIR TREE.
Gingko biloba **'Autumn Glory'**.
Gingko biloba **'Autumn Gold'**.
Gingko biloba **'Fairmount'**.
Gingko biloba **'Lakeview'**.
Gingko biloba **'Sentry'**.
Glaucotheca armata (S. Wats.) O. F. Cook. See **Brahea armata**.
Gleditsia inermis L. See **G. triacanthos** forma **inermis**.
Gleditsia triacanthos L. HONEY LOCUST.
Gleditsia triacanthos CVS. See **G. triacanthos** forma **inermis** CVS.
Gleditsia triacanthos forma **inermis** (L.) Zabel. Syn.: *G. inermis*. THORNLESS HONEY LOCUST.
Gleditsia triacanthos forma **inermis 'Imperial'**. IMPERIAL HONEY LOCUST.
Gleditsia triacanthos forma **inermis 'Moraine'**. MORAINE HONEY LOCUST.
Gleditsia triacanthos forma **inermis 'Ruby Lace'**. RUBY LACE HONEY LOCUST.
Gleditsia triacanthos forma **inermis 'Shademaster'**. SHADEMASTER HONEY LOCUST.
Gleditsia triacanthos forma **inermis 'Skyline'**. SKYLINE HONEY LOCUST.
Gleditsia triacanthos forma **inermis 'Sunburst'**. SUNBURST HONEY LOCUST.
GLORY-BOWER, BLEEDING-HEART: **Clerodendrum thomsoniae**.
   HARLEQUIN: **Clerodendrum trichotomum**.
GLORY VINE, CRIMSON: **Vitis coignetiae**.
Gnidia polystachya Bergius.
GOAT NUT: **Simmondsia chinensis**.
GOLD-DUST PLANT: **Aucuba japonica 'Variegata'**.
GOLDEN-BELLS, FORTUNE: **Forsythia suspensa** var. **fortunei**.
GOLDENBRUSH: **Haplopappus parishii**.
GOLDEN-CHAIN: **Laburnum alpinum**.
   WEEPING: **Laburnum alpinum 'Pendulum'**.
GOLDEN-CHAIN TREE: **Laburnum anagyroides**.
   LONG-CLUSTER: **Laburnum × watereri**.
GOLDEN FRAGRANCE: **Pittosporum napaulense**.
GOLDEN-SHOWER TREE: **Cassia fistula**.
GOLDEN VINE, BRAZILIAN: **Stigmaphyllon ciliatum**.
GOLD FLOWER: **Hypericum beanii**.
GOLD FLOWER: **Hypericum calycinum**.
GOLD FLOWER: **Hypericum × moseranum**.
Gonospermum canariense Less.
GOOD-LUCK PLANT, HAWAIIAN: **Cordyline terminalis**.
GOOSEBERRY, CANYON: **Ribes menziesii**.
   CHINESE: **Actinidia chinensis**.
   FUCHSIA-FLOWERED: **Ribes speciosum**.
   SIERRA: **Ribes roezlii**.
GORDONIA: **Franklinia alatamaha**.
Gordonia alatamaha (Marsh.) Sarg. See **Franklinia alatamaha**.
Gordonia axillaris (Roxb. ex Ker-Gawl.) D. Dietrich.
Gordonia lasianthus (L.) Ellis. LOBLOLLY BAY. BLACK LAUREL.
GORSE: **Ulex europaeus**.
GORSE, SPANISH: **Genista hispanica**.
Gossypium sturtianum Willis. Syn.: *G. sturtii*. STURT'S DESERT ROSE.
Gossypium sturtii F. J. Muell. See **G. sturtianum**.
GRANADILLA, PURPLE: **Passiflora edulis**.
   RED: **Passiflora coccinea**.
   YELLOW: **Passiflora laurifolia**.
GRAPE, CALIFORNIA WILD: **Vitis californica**.
   CAPE: **Rhoicissus capensis**.
   COMPACT OREGON: **Mahonia aquifolium 'Compacta'**.
   EUROPEAN: **Vitis vinifera**.
   EVERGREEN: **Rhoicissus capensis**.
   OREGON: **Mahonia aquifolium**.
   TINTED: **Vitis vinifera 'Purpurea'**.
   WINE: **Vitis vinifera**.
GRAPEFRUIT: **Citrus × paradisi**.
   MARSH SEEDLESS: **Citrus × paradisi 'Marsh'**.
GRASS, GIANT REED: **Arundo donax**.
   PAMPAS: **Cortaderia selloana**.
   ZEBRA: **Miscanthus sinensis 'Zebrina'**.
GRASS TREE: **Xanthorrhoea arborea**.
GRASS TREE: **Xanthorrhoea quadrangulata**.
Grayia spinosa (Hook.) Moq.
GREASEWOOD: **Adenostoma fasciculatum**.
GREENWEED, DYER'S: **Genista tinctoria**.
GREVILLEA, KRG HYBRID: **Grevillea 'Montezuma's Cloak'**.
   ROSEMARY: **Grevillea rosmarinifolia**.
   WOOLLY: **Grevillea lanigera**.
Grevillea aquifolium Lindl.
Grevillea banksii R. Br. Plants called *G. banksii* var. *forsteri* belong here.
Grevillea banksii var. *forsteri*. Listed name for **G. banksii**.
Grevillea **'Canberra'**.
Grevillea **'Constance'**.
Grevillea hilliana F. J. Muell.
Grevillea juniperina R. Br.
Grevillea juniperina **'Rosea'**.
Grevillea lanigera A. Cunn. ex R. Br. WOOLLY GREVILLEA.
Grevillea leucopteris Meissn.
Grevillea **'Montezuma's Cloak'**. KRG HYBRID GREVILLEA.
Grevillea **'Noell'**.
Grevillea obliquistigma C. A. Gardner.
Grevillea obtusifolia Meissn.
Grevillea petrophiloides Meissn.
Grevillea **'Red Top'**.
Grevillea robusta A. Cunn. ex R. Br. SILK OAK.
Grevillea rosmarinifolia A. Cunn. ROSEMARY GREVILLEA.
Grevillea thelemanniana Endl. HUMMINGBIRD-BUSH. SPIDER NET.
Grevillea tridentifera (Endl.) Meissn.
Grevillea victoriae F. J. Muell.
Grewia caffra Meissn. LAVENDER STAR PLANT. LAVENDER STAR FLOWER.
Greyia radlkoferi Szysz.
Greyia sutherlandii Hook. & Harv.

Griselinia littoralis Raoul.
Griselinia littoralis 'Variegata'.
Griselinia lucida G. Forst.
Griselinia lucida 'Variegata'.
Griselinia macrophylla?
GROUNDSEL, VELVET: **Senecio petasitis**.
GUAVA: **Psidium guajava**.
    CHILEAN: **Ugni molinae**.
    COOLIDGE PINEAPPLE: **Feijoa sellowiana 'Coolidgei'**.
    LEMON: **Psidium guajava**.
    PINEAPPLE: **Feijoa sellowiana**.
    STRAWBERRY: **Psidium cattleianum**.
    YELLOW STRAWBERRY: **Psidium cattleianum 'Lucidum'**.
GUINEA PLANT, GOLD: **Hibbertia scandens**.
GUM, AMERICAN SWEET: **Liquidambar styraciflua**.
    APPLE: **Eucalyptus clavigera**.
    ARCHER'S ALPINE: **Eucalyptus archeri**.
    BLACK: **Nyssa sylvatica**.
    BLAKELY'S RED: **Eucalyptus blakelyi**.
    BLUE: **Eucalyptus globulus**.
    BLUE-LEAVED: **Eucalyptus crucis**.
    BLUE WEEPING: **Eucalyptus sepulcralis**.
    BUSHY BLUE: **Eucalyptus globulus 'Compacta'**.
    CABBAGE: **Eucalyptus pauciflora**.
    CANDLE-BARK: **Eucalyptus rubida**.
    CHINESE SWEET: **Liquidambar formosana**.
    CIDER: **Eucalyptus gunnii**.
    CORAL: **Eucalyptus torquata**.
    COTTON: **Nyssa aquatica**.
    CUP: **Eucalyptus cosmophylla**.
    DEANE'S: **Eucalyptus deanei**.
    DESERT: **Eucalyptus rudis**.
    DWARF SUGAR: **Eucalyptus cladocalyx 'Nana'**.
    DWARF TASMANIAN BLUE: **Eucalyptus globulus 'Compacta'**.
    DWYER'S MALLEE: **Eucalyptus dwyeri**.
    FOREST RED: **Eucalyptus tereticornis**.
    FORMOSAN SWEET: **Liquidambar formosana**.
    FUCHSIA: **Eucalyptus forrestiana**.
    GHOST: **Eucalyptus pauciflora**.
    GIANT: **Eucalyptus regnans**.
    GRAY: **Eucalyptus punctata**.
    JOHNSTON'S: **Eucalyptus johnstonii**.
    KARRI: **Eucalyptus diversicolor**.
    LEMON: **Eucalyptus citriodora**.
    MAIDEN'S: **Eucalyptus maidenii**.
    MANNA: **Eucalyptus viminalis**.
    MINDANAO: **Eucalyptus deglupta**.
    MOUNTAIN: **Eucalyptus dalrympleana**.
    OMEO ROUND-LEAVED: **Eucalyptus neglecta**.
    ORIENTAL SWEET: **Liquidambar orientalis**.
    RED: **Eucalyptus calophylla**.
    RED: **Eucalyptus camaldulensis**.
    RED: **Liquidambar styraciflua**.
    RED CAP: **Eucalyptus erythrocorys**.
    RED-FLOWERING: **Eucalyptus ficifolia**.
    RED-SPOTTED: **Eucalyptus mannifera** subsp. **maculosa**.
    RIBBON: **Eucalyptus viminalis**.
    RIVER RED: **Eucalyptus camaldulensis**.
    ROSE: **Eucalyptus grandis**.
    ROUND-LEAVED SNOW: **Eucalyptus perriniana**.
    SALMON: **Eucalyptus salmonophloia**.
    SCARLET-FLOWERING: **Eucalyptus ficifolia**.
    SCARLET PEAR: **Eucalyptus stoatei**.
    SILVER-DOLLAR: **Eucalyptus polyanthemos**.
    SILVER MOUNTAIN: **Eucalyptus pulverulenta**.
    SMALL-LEAVED: **Eucalyptus parvifolia**.
    SNAPPY: **Eucalyptus racemosa**.
    SNOW: **Eucalyptus niphophila**.
    SNOW: **Eucalyptus pauciflora**.
    SOUR: **Nyssa sylvatica**.
    SPINNING: **Eucalyptus perriniana**.
    SPOTTED: **Eucalyptus maculata**.
    STEEDMAN'S: **Eucalyptus steedmanii**.
    STRICKLAND'S: **Eucalyptus stricklandii**.
    SUGAR: **Eucalyptus cladocalyx**.
    SWAMP: **Eucalyptus ovata**.
    SWAMP: **Eucalyptus rudis**.
    SWEET: **Liquidambar styraciflua**.
    SYDNEY BLUE: **Eucalyptus saligna**.
    TASMANIAN BLUE: **Eucalyptus globulus**.
    TASMANIAN SNOW: **Eucalyptus coccifera**.
    TINGIRINGI: **Eucalyptus glaucescens**.
    TUPELO: **Nyssa aquatica**.
    URN-FRUITED: **Eucalyptus urnigera**.
    VARNISH-LEAVED: **Eucalyptus vernicosa**.
    WHITE: **Eucalyptus viminalis**.
    YORK: **Eucalyptus loxophleba**.
GUMI: **Elaeagnus multiflora**.
GUNGURRU: **Eucalyptus caesia**.
Gunnera chilensis Lam. See **G. tinctoria**.
Gunnera dentata T. Kirk.
Gunnera manicata Lindl.
Gunnera tinctoria (Molina) Mirb. Syn.: *G. chilensis*.
Gymnocladus dioica (L.) K. Koch. KENTUCKY COFFEE TREE.

HACKBERRY: **Celtis occidentalis**.
    CHINESE: **Celtis sinensis**.
    EUROPEAN: **Celtis australis**.
    MEDITERRANEAN: **Celtis australis**.
    MISSISSIPPI: **Celtis laevigata**.
    WESTERN: **Celtis reticulata**.
HACKMATACK: **Larix laricina**.
HACKMATACK: **Larix occidentalis**.
HACKMATACK: **Populus balsamifera**.
HAKEA, SWEET: **Hakea suaveolens**.
Hakea acicularis R. Br.
Hakea arborescens R. Br.
Hakea cristata R. Br.
Hakea erinacea Meissn.
Hakea laurina R. Br. SEA-URCHIN TREE. PINCUSHION TREE.
Hakea multilineata Meissn.
Hakea petiolaris Meissn.
Hakea petiolaris 'Erecta'.
Hakea recurva Meissn.
Hakea ruscifolia Labill.
Hakea salicifolia (Vent.) B. L. Burtt. Syn.: *H. saligna*.
Hakea saligna (Andr.) Knight. See **H. salicifolia**.
Hakea sericea Schrad. & Wendl.
Hakea suaveolens R. Br. NEEDLE BUSH. SWEET HAKEA.
Hakea subsulcata Meissn.
Hakea victoriae Meissn.
Halesia carolina L. Syn.: *H. tetraptera*. SILVER-BELL. SNOWDROP TREE. WILD OLIVE.
Halesia monticola (Rehd.) Sarg. MOUNTAIN SILVER-BELL.
Halesia tetraptera Ellis. See **H. carolina**.

× *Halimiocistus sahucii* (Coste & Soulié) Janchen. [*Cistus salviifolius* × *Halimium umbellatum*].
Halimium lasianthum (Lam.) Spach. Syn.: *Helianthemum lasianthum*.
Halimium lasianthum subsp. **formosum** (Curtis) Heywood. Syn.: *Cistus formosus*.
Halimium ocymoides (Lam.) Willk. & J. Lange. Syn.: *Helianthemum ocymoides*.
Halimium umbellatum (L.) Spach.
Hamamelis 'Fire Charm'. See *H.* × *intermedia* 'Fire Charm'.
Hamamelis × *intermedia* Rehd. [*H. japonica* × *H. mollis*].
Hamamelis × *intermedia* 'Fire Charm'.
Hamamelis × *intermedia* 'Jelena'.
Hamamelis × *intermedia* 'Ruby Glow'.
Hamamelis japonica Siebold & Zucc. JAPANESE WITCH HAZEL.
Hamamelis mollis D. Oliver. CHINESE WITCH HAZEL.
Hamamelis 'Ruby Glow'. See *H.* × *intermedia* 'Ruby Glow'.
Hamamelis vernalis Sarg. SPRING WITCH HAZEL.
Hamamelis virginiana L. WITCH HAZEL.
Hamelia patens Jacq. ORANGE-RED FLOWER. SCARLET BUSH.
HAND PLANT, MEXICAN: **Chiranthodendron pentadactylon**.
Haplopappus canus (Gray) S. F. Blake.
Haplopappus cooperi (Gray) Hall.
Haplopappus parishii (Greene) S. F. Blake. GOLDENBRUSH.
Haplopappus squarrosus Hook. & Arn.
Haplopappus venetus (HBK) S. F. Blake subsp. **vernonioides** (Nutt.) Hall.
Hardenbergia comptoniana (Andr.) Benth. Syn.: *Kennedia comptoniana*. VINE LILAC.
Hardenbergia monophylla Benth. See **H. violacea**.
Hardenbergia violacea (Schneev.) Stearn. Syn.: *H. monophylla*. HARDY VIOLET VINE LILAC.
HARDHACK, GOLDEN: **Potentilla fruticosa**.
HARDTACK: **Cercocarpus betuloides**.
Harpephyllum caffrum Bernh. ex C. F. Krauss. KAFFIR PLUM.
Harpullia arborea (Blanco) Radlk.
HAT PLANT, CHINESE: **Holmskioldia sanguinea**.
Hauya cornigera. Listed name for **H. elegans**.
Hauya elegans DC. Syn.: *H. ruacophila*. SUPORA SHRUB.
Hauya ruacophila J. D. Sm. & Rose. See **H. elegans**.
HAW, BLACK: **Viburnum prunifolium**.
  BLUE: **Viburnum rufidulum**.
  SOUTHERN BLACK: **Viburnum rufidulum**.
  SWEET: **Viburnum prunifolium**.
HAWTHORN, BLACK: **Crataegus douglasii**.
  CARRIÈRE: **Crataegus** × **lavallei**.
  CHINESE: **Crataegus pinnatifida**.
  DOUBLE PINK: **Crataegus laevigata** 'Punicea Flore Pleno'.
  DOUBLE WHITE ENGLISH: **Crataegus laevigata** 'Plena'.
  ENGLISH: **Crataegus laevigata**.
  ENGLISH: **Crataegus monogyna**.
  INDIAN: **Rhaphiolepis indica**.
  LARGE CHINESE: **Crataegus pinnatifida** var. **major**.
  MEXICAN: **Crataegus pubescens**.
  PAUL'S SCARLET: **Crataegus laevigata** 'Coccinea Flore Pleno'.
  SINGLE PINK ENGLISH: **Crataegus laevigata** 'Rosea'.
  SINGLE-SEED: **Crataegus monogyna**.
  YEDDO: **Rhaphiolepis umbellata**.
HAZEL, BUTTERCUP WINTER: **Corylopsis pauciflora**.
  CHINESE WITCH: **Hamamelis mollis**.
  JAPANESE WITCH: **Hamamelis japonica**.
  SPIKE WINTER: **Corylopsis spicata**.
  SPRING WITCH: **Hamamelis vernalis**.
  TURKISH: **Corylus colurna**.
  WINTER: **Corylopsis pauciflora**.
  WINTER: **Corylopsis spicata**.
  WITCH: **Hamamelis virginiana**.
HAZELNUT, BEAKED: **Corylus cornuta**.
  EUROPEAN: **Corylus avellana**.
  TURKISH: **Corylus colurna**.
  WESTERN: **Corylus cornuta** var. **californica**.
HEATH: also see HEATHER. Heath and heather are used interchangeably.
HEATH, CORNISH: **Erica vagans**.
  CROSS-LEAVED: **Erica tetralix**.
  DORSET: **Erica ciliaris**.
  IRISH: **Daboecia cantabrica**.
  IRISH BELL: **Daboecia cantabrica**.
  MEDITERRANEAN: **Erica erigena**.
  MEDITERRANEAN: **Erica herbacea**.
  SOUTHERN: **Erica australis**.
  SPANISH: **Erica lusitanica**.
  SPIKE: **Bruckenthalia spiculifolia**.
  TREE: **Erica arborea**.
  WINTER: **Erica herbacea**.
HEATHER: This list includes cv. names and common names to help place cvs. in the appropriate species.
  ABERDEEN: **Calluna vulgaris** 'Aberdeen'.
  ALBA: **Daboecia cantabrica** 'Alba'.
  ALBA: **Erica cinerea** 'Alba'.
  ALBA: **Erica herbacea** 'Alba'.
  ALBA: **Erica tetralix** 'Alba'.
  ALBA: **Erica vagans** 'Alba'.
  ALBA MINOR: **Erica cinerea** 'Alba Minor'.
  ALBA MOLLIS: **Erica tetralix** 'Alba Mollis'.
  ALBA PLENA: **Calluna vulgaris** 'Alba Plena'.
  ALPINA: **Erica arborea** 'Alpina'.
  ALPORT: **Calluna vulgaris** 'Alportii'.
  ARTHUR JOHNSON: **Erica** × **darleyensis** 'Arthur Johnson'.
  ATROPURPUREA: **Daboecia cantabrica** 'Atropurpurea'.
  ATRORUBENS: **Erica cinerea** 'Atrorubens'.
  ATROSANGUINEA: **Erica cinerea** 'Atrosanguinea'.
  AUREA: **Erica herbacea** 'Aurea'.
  AUREIFOLIA: **Calluna vulgaris** 'Hammondii Aureifolia'.
  AUTUMN GLOW: **Calluna vulgaris** 'Autumn Glow'.
  BARNETT ANLEY: **Calluna vulgaris** 'Barnett Anley'.
  BELL: **Erica cinerea**.
  BIRCH GLOW: **Erica vagans** 'Birch Glow'.
  BLACK-EYED: **Erica canaliculata**.
  BOSCAWEN'S BLACK-EYED: **Erica canaliculata** 'Boscaweniana'.
  BRIGHTNESS: **Erica erigena** 'Brightness'.
  C. D. EASON: **Erica cinerea** 'C. D. Eason'.
  C. D. FORD: **Erica cinerea** 'C. D. Ford'.
  C. W. NIX: **Calluna vulgaris** 'C. W. Nix'.
  CARLTON: **Calluna vulgaris** 'Carlton'.
  CECILIA M. BEALE: **Erica herbacea** 'Cecilia M. Beale'.
  CHRISTMAS: **Erica canaliculata** 'Rosea'.
  COLLIGAN BRIDGE: **Erica cinerea** 'Colligan Bridge'.
  CONNIE UNDERWOOD: **Erica tetralix** 'Con Underwood'.
  CONSTANCE UNDERWOOD: **Erica tetralix** 'Con Underwood'.
  CON UNDERWOOD: **Erica tetralix** 'Con Underwood'.
  CORSICAN: **Erica terminalis**.
  COUNTY WICKLOW: **Calluna vulgaris** 'County Wicklow'.
  CRISPA: **Calluna vulgaris** 'Crispa'.
  CUPREA: **Calluna vulgaris** 'Cuprea'.
  CYNDY: **Calluna vulgaris** 'Cyndy'.
  DAINTY BESS: **Calluna vulgaris** 'Dainty Bess'.
  DARLEY DALE: **Erica** × **darleyensis** 'Darley Dale'.
  DARLEYENSIS: **Calluna vulgaris** 'Darleyensis'.
  DARLEYENSIS: **Erica** × **darleyensis**.
  DARLEYENSIS: **Erica tetralix** 'Darleyensis'.
  DAVID EASON: **Calluna vulgaris** 'David Eason'.

HEATHER, continued
  DAWN: **Erica × watsonii 'Dawn'**.
  DIANA HORNIBROOK: **Erica vagans 'Diana Hornibrook'**.
  DOMINO: **Erica cinerea 'Domino'**.
  DURFORD: **Calluna vulgaris 'Durfordii'**.
  DWARF: **Erica erigena 'Maxima'**.
  EDEN VALLEY: **Erica cinerea 'Eden Valley'**.
  EILEEN PORTER: **Erica herbacea 'Eileen Porter'**.
  ELATA: **Calluna vulgaris 'Elata'**.
  ELEGANTISSIMA: **Calluna vulgaris 'Elegantissima'**.
  ELSE FRYE: **Calluna vulgaris 'Else Frye'**.
  ELSIE PURNELL: **Calluna vulgaris 'Elsie Purnell'**.
  EVERBLOOMING FRENCH: **Erica doliiformis**.
  FALSE: **Cuphea hyssopifolia**.
  FELIX FAURE: **Erica 'Felix Faure'**.
  FLORE PLENO: **Calluna vulgaris 'Flore Pleno'**.
  FOX: **Calluna vulgaris 'Foxii'**.
  FOXHOLLOW MAHOGANY: **Erica cinerea 'Foxhollow Mahogany'**.
  FOXHOLLOW WANDERER: **Calluna vulgaris 'Foxhollow Wanderer'**.
  FOXII NANA: **Calluna vulgaris 'Foxii Nana'**.
  FRENCH: **Erica hyemalis**.
  GEORGE RENDALL: **Erica × darleyensis 'George Rendall'**.
  GEORGE UNDERWOOD: **Erica vagans 'George Underwood'**.
  GOLDEN: **Calluna vulgaris 'Aurea'**.
  GOLDEN DROP: **Erica cinerea 'Golden Drop'**.
  GOLDEN HUE: **Erica cinerea 'Golden Hue'**.
  GOLDSWORTH CRIMSON: **Calluna vulgaris 'Goldsworth Crimson'**.
  GWEN: **Erica × watsonii 'Gwen'**.
  HAMMOND AUREIFOLIA: **Calluna vulgaris 'Hammondii Aureifolia'**.
  H. E. BEALE: **Calluna vulgaris 'H. E. Beale'**.
  HIBERNICA: **Calluna vulgaris 'Hibernica'**.
  HIRSUTA: **Calluna vulgaris 'Hirsuta'**.
  HIRSUTA ALBIFLORA: **Calluna vulgaris 'Hirsuta Albiflora'**.
  J. H. HAMILTON: **Calluna vulgaris 'J. H. Hamilton'**.
  JAMES BACKHOUSE: **Erica herbacea 'James Backhouse'**.
  JOAN SPARKES: **Calluna vulgaris 'Joan Sparkes'**.
  JOHN MCLAREN: **Erica mammosa 'John McLaren'**.
  JOHNSON: **Calluna vulgaris 'Hiemalis'**.
  JUBILEE: **Erica mammosa 'Jubilee'**.
  KING GEORGE SPRING: **Erica herbacea 'King George'**.
  KUPHALDT: **Calluna vulgaris 'Kuphaldtii'**.
  LAVENDER SCOTCH: **Calluna vulgaris 'Serlei Purpurea'**.
  LYONESSE: **Erica vagans 'Lyonesse'**.
  MAIR'S VARIETY: **Calluna vulgaris 'Mair's Variety'**.
  MAXIMA: **Erica erigena 'Maxima'**.
  MEDITERRANEAN HYBRID: **Erica × darleyensis**.
  MOLLIS: **Erica tetralix 'Mollis'**.
  MOUNTAIN: **Phyllodoce empetriformis**.
  MOUNTAIN: **Phyllodoce glanduliflora**.
  MR. ROBERT: **Erica australis 'Mr. Robert'**.
  MRS. C. H. GILL: **Erica ciliaris 'Mrs. C. H. Gill'**.
  MRS. D. F. MAXWELL: **Erica vagans 'Mrs. D. F. Maxwell'**.
  MRS. PAT: **Calluna vulgaris 'Mrs. Pat'**.
  MRS. RONALD GRAY: **Calluna vulgaris 'Mrs. Ronald Gray'**.
  MUIRHEAD: **Cassiope 'Muirhead'**.
  MULLION: **Calluna vulgaris 'Mullion'**.
  NANA: **Calluna vulgaris 'Nana'**.
  NANA: **Erica vagans 'Nana'**.
  NANA COMPACTA: **Calluna vulgaris 'Nana Compacta'**.
  PALLIDA: **Daboecia cantabrica 'Pallida'**.
  PENHALE: **Calluna vulgaris 'Penhale'**.
  PINK FRENCH: **Erica 'Felix Faure'**.
  PINK SCOTCH: **Calluna vulgaris 'H. E. Beale'**.
  PLENA: **Erica mackaiana 'Plena'**.
  PRAECOX RUBRA: **Erica herbacea 'Praecox Rubra'**.
  PRAEGER: **Daboecia cantabrica 'Praegerae'**.
  PRAEGER: **Erica × praegeri**.
  P. S. PATRICK: **Erica cinerea 'P. S. Patrick'**.
  PYGMAEA: **Calluna vulgaris 'Pygmaea'**.
  RAINBOW: **Calluna vulgaris 'Rainbow'**.
  RED EVERBLOOMING: **Erica doliiformis**.
  RIGIDA: **Calluna vulgaris 'Rigida'**.
  RIVERSLEA: **Erica australis 'Riverslea'**.
  ROBERT CHAPMAN: **Calluna vulgaris 'Robert Chapman'**.
  ROMA: **Calluna vulgaris 'Roma'**.
  ROSE QUEEN: **Erica cinerea 'Rose Queen'**.
  ROSEA: **Daboecia cantabrica 'Pallida'**.
  ROSEA: **Erica cinerea 'Rosea'**.
  RUBY GLOW: **Erica herbacea 'Ruby Glow'**.
  RUTH SPARKES: **Calluna vulgaris 'Ruth Sparkes'**.
  SCOTCH: **Calluna vulgaris**.
  SERLEI: **Calluna vulgaris 'Serlei'**.
  SERLEI AUREA: **Calluna vulgaris 'Serlei Aurea'**.
  SERLEI PURPUREA: **Calluna vulgaris 'Serlei Purpurea'**.
  SERLEI RUBRA: **Calluna vulgaris 'Serlei Rubra'**.
  SHERWOOD CREEPING: **Erica herbacea 'Sherwood Creeping'**.
  SHERWOOD EARLY RED: **Erica herbacea 'Sherwood Early Red'**.
  SHERWOODII: **Erica herbacea 'Sherwoodii'**.
  SILBERSCHMELZE: **Erica × darleyensis 'Silberschmelze'**.
  SISTER ANNE: **Calluna vulgaris 'Sister Anne'**.
  SMITH'S MINIMA: **Calluna vulgaris 'Smith's Minima'**.
  SNOW QUEEN: **Erica herbacea 'Snow Queen'**.
  SPITFIRE: **Calluna vulgaris 'Spitfire'**.
  SPRINGWOOD PINK: **Erica herbacea 'Springwood Pink'**.
  SPRINGWOOD WHITE: **Erica herbacea 'Springwood White'**.
  ST. KEVERNE: **Erica vagans 'St. Keverne'**.
  ST. NICK: **Calluna vulgaris 'St. Nick'**.
  STARTLER: **Erica herbacea 'Startler'**.
  STOBOROUGH: **Erica ciliaris 'Stoborough'**.
  SUPERBA: **Erica erigena 'Superba'**.
  TENUIS: **Calluna vulgaris 'Tenuis'**.
  TIB: **Calluna vulgaris 'Tib'**.
  TOM THUMB: **Calluna vulgaris 'Tom Thumb'**.
  TOMENTOSA: **Calluna vulgaris 'Tomentosa'**.
  VARIEGATA: **Erica regia 'Variegata'**.
  VIOLACEA: **Erica cinerea 'Violacea'**.
  VIVELL: **Erica herbacea 'Vivellii'**.
  W. T. RACKLIFF: **Erica erigena 'W. T. Rackliff'**.
  WHITE SCOTCH: **Calluna vulgaris 'Alba'**.
  WHITE-TIPPED WINTER: **Erica hyemalis**.
  WINTER BEAUTY: **Erica herbacea 'Winter Beauty'**.
  WYCH: **Erica ciliaris 'Wych'**.
HEBE, BOX-LEAF: **Hebe odora**.
Hebe × andersonii (Lindl. & Paxt.) Cockayne. [? *H. salicifolia × H. speciosa*]. Syn.: *Veronica × andersonii*.
Hebe 'Autumn Glory'.
Hebe 'Barnettii'.
Hebe × bidwillii (Hook.) Allan. See **Parahebe × bidwillii**.
Hebe buchananii (Hook. f.) Cockayne & Allan.
Hebe buchananii 'Minor'.
Hebe buxifolia (Benth.) Cockayne & Allan. Syn.: *Veronica buxifolia*. Probably not cult. Plants so called are **H. odora**.
Hebe canescens Hort. See **Parahebe canescens**.
Hebe × carnea (J. B. Armstr.) Hillier.
Hebe × carnea 'Carnea'.
Hebe catarractae (G. Forst.) Allan. See **Parahebe catarractae**.
Hebe chathamica (Buchanan) Cockayne & Allan. Syn.: *Veronica chathamica*.
Hebe 'Christensenii'.
Hebe ciliolata (Hook. f.) Cockayne & Allan.

Hebe 'Co-ed'.
Hebe cupressoides (Hook. f.) Cockayne & Allan.
Hebe cupressoides 'Nana'.
Hebe decussata. Listed name for **H. elliptica**.
Hebe 'Desilor'.
Hebe divaricata (Cheeseman) Cockayne & Allan. Some plants called **H. menziesii** and *Veronica menziesii* belong here.
Hebe 'Duet'.
Hebe elliptica (G. Forst.) Penn. Syn.: *Veronica decussata. V. elliptica.*
Hebe elliptica 'Variegata'.
Hebe epacridea (Hook. f.) Cockayne & Allan.
Hebe 'Evansii'?
Hebe 'Evansii Rubra'?
Hebe × franciscana (Eastw.) Souster. [**H. elliptica × H. speciosa**].
Hebe × franciscana 'Lavender Queen'.
Hebe gibbsii (T. Kirk) Cockayne & Allan.
Hebe glaucophylla (Cockayne) Cockayne. Syn.: *Veronica glaucophylla.*
Hebe hectori (Hook. f.) Cockayne & Allan.
Hebe hulkeana (F. J. Muell.) Cockayne & Allan.
Hebe imperialis (Boucharlat ex Planch.) Cockayne. See **Hebe speciosa** 'Imperialis.'
Hebe laevis (Benth.) Cockayne & Allan. See **H. venustula**.
Hebe 'Lavender Queen'. See **H. × franciscana** 'Lavender Queen'.
Hebe lyallii (Hook. f.) Allan. See **Parahebe lyallii**.
Hebe lycopodioides (Hook. f.) Cockayne & Allan.
Hebe 'MacEwanii'. [**H. pimeleoides × H. sp.**].
Hebe menziesii (Benth.) Cockayne & Allan. Syn.: *Veronica menziesii.* Plants called *Veronica menziesii* and **Hebe menziesii** are **H. divaricata**.
Hebe montana J. B. Armstr. See **H. subalpina**.
Hebe odora (Hook. f.) Cockayne. BOX-LEAF HEBE. Plants called **H. buxifolia** and *Veronica buxifolia* belong here.
Hebe 'Patty's Purple'.
Hebe pimeleoides (Hook. f.) Cockayne & Allan.
Hebe pinguifolia (Hook. f.) Cockayne & Allan.
Hebe pinguifolia 'Pagei'.
Hebe propinqua (Cheeseman) Cockayne & Allan.
Hebe recurva Simps. & J. S. Thoms.
Hebe 'Reevesii'?
Hebe 'Rubra'. See **H. speciosa** 'Rubra'.
Hebe salicifolia (G. Forst.) Penn. Syn.: *Veronica salicifolia.*
Hebe selaginoides?
Hebe speciosa (R. Cunn. ex A. Cunn.) Cockayne & Allan. Syn.: *Veronica speciosa.*
Hebe speciosa 'Imperialis'. Syn.: *H. imperialis. Veronica imperialis.*
Hebe speciosa 'Rubra'.
Hebe subalpina (Cockayne) Cockayne & Allan. Syn.: *H. montana.*
Hebe tetragona (Hook.) Cockayne & Allan.
Hebe topiaria L. B. Moore.
Hebe traversii (Hook. f.) Cockayne & Allan. Syn.: *Veronica traversii.*
Hebe venustula (Colenso) L. B. Moore. Syn.: *H. laevis.*
Hebe 'Youngii'.
Hedera algeriensis Hibb. See **H. canariensis**.
Hedera azorica Carrière. See **H. canariensis** 'Azorica'.
Hedera canariensis Willd. Syn.: *H. algeriensis. H. maderensis.* ALGERIAN IVY.
Hedera canariensis 'Azorica'. Syn.: *H. azorica.*
Hedera canariensis 'Striata'.
Hedera canariensis 'Variegata'. VARIEGATED ALGERIAN IVY. GLOIRE-DE-MARENGO IVY.
Hedera canariensis 'Variegata Arborescens'. GHOST TREE IVY. SOLEDAD TREE IVY.
Hedera colchica (K. Koch) K. Koch. Syn.: *H. cordifolia. H. regneriana.* COLCHIS IVY. PERSIAN IVY.
Hedera colchica 'Dentata'.
Hedera colchica 'Dentata Variegata'.
Hedera 'Conglomerata'. See **H. helix** 'Conglomerata'.
Hedera cordifolia Hibb. See **H. colchica**.
Hedera 'Dentata Aurea'. See **H. colchica** 'Dentata Variegata'.
Hedera 'Glacier'. See **H. helix** 'Glacier'.
Hedera 'Hahn's'. See **H. helix** 'Hahn's Self-branching'.
Hedera helix L. ENGLISH IVY.
Hedera helix 'Arborescens'.
Hedera helix 'Argenteo-variegata'. SILVERLEAF IVY.
Hedera helix 'Baltica'.
Hedera helix 'Caenwoodiana'.
Hedera helix 'California Gold'.
Hedera helix 'Conglomerata'.
Hedera helix 'Emerald Globe'.
Hedera helix 'Glacier'.
Hedera helix 'Gold Heart'.
Hedera helix 'Hahn's'. See **H. helix** 'Hahn's Self-branching'.
Hedera helix 'Hahn's Self-branching'. HAHN'S IVY.
Hedera helix 'Manda's Crested'.
Hedera helix 'Merion Beauty'.
Hedera helix 'Needlepoint'.
Hedera helix 'Pedata'. BIRD'S-FOOT IVY.
Hedera helix 'Pixie'.
Hedera helix 'Purpurea'.
Hedera helix 'Rochester'.
Hedera helix 'Sagittifolia'. ARROWLEAF IVY.
Hedera helix 'Sherwood Creeper'.
Hedera helix 'Thorndale'.
Hedera maderensis K. Koch. See **H. canariensis**.
Hedera 'Merion Beauty'. See **H. helix** 'Merion Beauty'.
Hedera regnerana Hibb. See **H. colchica**.
Hedera rhombea (Miq.) Bean. JAPANESE IVY.
Hedera 'Rochester'. See **H. helix** 'Rochester'.
Hedera 'Striata'. See **H. canariensis** 'Striata'.
Hedychium coronarium J. Koenig. BUTTERFLY GINGER. GARLAND FLOWER.
Hedychium gardneranum Roscoe. KAHILI. KAHILI GINGER.
Hedyscepe canterburyana (C. Moore & F. J. Muell.) H. Wendl. & Drude. UMBRELLA PALM.
Heimerliodendron brunonianum (Endl.) Skottsberg. See **Pisonia umbellifera**.
Helianthemum 'American Beauty'.
Helianthemum chamaecistus Mill. See **H. nummularium**.
Helianthemum lasianthum (Lam.) Pers. See **Halimium lasianthum**.
Helianthemum nummularium (L.) Mill. Syn.: *H. chamaecistus.* SUN ROSE.
Helianthemum ocymoides (Lam.) Pers. See **Halimium ocymoides**.
Helianthemum 'Rose Glory'.
Helianthemum scoparium Nutt. RUSH ROSE.
Helianthemum scoparium var. *vulgare* Jeps.
Helichrysum petiolatum (L.) DC. COTTONLEAF. LICORICE PLANT.
HELIOTROPE: **Heliotropium arborescens**.
Heliotropium arborescens L. Syn.: *H. peruvianum.* HELIOTROPE.
Heliotropium arborescens 'Black Beauty'.
Heliotropium arborescens 'Black Prince'.
Heliotropium arborescens 'Lilac Queen'.

Heliotropium arborescens 'Royal Fragrance'.
Heliotropium peruvianum L. See **H. arborescens**.
HELLEBORE: **Helleborus foetidus**.
   CORSICAN: **Helleborus lividus** subsp. **corsicus**.
Helleborus corsicus Willd. See **H. lividus** subsp. **corsicus**.
Helleborus foetidus L. HELLEBORE.
Helleborus lividus Ait.
Helleborus lividus subsp. **corsicus** (Willd.) Tutin. Syn.: *H. corsicus*. CORSICAN HELLEBORE.
Helleborus niger L. CHRISTMAS ROSE.
Helleborus orientalis Lam. LENTEN ROSE.
HELMET SHRUB: **Dyschoriste thunbergiiflora**.
HEMLOCK, CANADIAN: **Tsuga canadensis**.
   CANADIAN WEEPING: **Tsuga canadensis** 'Pendula'.
   CAROLINA: **Tsuga caroliniana**.
   COAST: **Tsuga heterophylla**.
   EASTERN: **Tsuga canadensis**.
   JAPANESE: **Tsuga diversifolia**.
   MOUNTAIN: **Tsuga mertensiana**.
   SARGENT WEEPING: **Tsuga canadensis** 'Pendula'.
   WESTERN: **Tsuga heterophylla**.
HEMP, AFRICAN: **Sparmannia africana**.
   CEYLON BOWSTRING: **Sansevieria zeylanica**.
HERALD'S-TRUMPET: **Beaumontia grandiflora**.
HERB-OF-GRACE: **Ruta graveolens**.
HERCULES'-CLUB: **Aralia spinosa**.
Hermannia verticillata (L.) K. Schum. Syn.: *Mahernia verticillata*. HONEYBELLS.
Hesperaloe parviflora (Torr.) Coult. RED YUCCA.
Heterocentron elegans (Schlechtend.) Kuntze. Syn.: *Schizocentron elegans*. SPANISH-SHAWL.
Heteromeles arbutifolia (Ait.) M. J. Roemer. Syn.: *Photinia arbutifolia*. CALIFORNIA HOLLY. CHRISTMAS BERRY. TOYON.
Heteromeles arbutifolia var. **macrocarpa** (Munz) Munz. Syn.: *Photinia arbutifolia* var. *macrocarpa*.
Heteropteris glabra Hook. & Arn.
Heteropteris syringifolia Griseb.
Hibbertia cuneiformis (Labill.) Sm. Syn.: *Candollea cuneiformis*.
Hibbertia obtusifolia DC.
Hibbertia scandens (Willd.) Dryander. Syn.: *H. volubilis*. SNAKE VINE. GOLD GUINEA PLANT.
Hibbertia volubilis (Andr.) Vent. See **H. scandens**.
HIBISCUS, BLUE: **Alyogyne huegelii**.
   FRINGED: **Hibiscus schizopetalus**.
Hibiscus acetosella Welw. ex Hiern. Syn.: *H. eetveldeanus*.
Hibiscus eetveldeanus De Wild. & T. Durand. See **H. acetosella**.
Hibiscus huegelii Endl. See **Alyogyne huegelii**.
Hibiscus 'Mrs. Orpha'.
Hibiscus mutabilis L. CONFEDERATE ROSE.
Hibiscus mutabilis 'Roseus'.
Hibiscus mutabilis 'Rubrus'.
Hibiscus 'Pompon'. Listed name, may be **H. syriacus** 'Pompon Rouge'.
Hibiscus rosa-sinensis L. ROSE-OF-CHINA.
Hibiscus rosa-sinensis 'Agnes Gault'.
Hibiscus rosa-sinensis 'Aloha'.
Hibiscus rosa-sinensis 'American Beauty'.
Hibiscus rosa-sinensis 'Apricot'.
Hibiscus rosa-sinensis 'Brilliant'.
Hibiscus rosa-sinensis 'Butterball'.
Hibiscus rosa-sinensis 'Buttercup'.
Hibiscus rosa-sinensis 'Butterfly'.
Hibiscus rosa-sinensis 'California Gold'.
Hibiscus rosa-sinensis 'Cecile'.
Hibiscus rosa-sinensis 'Cile Tinney'.
Hibiscus rosa-sinensis 'Cooperi'.
Hibiscus rosa-sinensis 'Coral Beauty'.
Hibiscus rosa-sinensis 'Crown of Bohemia'.
Hibiscus rosa-sinensis 'Delight'.
Hibiscus rosa-sinensis 'Diamond Head'.
Hibiscus rosa-sinensis 'Ecstasy'.
Hibiscus rosa-sinensis 'Fiesta'.
Hibiscus rosa-sinensis 'Flamingo Plume'.
Hibiscus rosa-sinensis 'Florida Sunset'.
Hibiscus rosa-sinensis 'Fullmoon'.
Hibiscus rosa-sinensis 'General Marshall'.
Hibiscus rosa-sinensis 'Golden Dust'.
Hibiscus rosa-sinensis 'Golden Harvest'.
Hibiscus rosa-sinensis 'Golden Isle'.
Hibiscus rosa-sinensis 'Hula Girl'.
Hibiscus rosa-sinensis 'Jane Cowell'.
Hibiscus rosa-sinensis 'Kamapuaa'.
Hibiscus rosa-sinensis 'Kate Sessions'.
Hibiscus rosa-sinensis 'Kauai'.
Hibiscus rosa-sinensis 'Kawailoa Beauty'.
Hibiscus rosa-sinensis 'Kona'.
Hibiscus rosa-sinensis 'Kona Improved'.
Hibiscus rosa-sinensis 'Luna'.
Hibiscus rosa-sinensis 'Misty Isle'.
Hibiscus rosa-sinensis 'Oahu'.
Hibiscus rosa-sinensis 'Paradise Moon'.
Hibiscus rosa-sinensis 'Pink Wings'.
Hibiscus rosa-sinensis 'President'.
Hibiscus rosa-sinensis 'Psyche'.
Hibiscus rosa-sinensis 'Radiane'.
Hibiscus rosa-sinensis 'Red Dragon'.
Hibiscus rosa-sinensis 'Red Monarch'.
Hibiscus rosa-sinensis 'Ross Estey'.
Hibiscus rosa-sinensis 'Ruby Red'.
Hibiscus rosa-sinensis 'San Chaba'.
Hibiscus rosa-sinensis 'Santa Ana'.
Hibiscus rosa-sinensis 'Scarlet Giant'.
Hibiscus rosa-sinensis 'Sensation'.
Hibiscus rosa-sinensis 'Stop Light'.
Hibiscus rosa-sinensis 'Sundown'.
Hibiscus rosa-sinensis 'Tangerine'.
Hibiscus rosa-sinensis 'Texas Star'.
Hibiscus rosa-sinensis 'The Bride'.
Hibiscus rosa-sinensis 'Vulcan'.
Hibiscus rosa-sinensis 'White Wings'.
Hibiscus 'Roseus'. See **H. mutabilis** 'Roseus'.
Hibiscus 'San Diego Red'. See **H. rosa-sinensis** 'Brilliant'.
Hibiscus schizopetalus (M. T. Mast.) Hook. f. FRINGED HIBISCUS.
Hibiscus scottii Balf. f.
Hibiscus 'Spotlight'?
Hibiscus syriacus L. Syn.: *Althaea syriaca*. SHRUB ALTHEA. ROSE-OF-SHARON.
Hibiscus syriacus 'Albus'. WHITE ROSE-OF-SHARON.
Hibiscus syriacus 'Amplissimus'.
Hibiscus syriacus 'Anemonaeflorus'.
Hibiscus syriacus 'Ardens'. PURPLE ROSE-OF-SHARON.
Hibiscus syriacus 'Blue Bird'.
Hibiscus syriacus 'Boule de Feu'.
Hibiscus syriacus 'Coelestis'.
Hibiscus syriacus 'Collie Mullens'.
Hibiscus syriacus 'Hamabo'.
Hibiscus syriacus 'Jeanne d'Arc'.
Hibiscus syriacus 'Lady Stanley'.
Hibiscus syriacus 'Lucy'.
Hibiscus syriacus 'Paeoniflorus'.

Hibiscus syriacus 'Pink Delight'.
Hibiscus syriacus 'Pompon Rouge'.
Hibiscus syriacus 'Pulcherrimus'. PINK ROSE-OF-SHARON.
Hibiscus syriacus 'Purpureus'.
Hibiscus syriacus 'Red Heart'.
Hibiscus syriacus 'Rose Pink'.
Hibiscus syriacus 'Woodbridge'.
Hibiscus tiliaceus L. MAHOE.
Hibiscus 'Vice President'?
Hibiscus wrayae Lindl. See *Alyogyne huegelii*.
HIMALAYA BERRY: **Rubus procerus**.
HINDU ROPE PLANT: **Hoya carnosa 'Krinkle Kurl'**.
HINOKI: **Chamaecyparis obtusa**.
   DWARF GOLDEN: **Chamaecyparis obtusa 'Nana Aurea'**.
   GOLDEN: **Chamaecyparis obtusa 'Aurea'**.
HIRYU: **Poncirus trifoliata 'Monstrosa'**.
Hoheria glabrata Sprague & Summerhayes. MOUNTAIN RIBBONWOOD. LACEBARK.
Hoheria populnea A. Cunn. NEW ZEALAND LACEBARK.
Hoheria sexstylosa Colenso. RIBBONWOOD.
HOLLY, AMERICAN: **Ilex opaca**.
   BELGIAN: **Ilex × altaclarensis 'Belgica'**.
   BOX-LEAVED: **Ilex crenata**.
   BURFORD: **Ilex cornuta 'Burfordii'**.
   CALIFORNIA: **Heteromeles arbutifolia**.
   CHINESE: **Ilex cornuta**.
   CHINESE: **Osmanthus heterophyllus**.
   CONVEX: **Ilex crenata 'Convexa'**.
   COSTA RICAN: **Olmediella betschlerana**.
   DESERT: **Artiplex hymenelytra**.
   DUTCH: **Ilex aquifolium 'J. C. van Tol'**.
   DWARF BURFORD: **Ilex cornuta 'Dwarf Burford'**.
   DWARF CHINESE: **Ilex cornuta 'Rotunda'**.
   DWARF JAPANESE: **Ilex crenata 'Compacta'**.
   ENGLISH: **Ilex aquifolium**.
   FALSE: **Osmanthus heterophyllus**.
   GUATEMALAN: **Olmediella betschlerana**.
   HEDGEHOG: **Ilex aquifolium 'Ferox'**.
   HELLER'S JAPANESE: **Ilex crenata 'Helleri'**.
   HIGHCLERE: **Ilex × altaclarensis 'Altaclarensis'**.
   JAPANESE: **Ilex crenata**.
   KASHI: **Ilex chinensis**.
   LUSTER-LEAF: **Ilex latifolia**.
   MADEIRA: **Ilex perado**.
   MINIATURE: **Malpighia coccigera**.
   PERRY'S WEEPING ENGLISH: **Ilex aquifolium 'Argenteo-marginata Pendula'**.
   PORCUPINE: **Ilex aquifolium 'Ferox'**.
   PUERTO RICAN: **Olmediella betschlerana**.
   SILVER HEDGEHOG: **Ilex aquifolium 'Ferox Argentea'**.
   SILVER-VARIEGATED ENGLISH: **Ilex aquifolium 'Argenteo-marginata'**.
   SINGAPORE: **Malpighia coccigera**.
   SUMMER: **Comarostaphylis diversifolia**.
   VARIEGATED ENGLISH: **Ilex aquifolium 'Aureo-marginata'**.
   WEST INDIAN: **Leea coccinea**.
   WILSON: **Ilex × altaclarensis 'Wilsonii'**.
HOLLY-GRAPE, CALIFORNIA: **Mahonia pinnata**.
   CHINESE: **Mahonia lomariifolia**.
   OREGON: **Mahonia aquifolium**.
Holmskioldia sanguinea Retzius. CHINESE HAT PLANT.
Holodiscus discolor (Pursh) Maxim. CREAMBUSH. OCEAN SPRAY.
Homalanthus populifolius R. Graham. QUEENSLAND POPLAR.
Homalocladium platycladum (F. J. Muell.) L. H. Bailey. Syn.: *Muehlenbeckia platyclada*. CENTIPEDE PLANT. RIBBONBUSH. TAPEWORM PLANT.

HONEYBELLS: **Hermannia verticillata**.
HONEYBUSH: **Melianthus major**.
HONEY LOCUST: See LOCUST, HONEY.
HONEY PLANT: **Hoya carnosa**.
HONEYSUCKLE, BOX: **Lonicera nitida**.
   CALIFORNIA: **Lonicera hispidula**.
   CAPE: **Tecomaria capensis**.
   CHINESE: **Lonicera japonica var. chinensis**.
   CORAL: **Lonicera heckrottii**.
   CORAL: **Lonicera sempervirens**.
   DUTCH: **Lonicera periclymenum var. belgica**.
   GIANT BURMESE: **Lonicera hildebrandiana**.
   GOLD-FLAME: **Lonicera heckrottii**.
   GOLD-NET: **Lonicera japonica 'Aureo-reticulata'**.
   HAIRY: **Lonicera hispidula var. vacillans**.
   HALL'S JAPANESE: **Lonicera japonica 'Halliana'**.
   JAPANESE: **Lonicera japonica**.
   PINK TATARIAN: **Lonicera tatarica 'Rosea'**.
   PRIVET: **Lonicera pileata**.
   PURPLE-LEAVED: **Lonicera japonica 'Purpurea'**.
   RED TATARIAN: **Lonicera tatarica 'Siberica'**.
   SOUTH AFRICAN: **Turraea obtusifolia**.
   TATARIAN: **Lonicera tatarica**.
   TRUMPET: **Lonicera sempervirens**.
   WHITE TATARIAN: **Lonicera tatarica 'Alba'**.
   WILD: **Lonicera subspicata**.
   WINTER: **Lonicera fragrantissima**.
   YELLOW CAPE: **Tecomaria capensis 'Aurea'**.
HOPBUSH: **Dodonaea viscosa**.
HOPSEED, CLAMMY: **Dodonaea viscosa**.
HOPSEED, RED: **Dodonaea viscosa**.
HOPSEED BUSH: **Dodonaea viscosa**.
HOPSEED TREE: **Dodonaea viscosa 'Purpurea'**.
HORNBEAM, AMERICAN: **Carpinus caroliniana**.
   AMERICAN HOP: **Ostrya virginiana**.
   EUROPEAN: **Carpinus betulus**.
   EUROPEAN HOP: **Ostrya carpinifolia**.
   JAPANESE: **Carpinus japonica**.
HORNCONE, MEXICAN: **Ceratozamia mexicana**.
HORSECHESTNUT: See CHESTNUT, HORSE.
HORSETAIL TREE: **Casuarina cunninghamiana**.
HORSETAIL TREE: **Casuarina stricta**.
HORTENSIA: **Hydrangea macrophylla**.
HOTTENTOT-BREAD: **Dioscorea elephantipes**.
Hovenia dulcis Thunb. JAPANESE RAISIN TREE.
Howea belmoreana (C. Moore & F. J. Muell.) Becc. Syn.: *Kentia belmoreana*. SENTRY PALM. BELMORE SENTRY PALM.
Howea forsterana (C. Moore & F. J. Muell.) Becc. *Kentia forsterana*. PARADISE PALM. FORSTER SENTRY PALM. SENTRY PALM.
Hoya bella Hook. MINIATURE WAX PLANT.
Hoya carnosa (L. f.) R. Br. Syn.: *H. motoskei*. HONEY PLANT. WAX PLANT.
Hoya carnosa 'Compacta'.
Hoya carnosa 'Exotica'.
Hoya carnosa 'Green Curls'. See *H. carnosa* 'Krinkle Kurl'.
Hoya carnosa 'Krinkle Kurl'. HINDU ROPE PLANT.
Hoya carnosa 'Variegata'. VARIEGATED WAX PLANT.
Hoya cinnamomifolia Hook.
Hoya imperialis Lindl.
Hoya keysii F. M. Bailey.
Hoya longifolia Wallich.
Hoya longifolia var. *shepherdii* (Hook.) N. E. Br. Syn.: *H. shepherdii*.

Hoya motoskei Teysm. & Binnend. See **H. carnosa**.
Hoya purpurea-fusca Hook.
Hoya shepherdii Hook. See **H. longifolia** var. **shepherdii**.
Hoya 'Silver Pink'. See **H. purpurea-fusca**.
HUCKLEBERRY, CALIFORNIA: **Vaccinium ovatum**.
    EVERGREEN: **Vaccinium ovatum**.
    RED: **Vaccinium parvifolium**.
HUMMINGBIRD BUSH: **Grevillea thelemanniana**.
HYACINTH SHRUB: **Xanthoceras sorbifolium**.
Hybophrynium braunianum K. Schum. Syn.: *Bamburanta arnoldiana*.
HYDRANGEA, BIG-LEAF: **Hydrangea macrophylla**.
    CLIMBING: **Decumaria barbara**.
    CLIMBING: **Hydrangea anomala** subsp. **petiolaris**.
    FRENCH: **Hydrangea macrophylla**.
    GARDEN: **Hydrangea macrophylla**.
    OAKLEAF: **Hydrangea quercifolia**.
    PEEGEE: **Hydrangea paniculata 'Grandiflora'**.
    SMOOTH: **Hydrangea arborescens**.
    SNOWHILL: **Hydrangea arborescens 'Grandiflora'**.
    VINE, JAPANESE: **Schizophragma hydrangeoides**.
Hydrangea acuminata Siebold & Zucc. See **H. macrophylla** subsp. **serrata**.
Hydrangea acuminata 'Pink Beauty'. See **H. macrophylla 'Pink Beauty'**.
Hydrangea acuminata 'Preziosa'. See **H. macrophylla 'Preziosa'**.
Hydrangea anomala D. Don. subsp. ***petiolaris*** (Siebold & Zucc.) McClint. Syn.: *H. petiolaris*. CLIMBING HYDRANGEA.
Hydrangea arborescens L. SMOOTH HYDRANGEA.
Hydrangea arborescens 'Grandiflora'. SNOWHILL HYDRANGEA.
Hydrangea hortensia Siebold. See **H. macrophylla**.
Hydrangea 'Kuhnert'. See **H. macrophylla 'Kuhnert'**.
Hydrangea macrophylla (Thunb.) Ser. Syn.: *H. hortensia. H. opuloides. H. otaksa*. HORTENSIA. FRENCH HYDRANGEA. GARDEN HYDRANGEA. BIG-LEAF HYDRANGEA.
***Hydrangea macrophylla* 'All Summer Beauty'**.
***Hydrangea macrophylla* 'Altona'**.
***Hydrangea macrophylla* 'Ami Pasquier'**.
***Hydrangea macrophylla* 'Blue Prince'**.
***Hydrangea macrophylla* 'Blue Wave'**.
***Hydrangea macrophylla* 'Bouffant'**.
***Hydrangea macrophylla* 'Celeste'**.
***Hydrangea macrophylla* 'Charm'**.
***Hydrangea macrophylla* 'Chautard'**.
***Hydrangea macrophylla* 'Daphne'**.
***Hydrangea macrophylla* 'Domotoi'**.
***Hydrangea macrophylla* 'Engel's White'**.
***Hydrangea macrophylla* 'Europa'**.
***Hydrangea macrophylla* 'Europa Rose'**.
***Hydrangea macrophylla* 'Gertrude Glahn'**.
***Hydrangea macrophylla* 'Giant White'**.
***Hydrangea macrophylla* 'Goliath'**.
***Hydrangea macrophylla* 'Hamburg'**.
***Hydrangea macrophylla* 'Hamburg Scarlet'**.
***Hydrangea macrophylla* 'Hollandia'**.
Hydrangea macrophylla hortensis. Listed name for **H. macrophylla**.
Hydrangea macrophylla hortensis compacta. Listed name for **H. macrophylla**.
***Hydrangea macrophylla* 'Kuhnert'**.
***Hydrangea macrophylla* 'Merrit's Beauty'**.
***Hydrangea macrophylla* 'Merveille'**.
***Hydrangea macrophylla* 'Mme. E. Moulliere'**.
***Hydrangea macrophylla* 'Nikko'**.
***Hydrangea macrophylla* 'Nikko Blue'**.
***Hydrangea macrophylla* 'Pink Beauty'**.
***Hydrangea macrophylla* 'Preziosa'**.
***Hydrangea macrophylla* 'Red Clackamas'**.
***Hydrangea macrophylla* 'Revelation'**.
***Hydrangea macrophylla* 'Rouget de Lisle'**.
Hydrangea macrophylla subsp. ***serrata*** (Thunb.) Makino. Syn.: *H. acuminata*.
***Hydrangea macrophylla* 'Strafford'**.
***Hydrangea macrophylla* 'Tricolor'**.
***Hydrangea macrophylla* 'Triomphe'**.
***Hydrangea macrophylla* 'Trophy'**.
***Hydrangea macrophylla* 'Variegata'**.
Hydrangea opuloides Lam. See **H. macrophylla**.
Hydrangea otaksa Siebold & Zucc. See **H. macrophylla**.
Hydrangea paniculata Siebold.
***Hydrangea paniculata* 'Grandiflora'**. PEEGEE HYDRANGEA.
Hydrangea petiolaris Siebold & Zucc. See **H. anomala** subsp. **petiolaris**.
Hydrangea quercifolia Bartr. OAKLEAF HYDRANGEA.
Hymenosporum flavum (Hook.) F. J. Muell. SWEETSHADE.
Hyophorbe verschaffeltii H. Wendl. Syn.: *Mascarena verschaffeltii*. SPINDLE PALM.
Hypericum androsaemum L. TUTSAN.
Hypericum aureum Bartr. See **H. frondosum**.
Hypericum beanii N. Robs. Syn.: *H. patulum* var. *henryi*. GOLD FLOWER. HENRY ST. JOHN'S-WORT.
Hypericum calycinum L. AARON'S-BEARD. CREEPING ST. JOHN'S-WORT. GOLD FLOWER. ROSE-OF-SHARON.
Hypericum coris L.
Hypericum dyeri Rehd.
Hypericum empetrifolium Willd.
Hypericum forrestii (Chitt.) N. Robs. Syn.: *H. patulum* var. *forrestii*.
Hypericum frondosum Michx. Syn.: *H. aureum*.
Hypericum henryi Lév. & Vaniot. See **H. patulum**.
Hypericum 'Hidcote'.
Hypericum hookeranum Wight & Arn.
Hypericum kalmianum L.
Hypericum kouytchense Lév.
***Hypericum kouytchense* 'Sun Gold'**.
Hypericum lanceolatum Lam. See **H. revolutum**.
Hypericum leschenaultii Choisy.
Hypericum linarioides Bosse. Syn.: *H. repens*.
Hypericum × moseranum André. [*H. patulum* × *H. calycinum*]. GOLD FLOWER.
Hypericum olympicum L.
Hypericum patulum Thunb. Syn.: *H. henryi*.
Hypericum patulum var. *forrestii* Chitt. See **H. forrestii**.
Hypericum patulum var. *henryi* Bean. See **H. beanii**.
Hypericum patulum var. *uralum* (Buch.-Ham. ex D. Don) Koehne. See **H. uralum**.
Hypericum repens of authors, not L. See **H. linarioides**.
Hypericum revolutum Vahl. Syn.: *H. lanceolatum*.
Hypericum 'Rowallane'.
Hypericum 'Sun Gold'. See **H. kouytchense 'Sun Gold'**.
Hypericum uralum Buch.-Ham. ex D. Don. Syn.: *H. patulum* var. *uralum*.
Hypoestes phyllostachys Baker. PINK POLKA-DOT PLANT. FRECKLE-FACE. Some plants called **H. sanguinolenta** belong here.
Hypoestes sanguinolenta (Van Houtte) Hook. f. May not be cult. Plants so called may be **H. phyllostachys**.
Hyptis emoryi Torr. DESERT LAVENDER.

I

Iberis semperflorens L. CANDYTUFT.
Iberis sempervirens L. EDGING CANDYTUFT.
Iberis sempervirens 'Camlaensis'.
Iberis sempervirens 'Little Cushion'.
Iberis sempervirens 'Little Gem'.
Iberis sempervirens 'Purity'.
Iberis sempervirens 'Snowflake'.
Idesia polycarpa Maxim.
IDRIA: **Fouquieria columnaris**.
Idria columnaris Kell. See **Fouquieria columnaris**.
Ilex × *altaclarensis* (Dallim.) Rehd. [*I. aquifolium* × *I. perado*].
Ilex × *altaclarensis* 'Altaclarensis'. HIGHCLERE HOLLY.
Ilex × *altaclarensis* 'Belgica'. BELGIAN HOLLY.
Ilex × *altaclarensis* 'Camelliifolia'.
Ilex × *altaclarensis* 'Eldridge'.
Ilex × *altaclarensis* 'J. C. van Tol'. See *I. aquifolium* 'J. C. van Tol'.
Ilex × *altaclarensis* 'Wilsonii'. WILSON HOLLY. Not to be confused with *I. wilsonii*.
Ilex aquifolium L. ENGLISH HOLLY.
Ilex aquifolium 'Albo-marginata'.
Ilex aquifolium 'Altaclarensis'. See *I.* × *altaclarensis*.
Ilex aquifolium 'Angustifolia'.
Ilex aquifolium 'Argentea Medio-picta'. Collective name for plants with leaves centrally blotched with creamy white.
Ilex aquifolium 'Argentea Regina'.
Ilex aquifolium 'Argenteo-marginata'. SILVER-VARIEGATED ENGLISH HOLLY. Collective name for plants with silver-margined leaves.
Ilex aquifolium 'Argenteo-marginata Pendula'. PERRY'S WEEPING ENGLISH HOLLY.
Ilex aquifolium 'Aurea Medio-picta'. Collective name for plants with leaves centrally blotched with yellow.
Ilex aquifolium 'Aurea Regina'.
Ilex aquifolium 'Aureo-marginata'. VARIEGATED ENGLISH HOLLY. Collective name for plants with yellow margined leaves.
Ilex aquifolium 'Bacciflava'.
Ilex aquifolium 'Bailey's Pride'.
Ilex aquifolium 'Balkans'.
Ilex aquifolium 'Big Bull'.
Ilex aquifolium 'Boulder Creek'.
Ilex aquifolium 'Brilliant'. See *I.* × *aquipernyi* 'Brilliant'.
Ilex aquifolium 'Camelliifolia'. See *I.* × *altaclarensis* 'Camelliifolia'.
Ilex aquifolium 'Chambers'.
Ilex aquifolium 'Ciliata Major'.
Ilex aquifolium 'Deluxe'.
Ilex aquifolium 'Femina'?
Ilex aquifolium 'Ferox'. HEDGEHOG HOLLY. PORCUPINE HOLLY.
Ilex aquifolium 'Ferox Argentea'. SILVER HEDGEHOG HOLLY.
Ilex aquifolium 'Ferox Argentea Marginata'. Listed name. May be *I. aquifolium* 'Ferox Argentea'.
Ilex aquifolium 'Ferox Aurea'.
Ilex aquifolium 'Fertilis'?
Ilex aquifolium 'Fructu Luteo'. See *I. aquifolium* 'Bacciflava'.
Ilex aquifolium 'Gold Edge'?
Ilex aquifolium 'Golden King'. See *I. aquifolium* 'Aurea Regina'.
Ilex aquifolium 'Golden Milkmaid'. See *I. aquifolium* 'Aurea Medio-picta'.
Ilex aquifolium 'Golden Queen'. See *I. aquifolium* 'Aurea Regina'.
Ilex aquifolium 'Golden Variegated'?
Ilex aquifolium 'James G. Esson'. See *I.* × *aquipernyi* 'James G. Esson'.
Ilex aquifolium 'J. C. van Tol'. DUTCH HOLLY.
Ilex aquifolium 'Lily Gold'.
Ilex aquifolium 'Little Bull'.
Ilex aquifolium 'Moonlight'.
Ilex aquifolium 'Pendula'.
Ilex aquifolium 'Pinto'.
Ilex aquifolium 'Pyramidalis'.
Ilex aquifolium 'Rederly'.
Ilex aquifolium 'San Gabriel'.
Ilex aquifolium 'Santa Ana'.
Ilex aquifolium 'Silver King'. See *I. aquifolium* 'Argentea Regina'.
Ilex aquifolium 'Silver Milkmaid'. See *I. aquifolium* 'Argentea Medio-picta'.
Ilex aquifolium 'Silver Queen'. See *I. aquifolium* 'Argentea Regina'.
Ilex aquifolium 'Silver Star'.
Ilex aquifolium 'Silver Variegated'?
Ilex aquifolium 'Sparkler'.
Ilex aquifolium 'Teufel's Deluxe'. See *I. aquifolium* 'Deluxe'.
Ilex aquifolium 'Teufel's Hybrid'.
Ilex aquifolium 'Variegated'?
Ilex aquifolium 'Zero'.
Ilex × *aquipernyi* Gable ex W. B. Clarke. [*I. aquifolium* × *I. pernyi*].
Ilex × *aquipernyi* 'Brilliant'.
Ilex × *aquipernyi* 'James G. Esson'.
Ilex 'Argenteo-marginata'. See *I. aquifolium* 'Argenteo-marginata'.
Ilex × *attenuata* Ashe. [*I. opaca* × *I. cassine*].
Ilex × *attenuata* 'East Palatka'.
Ilex × *attenuata* 'Foster No. 2'.
Ilex 'Biloxi'. See *I. crenata* 'Biloxi'.
Ilex bioritensis Hayata. Syn.: *I. pernyi* var. *veitchii*.
Ilex 'Brilliant'. See *I.* × *aquipernyi* 'Brilliant'.
Ilex 'Burfordii'. See *I. cornuta* 'Burfordii'.
Ilex 'Burfordii Compacta'. See *I. cornuta* 'Dwarf Burford'.
Ilex 'Burfordii Nana'. See *I. cornuta* 'Dwarf Burford'.
Ilex 'Camelliifolia'. See *I.* × *altaclarensis* 'Camelliifolia'.
Ilex 'Chambers'. See *I. aquifolium* 'Chambers'.
Ilex chinensis Sims. Syn.: *I. purpurea* var. *oldhamii*. KASHI HOLLY.
Ilex ciliospinosa Loes.
Ilex 'Convexa Bullata'. See *I. crenata* 'Convexa'.
Ilex cornuta Lindl. & Paxt. CHINESE HOLLY.
Ilex cornuta 'Azusa'.
Ilex cornuta 'Berries Jubilee'.
Ilex cornuta 'Burfordii'. BURFORD HOLLY.
Ilex cornuta 'Burfordii Compacta'. See *I. cornuta* 'Dwarf Burford'.
Ilex cornuta 'Burfordii Nana'. See *I. cornuta* 'Dwarf Burford'.
Ilex cornuta 'Dazzler'.
Ilex cornuta 'D'Or'.
Ilex cornuta 'Dwarf Burford'. DWARF BURFORD HOLLY.
Ilex cornuta 'Giant Beauty'.
Ilex cornuta 'Rotunda'. DWARF CHINESE HOLLY.
Ilex cornuta 'Walder'.

Ilex cornuta 'Willowleaf'.
Ilex crenata Thunb. BOX-LEAVED HOLLY. JAPANESE HOLLY.
Ilex crenata 'Biloxi'.
Ilex crenata 'Bullata'. See *I. crenata* 'Convexa'.
Ilex crenata 'Compacta'. DWARF JAPANESE HOLLY.
Ilex crenata 'Convexa'.
Ilex crenata 'Glory'.
Ilex crenata 'Green Island'.
Ilex crenata 'Green Thumb'.
Ilex crenata 'Helleri'.
Ilex crenata 'Hetzii'.
Ilex crenata 'Latifolia'.
Ilex crenata 'Mariesii'.
Ilex crenata 'Microphylla'.
Ilex crenata 'Rotundifolia'.
Ilex 'Dazzler'. See *I. cornuta* 'Dazzler'.
Ilex 'Dwarf Burford'. See *I. cornuta* 'Dwarf Burford'.
Ilex 'East Palatka'. See *I.* × *attenuata* 'East Palatka'.
Ilex 'Foster No. 2'. See *I.* × *attenuata* 'Foster No. 2'.
Ilex glabra (L.) Gray. INKBERRY.
Ilex 'Green Island'. See *I. crenata* 'Green Island'.
Ilex integra Thunb.
Ilex 'John T. Morris'. [*I. cornuta* 'Burfordii' × *I. pernyi*].
Ilex latifolia Thunb. LUSTER-LEAF HOLLY. TARAJO.
Ilex 'Lydia Morris'. [*I. cornuta* 'Burfordii' × *I. pernyi*].
Ilex 'Nellie R. Stevens'. [*I. aquifolium* × *I. cornuta*].
Ilex opaca Ait. AMERICAN HOLLY.
Ilex opaca 'Brilliantissima'. See *I. opaca* 'George E. Hart'.
Ilex opaca 'East Palatka'. See *I.* × *attenuata* 'East Palatka'.
Ilex opaca 'George E. Hart'.
Ilex opaca 'Greenleaf'.
Ilex opaca 'Howard'.
Ilex opaca 'Mrs. Sarver'. See *I. opaca* 'Rosalind Sarver'.
Ilex opaca 'Rosalind Sarver'.
Ilex pedunculosa Miq.
Ilex perado Ait. MADEIRA HOLLY.
Ilex pernyi Franch.
Ilex pernyi 'Brilliant'. See *I.* × *aquipernyi* 'Brilliant'.
Ilex pernyi var. *veitchii* Bean ex Rehd. See *I. bioritensis*.
Ilex purpurea Hassk. var. *oldhamii* (Miq.) Loes. See *I. chinensis*.
Ilex rotunda Thunb.
Ilex 'San Jose'. [*I. altaclarensis* 'Wilsonii' × *I. sikkimensis* Kurz].
Ilex 'San Jose Hybrid'. See *I.* 'San Jose'.
Ilex serrata Thunb. JAPANESE WINTERBERRY.
Ilex serrata 'Sieboldii'. Syn.: *I. sieboldii*.
Ilex sieboldii Miq. See *I. serrata* 'Sieboldii'.
Ilex 'Van Tol'. See *I. aquifolium* 'J. C. van Tol'.
Ilex vomitoria Ait. YAUPON. CASSINE.
Ilex vomitoria 'City of Houston'. See *I. vomitoria* 'Pride of Houston'.
Ilex vomitoria 'Nana'. DWARF YAUPON.
Ilex vomitoria 'Pride of Houston'.
Ilex vomitoria 'Pride of Texas'.
Ilex vomitoria 'Stokes' Dwarf'.
Ilex wilsonii Loes. Not to be confused with *I.* × *altaclarensis* 'Wilsonii'.
Illicium floridanum Ellis. PURPLE ANISE.
ILLYARIE: **Eucalyptus erythrocorys**.
Impatiens oliveri C. H. Wright ex W. Wats. See *I. sodenii*.
Impatiens sodenii Engler & Warb. ex Engler. Syn.: *I. oliveri*. OLIVER'S SNAPWEED. POOR MAN'S RHODODENDRON. TOUCH-ME-NOT.

INCIENSO: **Encelia farinosa**.
INDIGO, FALSE: **Amorpha fruticosa**.
INDIGO BUSH: **Amorpha fruticosa**.
INKBERRY: **Ilex glabra**.
Iochroma cyaneum (Lindl.) M. L. Green. Syn.: *I. lanceolatum*. *I. tubulosum*. PURPLE TOBACCO.
Iochroma fuchsioides (HBK) Miers.
Iochroma lanceolatum (Miers) Miers. See *I. cyaneum*.
Iochroma tubulosum Benth. See *I. cyaneum*.
Ipomoea bolusiana Schinz.
Ipomoea fistulosa Mart. ex Choisy. MORNING-GLORY.
IRIS, AFRICAN: **Dietes iridioides**.
IRONBARK, LARGE-FRUITED WHITE: **Eucalyptus leucoxylon** var. **macrocarpa**.
  LEMON-SCENTED: **Eucalyptus staigerana**.
  MULGA: **Eucalyptus sideroxylon**.
  NARROW-LEAVED: **Eucalyptus crebra**.
  PINK: **Eucalyptus sideroxylon**.
  RED: **Eucalyptus sideroxylon**.
  WHITE: **Eucalyptus leucoxylon**.
IRON PLANT: **Aspidistra elatior**.
IRON TREE, NEW ZEALAND: **Metrosideros excelsus**.
  CRIMSON: **Metrosideros kermadecensis**.
IRONWOOD, CATALINA: **Lyonothamnus floribundus**.
  DESERT: **Olneya tesota**.
  FERN-LEAF CATALINA: **Lyonothamnus floribundus** subsp. **asplenifolius**.
  ISLAND: **Lyonothamnus floribundus** subsp. **asplenifolius**.
ISLAY: **Prunus ilicifolia**.
Isomeris arborea Nutt. Syn.: *Cleome isomeris*. BLADDERPOD. BLADDERBUSH. BURRO-FAT.
Isoplexis canariensis (L.) Lindl. ex G. Don. Syn.: *Digitalis canariensis*.
Isopogon anethifolius (Salisb.) Knight.
ISU TREE: **Distylium racemosum**.
Itea ilicifolia D. Oliver. HOLLY-LEAF SWEETSPIRE.
Itea japonica D. Oliver.
Itea virginiana L. SWEETSPIRE.
Itea yunnanensis Franch.
IVY, ALGERIAN: **Hedera canariensis**.
  AMERICAN: **Parthenocissus quinquefolia**.
  ARROWLEAF: **Hedera helix** 'Sagittifolia'.
  BIRD'S-FOOT: **Hedera helix** 'Pedata'.
  BOSTON: **Parthenocissus tricuspidata**.
  COLCHIS: **Hedera colchica**.
  ENGLISH: **Hedera helix**.
  FIVE-LEAVED: **Parthenocissus quinquefolia**.
  GERMAN: **Senecio mikanioides**.
  GHOST TREE: **Hedera canariensis** 'Variegata Arborescens'.
  GLOIRE-DE-MARENGO: **Hedera canariensis** 'Variegata'.
  GRAPE: **Cissus rhombifolia**.
  HAHN'S: **Hedera helix** 'Hahn's Self-branching'.
  JAPANESE: **Hedera rhombea**.
  JAPANESE: **Parthenocissus tricuspidata**.
  KANGAROO: **Cissus antarctica**.
  MINIATURE GRAPE: **Cissus striata**.
  PARLOR: **Philodendron scandens** subsp. **oxycardium**.
  PERSIAN: **Hedera colchica**.
  PERSIAN: **Hedera colchica** 'Dentata'.
  SILVERLEAF: **Hedera helix** 'Argenteo-variegata'.
  SOLEDAD TREE: **Hedera canariensis** 'Variegata Arborescens'.
  VARIEGATED ALGERIAN: **Hedera canariensis** 'Variegata'.
  VENEZUELA GRAPE: **Cissus rhombifolia**.
IVYBUSH: **Kalmia latifolia**.

J

Jacaranda acutifolia Humb. & Bonpl. GREEN EBONY. Some plants so called are **J. mimosifolia**.
Jacaranda acutifolia 'Alba'.
Jacaranda cuspidifolia Mart.
Jacaranda mimosifolia D. Don. Syn.: *J. ovalifolia*. Some plants so called may be **J. acutifolia**.
Jacaranda mimosifolia 'Alba'.
Jacaranda ovalifolia R. Br. See **J. mimosifolia**.
Jacobinia aurea (Schlechtend.) Hemsl. See **Justicia aurea**.
Jacobinia carnea (Lindl.) Nichols. See **Justicia carnea**.
Jacobinia ghiesbreghtiana (Lem.) Hemsl. See **Justicia ghiesbreghtiana**.
Jacobinia incana (Nees) Hemsl. See **Justicia leonardii**.
Jacobinia mohintlii (Nees) Hemsl. See **Justicia spicigera**.
Jacobinia pauciflora (Nees) Lindau. See **Justicia rizzinii**.
Jacobinia rosea. Listed name, perhaps for **Justicia carnea**.
Jacobinia spicigera (Schlechtend.) L. H. Bailey. See **Justicia spicigera**.
Jacobinia suberecta André. See **Dicliptera suberecta**.
JADE PLANT: **Crassula arborescens**.
JADE PLANT: **Crassula obliqua**.
JADE TREE: **Crassula ovata**.
JASMINE, ANGEL-WING: **Jasminum nitidum**.
  ARABIAN: **Jasminum sambac**.
  CAPE: **Gardenia augusta**.
  CHILEAN: **Mandevilla laxa**.
  DWARF: **Jasminum parkeri**.
  ITALIAN: **Jasminum humile**.
  ITALIAN: **Jasminum humile 'Revolutum'**.
  MADAGASCAR: **Stephanotis floribunda**.
  MYSTERY CAPE: **Gardenia augusta 'Mystery'**.
  PARKER: **Jasminum parkeri**.
  PINK: **Jasminum polyanthum**.
  PINWHEEL: **Jasminum gracillimum**.
  POET'S: **Jasminum officinale**.
  PRIMROSE: **Jasminum mesnyi**.
  SPANISH: **Jasminum grandiflorum**.
  STAR: **Jasminum gracillimum**.
  STAR: **Jasminum multiflorum**.
  STAR: **Trachelospermum jasminoides**.
  WINTER: **Jasminum nudiflorum**.
Jasminum absimile L. H. Bailey. See **J. leratii**.
Jasminum angulare Vahl. Some plants called **J. azoricum** belong here.
Jasminum azoricum L. Some plants so called are **J. angulare** or **J. fluminense**.
Jasminum beesianum Forrest & Diels.
Jasminum dichotomum Vahl.
Jasminum floridum Bunge.
Jasminum fluminense Vell. Some plants called **J. azoricum** belong here.
Jasminum fruticans L.
Jasminum gracillimum Hook. f. PINWHEEL JASMINE. STAR JASMINE.
Jasminum 'Grand Duke'. See **J. sambac 'Grand Duke'**.
Jasminum grandiflorum L. Syn.: *J. officinale* var. *grandiflorum*. SPANISH JASMINE.
Jasminum humile L. ITALIAN JASMINE.
Jasminum humile var. **kansuense** Kobuski.
Jasminum humile var. *revolutum* (Sims) Kobuski. See **J. humile 'Revolutum'**.
Jasminum humile 'Revolutum'. Syn.: *J. humile* var. *revolutum*. *J. revolutum*. ITALIAN JASMINE.
Jasminum kansuense. Listed name for **J. humile** var. **kansuense**.
Jasminum leratii Schlechter. Syn.: *J. absimile*. Some plants called **J. ligustrifolium, J. rigidum, J. simplicifolium** may be **J. leratii**.
Jasminum ligustrifolium Lam. Probably not cult. Plants so called are probably **J. leratii**.
Jasminum magnificum Lingelsh. Probably not cult. Plants so called are probably **J. nitidum**.
Jasminum 'Maid of Orleans'. See **J. sambac 'Maid of Orleans'**.
Jasminum mesnyi Hance. Syn.: *J. primulinum*. PRIMROSE JASMINE.
Jasminum multiflorum (Burm. f.) Andr. Syn.: *J. pubescens*. STAR JASMINE.
Jasminum nitidum Skan. ANGEL-WING JASMINE. Some plants called **J. magnificum** or **J. rigidum** belong here.
Jasminum nudiflorum Lindl. WINTER JASMINE.
Jasminum officinale L. POET'S JASMINE.
Jasminum officinale var. *grandiflorum* (L.) Stokes. See **J. grandiflorum**.
Jasminum parkeri Dunn. PARKER JASMINE. DWARF JASMINE.
Jasminum polyanthum Franch. PINK JASMINE.
Jasminum primulinum Hemsl. See **J. mesnyi**.
Jasminum pubescens Willd. See **J. multiflorum**.
Jasminum revolutum Sims. See **J. humile 'Revolutum'**.
Jasminum rex Dunn.
Jasminum rhynchospermum. Listed name for **Trachelospermum jasminoides**.
Jasminum rigidum Zenker. Probably not cult. Plants so called are probably **J. leratii** or **J. nitidum**.
Jasminum sambac (L.) Ait. ARABIAN JASMINE.
Jasminum sambac 'Grand Duke'.
Jasminum sambac 'Maid of Orleans'.
Jasminum simplicifolium G. Forst. May not be cult. Plants so called are probably **J. leratii**.
Jasminum × stephanense Lemoine. [*J. officinale* × *J. beesianum*].
Jatropha berlandieri Torr. See **J. cathartica**.
Jatropha cathartica Terán & Berl. Syn.: *J. berlandieri*.
JELLY-BEAN PLANT: **Sedum pachyphyllum**.
JELLY-BEANS: **Sedum pachyphyllum**.
JESSAMINE, CAROLINA: **Gelsemium sempervirens**.
  DAY-BLOOMING: **Cestrum diurnum**.
  NIGHT-BLOOMING: **Cestrum nocturnum**.
  ORANGE: **Murraya paniculata**.
  WILLOW-LEAVED: **Cestrum parqui**.
JETBEAD: **Rhodotypos scandens**.
JIM-BRUSH: **Ceanothus sorediatus**.
JOJOBA: **Simmondsia chinensis**.
JOSHUA TREE: **Yucca brevifolia**.
  JAEGER: **Yucca brevifolia** var. **jaegerana**.
Jovellana sinclairii (Hook.) Kränzlin.
Jubaea chilensis (Molina) Baill. Syn.: *J. spectabilis*. CHILEAN WINE PALM. MONKEY COCONUT. SYRUP PALM. WINE PALM.
Jubaea spectabilis HBK. See **J. chilensis**.
JUDAS TREE: **Cercis occidentalis**.
JUDAS TREE: **Cercis siliquastrum**.
  WHITE: **Cercis siliquastrum 'Alba'**.
Juglans californica S. Wats. SOUTHERN CALIFORNIA BLACK WALNUT.
Juglans cinerea L. BUTTERNUT.
Juglans hindsii (Jeps.) Jeps. CALIFORNIA BLACK WALNUT.
Juglans major (Torr.) Heller. NOGAL. ARIZONA WALNUT.

Juglans nigra L. BLACK WALNUT.
Juglans 'Paradox'. [*J. hindsii* × *J. regia*]. PARADOX WALNUT.
Juglans regia L. ENGLISH WALNUT. PERSIAN WALNUT.
JUJUBE: *Ziziphus jujuba*.
   CHINESE: *Ziziphus jujuba*.
JUNIPER, ALLIGATOR: *Juniperus deppeana*.
   ANDORRA: *Juniperus horizontalis* 'Plumosa'.
   ARMSTRONG: *Juniperus chinensis* 'Armstrongii'.
   ASHFORD: *Juniperus communis* 'Ashfordii'.
   BLUE: *Juniperus chinensis* 'Hetzii'.
   BLUE COLUMN CHINESE: *Juniperus chinensis* 'Columnaris Glauca'.
   BLUE HOLLYWOOD: *Juniperus chinensis* 'Kaizuka Blue'.
   BLUE PFITZER: *Juniperus chinensis* 'Pfitzerana Glauca'.
   BLUE SARGENT'S: *Juniperus chinensis* var. *sargentii* 'Glauca'.
   BLUE SHIMPAKU: *Juniperus chinensis* var. *sargentii* 'Glauca'.
     (This common name is sometimes misapplied to *J. chinensis* 'Blaauw'.)
   BLUE UPRIGHT: *Juniperus chinensis* 'Hetzii'.
   CALIFORNIA: *Juniperus californica*.
   CANADIAN: *Juniperus communis* var. *depressa*.
   CANAERT UPRIGHT: *Juniperus virginiana* 'Canaertii'.
   CANARY ISLAND: *Juniperus cedrus*.
   CHINESE: *Juniperus chinensis*.
   COMPACT PFITZER: *Juniperus chinensis* 'Pfitzerana Compacta'.
   CREEPING: *Juniperus horizontalis*.
   CREEPING: *Juniperus procumbens*.
   CYPRESS: *Juniperus sabina* 'Cupressifolia'.
   DESERT: *Juniperus osteosperma*.
   DWARF JAPGARDEN: *Juniperus procumbens* 'Nana'.
   DWARF SWEDISH: *Juniperus communis* 'Suecica Nana'.
   ENGLISH: *Juniperus communis*.
   FISHTAIL: *Juniperus squamata* 'Meyeri'.
   GOLDEN PFITZER: *Juniperus chinensis* 'Pfitzerana Aurea'.
   GOLDEN PLUME CHINESE: *Juniperus chinensis* 'Plumosa Aurea'.
   GRECIAN: *Juniperus excelsa*.
   GREEN COLUMN CHINESE: *Juniperus chinensis* 'Columnaris'.
   HETZ BLUE: *Juniperus chinensis* 'Hetzii'.
   HICKS: *Juniperus sabina* 'Hicksii'.
   HILL DUNDEE: *Juniperus virginana* 'Hillii'.
   HOLLYWOOD: *Juniperus chinensis* 'Kaizuka'.
   HORIZONTAL: *Juniperus communis* var. *montana*.
   IRISH: *Juniperus communis* 'Hibernica'.
   JAPGARDEN: *Juniperus procumbens*.
   KETELEER: *Juniperus chinensis* 'Keteleeri'.
   KOSTER: *Juniperus virginiana* 'Kosteri'.
   MEYER: *Juniperus squamata* 'Meyeri'.
   MEYER SINGLE-SEED: *Juniperus squamata* 'Meyeri'.
   MOUNTAIN: *Juniperus communis* var. *montana*.
   NICK'S COMPACT PFITZER: *Juniperus chinensis* 'Pfitzerana Compacta'.
   PFITZER: *Juniperus chinensis* 'Pfitzerana'.
   PLUME CHINESE: *Juniperus chinensis* 'Plumosa'.
   POLISH: *Juniperus communis* 'Cracovia'.
   PROSTRATE COMMON: *Juniperus communis* 'Prostrata'.
   ROCKY MOUNTAIN: *Juniperus scopulorum*.
   SAN JOSE CREEPING: *Juniperus chinensis* 'San Jose'.
   SARGENT: *Juniperus chinensis* var. *sargentii*.
   SAVIN: *Juniperus sabina*.
   SCALY-LEAVED NEPAL: *Juniperus squamata*.
   SHIMPAKU: *Juniperus chinensis* var. *sargentii*. (This common name is sometimes misapplied to cvs. of the 'Plumosa' group of *J. chinensis*.)
   SHORE: *Juniperus conferta*.
   SIERRA: *Juniperus occidentalis*.
   SILVER COLUMN: *Juniperus scopulorum* 'Hill's Silver'.
   SINGLE-SEED: *Juniperus squamata*.
   SPANISH: *Juniperus sabina* 'Tamariscifolia'.
   SWEDISH: *Juniperus communis* 'Suecica'.
   TAM: *Juniperus sabina* 'Tamariscifolia'.
   TAMARIX: *Juniperus sabina* 'Tamariscifolia'.
   TEAR DROP: *Juniperus chinensis* 'Blue Point'.
   TEXAS STAR: *Juniperus chinensis* 'Blue Vase'.
   TRIPARTITE: *Juniperus virginiana* 'Tripartita'.
   TWISTED CHINESE: *Juniperus chinensis* 'Kaizuka'.
   UPRIGHT COLUMN: *Juniperus communis* 'Kiyonoi'.
   UTAH: *Juniperus osteosperma*.
   VARIEGATED CHINESE: *Juniperus chinensis* 'Variegata'.
   VARIEGATED CREEPING: *Juniperus communis* 'Nana Aurea'.
   VARIEGATED HOLLYWOOD: *Juniperus chinensis* 'Kaizuka Variegated'.
   VARIEGATED HORIZONTAL: *Juniperus communis* 'Nana Aurea'.
   VON EHREN: *Juniperus sabina* 'Von Ehren'.
   WAUKEGAN: *Juniperus horizontalis* 'Douglasii'.
   WELCH ROCKY MOUNTAIN: *Juniperus scopulorum* 'Welchii'.
   WESTERN: *Juniperus occidentalis*.
   WHITE'S SILVER KING: *Juniperus scopulorum* 'Silver King'.
   WILTON CARPET: *Juniperus horizontalis* 'Wiltonii'.
Juniperus 'Arcadia'. See *J. sabina* 'Arcadia'.
Juniperus 'Armstrongii Aurea'. See *J. chinensis* 'Golden Armstrong'.
Juniperus 'Bar Harbor'. See *J. horizontalis* 'Bar Harbor'.
Juniperus 'Blue Point'. See *J. chinensis* 'Blue Point'.
Juniperus californica Carrière. CALIFORNIA JUNIPER.
Juniperus cedrus Webb & Berth. CANARY ISLAND JUNIPER.
Juniperus chinensis L. CHINESE JUNIPER.
Juniperus chinensis 'Alba'. Listed name. Plants so called may be *J. davurica* 'Expansa Variegata'.
Juniperus chinensis 'Ames'.
Juniperus chinensis 'Armstrongii'. ARMSTRONG JUNIPER.
Juniperus chinensis 'Aurea Drewii'?
Juniperus chinensis 'Aurea Gold Coast'. See *J. chinensis* 'Gold Coast'.
Juniperus chinensis 'Aureo-variegata'. See *J. chinensis* 'Plumosa Aureo-variegata'.
Juniperus chinensis 'Blaauw'.
Juniperus chinensis 'Blue Point'. TEAR-DROP JUNIPER.
Juniperus chinensis 'Blue Vase'. TEXAS STAR JUNIPER.
Juniperus chinensis 'Coastii'. See *J. chinensis* 'Gold Coast'.
Juniperus chinensis 'Coastii Aurea'. See *J. chinensis* 'Gold Coast'.
Juniperus chinensis 'Columnaris'. GREEN COLUMN CHINESE JUNIPER.
Juniperus chinensis 'Columnaris Blue'. See *J. chinensis* 'Columnaris Glauca'.
Juniperus chinensis 'Columnaris Glauca'. BLUE COLUMN CHINESE JUNIPER.
Juniperus chinensis 'Columnaris Mordigan'?
Juniperus chinensis 'Corymbosa'?
Juniperus chinensis 'Corymbosa Variegata'?
Juniperus chinensis 'Deigaardii'?
Juniperus chinensis 'Densa'.
Juniperus chinensis 'Densaerecta'. See *J. chinensis* 'Spartan'.
Juniperus chinensis 'Densaerecta Spartan'. See *J. chinensis* 'Spartan'.
Juniperus chinensis 'Densa Spartan'. See *J. chinensis* 'Spartan'.
Juniperus chinensis 'Expansa'. See *J. davurica* 'Expansa'.
Juniperus chinensis 'Expansa Aureo-spicata'. See *J. davurica* 'Expansa Aureo-spicata'.
Juniperus chinensis 'Expansa Variegata'. See *J. davurica* 'Expansa Variegata'.
Juniperus chinensis 'Fairview'.

Juniperus chinensis 'Foemina'.
Juniperus chinensis 'Fortunei'. See **J. chinensis 'Sheppardii'** or **J. chinensis 'Smithii'**.
Juniperus chinensis 'Fruitlandii'.
Juniperus chinensis 'Glauca'. Listed name. Plants so called may be any of several cvs.
Juniperus chinensis 'Glauca Hetzii'. See **J. chinensis 'Hetzii'**.
Juniperus chinensis 'Gold Coast'.
Juniperus chinensis 'Golden Armstrong'.
Juniperus chinensis 'Hetzii'. HETZ BLUE JUNIPER. BLUE UPRIGHT JUNIPER. BLUE JUNIPER.
Juniperus chinensis 'Hetzii Columnaris'. See **J. chinensis 'Fairview'**.
Juniperus chinensis 'Hibernica'. See **J. communis 'Hibernica'**.
Juniperus chinensis 'Iowa'.
Juniperus chinensis 'Japonica'. Some plants so called are **J. chinensis 'Veitchii', J. chinensis** var. ***sargentii, J. chinensis* 'Plumosa'** or **J. procumbens**. Some plants called *J. japonica* and *J. chinensis* 'Japonica Procumbens' may belong here.
Juniperus chinensis 'Japonica Procumbens'. Listed name. Plants so called may be **J. chinensis 'Japonica'** or **J. procumbens**.
Juniperus chinensis 'Japonica San Jose'. See **J. chinensis 'San Jose'**.
Juniperus chinensis 'Kaizuka'. HOLLYWOOD JUNIPER. TWISTED CHINESE JUNIPER.
Juniperus chinensis 'Kaizuka Blue'. BLUE HOLLYWOOD JUNIPER.
Juniperus chinensis 'Kaizuka Variegated'. VARIEGATED HOLLYWOOD JUNIPER.
Juniperus chinensis 'Keteleeri'. KETELEER JUNIPER.
Juniperus chinensis 'Maney'.
Juniperus chinensis 'Mint Julep'.
Juniperus chinensis 'Mordigan'?
Juniperus chinensis 'Mountbatten'.
Juniperus chinensis 'Neaboriensis'.
Juniperus chinensis 'Obelisk'.
Juniperus chinensis 'Oblonga'.
Juniperus chinensis 'Oklahoma'?
Juniperus chinensis 'Old Gold'.
Juniperus chinensis 'Parsonii'. See **J. davurica 'Expansa'**.
Junperus chinensis 'Parsonii Aureo-spicata'. See **J. davurica 'Expansa Aureo-spicata'**.
Juniperus chinensis 'Parsonii Variegata'. See **J. davurica 'Expansa Variegata'**.
Juniperus chinensis 'Pfitzer Blue Gold'.
Juniperus chinensis 'Pfitzerana'. PFITZER JUNIPER.
Juniperus chinensis 'Pfitzerana Aurea'. GOLDEN PFITZER JUNIPER.
Juniperus chinensis 'Pfitzerana Aurea Compacta'?
Juniperus chinensis 'Pfitzerana Aurea Nana'. See **J. chinensis 'Pfitzerana Aurea'**.
Juniperus chinensis 'Pfitzerana Blue'. See **J. chinensis 'Pfitzerana Glauca'**.
Juniperus chinensis 'Pfitzerana Blue Compacta'. See **J. chinensis 'Pfitzerana Glauca Compacta'**.
Juniperus chinensis 'Pfitzerana Compacta'. COMPACT PFITZER JUNIPER. NICK'S COMPACT PFITZER JUNIPER.
Juniperus chinensis 'Pfitzerana Glauca'. BLUE PFITZER JUNIPER.
Juniperus chinensis 'Pfitzerana Glauca Blue Gold'. See **J. chinensis 'Pfitzer Blue Gold'**.
Juniperus chinensis 'Pfitzerana Glauca Compacta'.
Juniperus chinensis 'Pfitzerana Green Compacta'. See **J. chinensis 'Pfitzerana Compacta'**.
Juniperus chinensis 'Pfitzerana Mordigan Aurea'?
Juniperus chinensis 'Pfitzerana Nana'?
Juniperus chinensis 'Pfitzerana Nelson Blue'?
Juniperus chinensis 'Pfitzerana Nick's Compacta'. See **J. chinensis 'Pfitzerana Compacta'**.

Juniperus chinensis 'Pfitzerana Old Gold'. See **J. chinensis 'Old Gold'**.
Juniperus chinensis 'Plumosa'. PLUME CHINESE JUNIPER. Some plants called **J. chinensis 'Japonica'** may belong here.
Juniperus chinensis 'Plumosa Albo-variegata'.
Juniperus chinensis 'Plumosa Aurea'. GOLDEN PLUME CHINESE JUNIPER. Some plants called *J. japonica* 'Bandai Sugi Aurea' may belong here.
Juniperus chinensis 'Plumosa Aureo-variegata'.
Juniperus chinensis 'Pyramidalis'. Some plants so called are **J. chinensis 'Stricta'**.
Juniperus chinensis 'Reevesiana'. See **J. chinensis 'Foemina'**.
Juniperus chinensis 'Richeson'.
Juniperus chinensis 'Robusta'. See **J. chinensis 'Robust Green'**.
Juniperus chinensis 'Robust Green'.
Juniperus chinensis 'San Jose'. SAN JOSE CREEPING JUNIPER.
Juniperus chinensis var. *sargentii* A. Henry. Syn.: *J. sargentii*. SARGENT JUNIPER. Some plants called *J. chinensis* 'Japonica' may belong here.
Juniperus chinensis var. *sargentii* 'Glauca'. BLUE SARGENT'S JUNIPER.
Juniperus chinensis var. *sargentii* 'Viridis'.
Juniperus chinensis 'Sea Green'?
Juniperus chinensis 'Sea Spray'.
Juniperus chinensis 'Sheppardii'. Some plants called *J. chinensis* 'Fortunei' belong here.
Juniperus chinensis 'Shimpaku'. See **J. chinensis** var. ***sargentii***. Some plants called *J. chinensis* 'Shimpaku' may belong to **J. chinensis 'Plumosa'**.
Juniperus chinensis 'Smithii'. Some plants called *J. chinensis* 'Fortunei' belong here.
Juniperus chinensis 'Spartan'.
Juniperus chinensis 'Stricta'. Plants in U.S.A. called **J. excelsa 'Stricta'** (spiny Greek juniper) and some called **J. chinensis 'Pyramidalis'** belong here.
Juniperus chinensis 'Torulosa'. See **J. chinensis 'Kaizuka'**.
Juniperus chinensis 'Torulosa Glauca'. See **J. chinensis 'Kaizuka Blue'**.
Juniperus chinensis 'Torulosa Variegata'. See **J. chinensis 'Kaizuka Variegata'**.
Juniperus chinensis 'Variegata'. VARIEGATED CHINESE JUNIPER.
Juniperus chinensis 'Veitchii'. Some plants called **J. chinensis 'Japonica'** belong here.
Juniperus chinensis 'Viridis'?
Juniperus chinensis 'Wilsonii'. See **J. squamata 'Wilsonii'**.
Juniperus chinensis 'Wintergreen'.
Juniperus 'Coastii Aurea'. See **J. chinensis 'Gold Coast'**.
Juniperus 'Columnaris'. See **J. chinensis 'Columnaris'** or **J. communis 'Columnaris'**.
Juniperus communis L. ENGLISH JUNIPER.
Juniperus communis var. *alpina* Gaudin. See **J. communis** var. **montana**.
Juniperus communis 'Ashfordii'. ASHFORD JUNIPER.
Juniperus communis var. *aureo-spica* Rehd. See **J. communis 'Depressa Aurea'**.
Juniperus communis 'Columnaris'.
Juniperus communis 'Compressa'.
Juniperus communis 'Cracovia'. POLISH JUNIPER.
Juniperus communis var. *depressa* Pursh. Syn.: *J. depressa*. CANADIAN JUNIPER.
Juniperus communis 'Depressa Aurea'.
Juniperus communis 'Depressa Dumosa'. See **J. communis 'Dumosa'**.
Juniperus communis var. *depressa* 'Vase'. See **J. communis 'Vase'**.
Juniperus communis 'Dumosa'.

Juniperus communis 'Effusa'.
Juniperus communis 'Fastigiata'. See **J. communis 'Hibernica'** or **J. communis 'Suecica'**.
Juniperus communis subsp. **hemisphaerica** (J. Presl & K. Presl) Nyman. Syn.: *J. communis* var. *saxatilis* (Pall.) Willd.
Juniperus communis 'Hibernica'. IRISH JUNIPER.
Juniperus communis 'Hibernica Fastigiata'. See **J. communis 'Hibernica'**.
Juniperus communis 'Hornibrookii'.
Juniperus communis 'Kiyonoi'. UPRIGHT COLUMN JUNIPER.
Juniperus communis var. **montana** Ait. Syn.: *J. communis* var. *saxatilis* of authors, not Pall. *J. communis* var. *nana*. *J. communis* var. *alpina*. *J. communis* var. *sibirica*. *J. sibirica*. MOUNTAIN JUNIPER. HORIZONTAL JUNIPER.
Juniperus communis var. *nana* (Willd.) Baumg. See **J. communis** var. **montana**.
Juniperus communis 'Nana Aurea'. VARIEGATED CREEPING JUNIPER. VARIEGATED HORIZONTAL JUNIPER.
Juniperus communis 'Prostrata'. PROSTRATE COMMON JUNIPER.
Juniperus communis var. *saxatilis* (Pall.) Willd. See **J. communis** subsp. **hemisphaerica**.
Juniperus communis var. *saxatilis* of authors, not (Pall.) Willd. See **J. communis** var. **montana**.
Juniperus communis 'Saxatilis Nana'. See **J. communis** var. **montana**.
Juniperus communis var. *saxatilis* 'Variegata'. Listed name, probably for **J. communis 'Nana Aurea'**.
Juniperus communis var. *sibirica* (Burgsd.) Rydb. See **J. communis** var. **montana**.
Juniperus communis 'Stricta'. See **J. communis 'Hibernica'**.
Juniperus communis 'Suecica'. SWEDISH JUNIPER.
Juniperus communis 'Suecica Nana'. DWARF SWEDISH JUNIPER.
Juniperus communis 'Variegata'. See **J. communis 'Nana Aurea'**.
Juniperus communis 'Vase'.
Juniperus 'Compacta'. Listed name. Plants so called may be **J. chinensis 'Armstrongii'** or **J. chinensis 'Pfitzerana Compacta'**.
Juniperus conferta Parl. Syn.: *J. litoralis*. SHORE JUNIPER.
Juniperus conferta 'Blue Pacific'.
Juniperus conferta 'Emerald Sea'.
Juniperus conferta litoralis. Listed name for **J. conferta**.
Juniperus 'Corymbosa'. Listed name. Plants so called are unidentified **J. chinensis** cvs.
Juniperus 'Corymbosa Variegata'. Listed name. Plants so called are **J. chinensis** cvs.
Juniperus davurica Pall.
Juniperus davurica 'Expansa'. Syn.: *J. chinensis* 'Expansa'. *J. chinensis* 'Parsonii'. *J. squamata* 'Parsonii'. *J. squamata* 'Expansa Parsonii'.
Juniperus davurica 'Expansa Aureo-spicata'. Syn.: *J. chinensis* 'Expansa Aureo-spicata'. *J. chinensis* 'Parsonii Aureo-spicata'. *J. squamata* 'Parsonii Aureo-spicata'.
Juniperus davurica 'Expansa Variegata'. Syn.: *J. chinensis* 'Expansa Variegata'. *J. chinensis* 'Parsonii Variegata'. *J. squamata* 'Albo-variegata'. *J. squamata* 'Parsonii Variegata'.
Juniperus davurica 'Parsonii'. See **J. davurica 'Expansa'**.
Juniperus deppeana Steud. Syn.: *J. pachyphlaea*. ALLIGATOR JUNIPER.
Juniperus depressa (Pursh) Raf. See **J. communis** var. **depressa**.
Juniperus 'Depressa Plumosa'. See **J. communis 'Dumosa'**.
Juniperus excelsa Bieb. GRECIAN JUNIPER.
Juniperus excelsa 'Stricta'. May not be cult. in U.S. Plants so called are **J. chinensis 'Stricta'**.
Juniperus fastigiata. Listed name for **J. communis 'Hibernica'**.
Juniperus 'Foemina'. See **J. chinensis 'Foemina'**.
Juniperus 'Fruitlandii'. See **J. chinensis 'Fruitlandii'**.
Juniperus 'Glauca'?
Juniperus 'Glauca Hetzii'. See **J. chinensis 'Hetzii'**.
Juniperus 'Hetzii'. See **J. chinensis 'Hetzii'**.
Juniperus 'Hetz's Columnaris'. See **J. chinensis 'Fairview'**.
Juniperus hibernica Lodd. See **J. communis 'Hibernica'**.
Juniperus hibernica 'Fastigiata'. See **J. communis 'Hibernica'**.
Juniperus horizontalis Moench. Syn.: *J. prostrata*. *J. virginiana* var. *prostrata*. CREEPING JUNIPER.
Juniperus horizontalis 'Admirabilis'.
Juniperus horizontalis 'Aunt Jemima'?
Juniperus horizontalis 'Bar Harbor'.
Juniperus horizontalis 'Blue Chip'.
Juniperus horizontalis 'Blue Horizon'.
Juniperus horizontalis 'Blue Mat'?
Juniperus horizontalis 'Blue Rug'. See **J. horizontalis 'Wiltonii'**.
Juniperus horizontalis 'Douglasii'. WAUKEGAN JUNIPER.
Juniperus horizontalis 'Emerald Spreader'.
Juniperus horizontalis 'Emerson'.
Juniperus horizontalis 'Emerson's Creeper'. See **J. horizontalis 'Emerson'**.
Juniperus horizontalis 'Eximius'?
Juniperus horizontalis 'Filicinus'.
Juniperus horizontalis 'Glauca'. Some plants so called are **J. horizontalis 'Wiltonii'**.
Juniperus horizontalis 'Gray Carpet'.
Juniperus horizontalis 'Green Bowers'.
Juniperus horizontalis 'Hughes'? May be **J. scopulorum 'Hughes'**.
Juniperus horizontalis 'Jade Spreader'.
Juniperus horizontalis 'Marcella'.
Juniperus horizontalis 'Marshall Creeper'. See **J. horizontalis 'Emerson'**.
Juniperus horizontalis 'Plumosa'. ANDORRA JUNIPER.
Juniperus horizontalis 'Plumosa Aunt Jemima'?
Juniperus horizontalis 'Plumosa Bar Harbor'. See **J. horizontalis 'Bar Harbor'**.
Juniperus horizontalis 'Plumosa Compacta'?
Juniperus horizontalis 'Plumosa Youngstown'. See **J. horizontalis 'Youngstown'**.
Juniperus horizontalis 'Prince of Wales'.
Juniperus horizontalis 'Prostrata'.
Juniperus horizontalis 'Turquoise Spreader'.
Juniperus horizontalis 'Variegata'.
Juniperus horizontalis 'Venusta'?
Juniperus horizontalis 'Webberi'.
Juniperus horizontalis 'Wiltonii'. WILTON CARPET JUNIPER. Some plants called **J. horizontalis 'Glauca'** belong here.
Juniperus horizontalis 'Youngstown'.
Juniperus horizontalis 'Yukon Belle'.
Juniperus 'Hornibrookii'. See **J. communis 'Hornibrookii'**.
Juniperus japonica Carrière. See **J. chinensis 'Japonica'**.
Juniperus japonica Hort., not Carrière. Listed name. Plants so called may be **J. procumbens** or **J. chinensis 'Japonica'**.
Juniperus japonica 'Bandai Sugi Aurea'. Listed name. May be **J. chinensis 'Plumosa Aurea'**.
Juniperus japonica 'Plumosa Aurea-variegata'. See **J. chinensis 'Plumosa Aureo-variegata'**.
Juniperus japonica 'San Jose'. See **J. chinensis 'San Jose'**.
Juniperus 'Keteleeri'. See **J. chinensis 'Keteleeri'**.
Juniperus litoralis Maxim. See **J. conferta**.
Juniperus 'Obelisk'. See **J. chinensis 'Obelisk'**.
Juniperus occidentalis Hook. SIERRA JUNIPER. WESTERN JUNIPER. WESTERN RED CEDAR.
Juniperus occidentalis 'Glauca'. See **J. occidentalis 'Sierra Silver'**.

Juniperus occidentalis 'Sierra Silver'.
Juniperus 'Old Gold'. See **J. chinensis 'Old Gold'**.
Juniperus osteosperma (Torr.) Little. Syn.: *J. utahensis*. UTAH JUNIPER. DESERT JUNIPER.
Juniperus pachyphlaea Torr. See **J. deppeana**.
Juniperus 'Pathfinder'. See **J. scopulorum 'Pathfinder'**.
Juniperus 'Pfitzerana'. See **J. chinensis 'Pfitzerana'**.
Juniperus 'Pfitzerana Aurea'. See **J. chinensis 'Pfitzerana Aurea'**.
Juniperus 'Pfitzerana Aurea Compacta'?
Juniperus 'Pfitzerana Compacta'. See **J. chinensis 'Pfitzerana Compacta'**.
Juniperus 'Pfitzerana Glauca'. See **J. chinensis 'Pfitzerana Glauca'**.
Juniperus 'Pfitzerana Nana'. See **J. chinensis 'Pfitzerana Nana'**.
Juniperus 'Pfitzerana Nick's Compacta'. See **J. chinensis 'Pfitzerana Compacta'**.
Juniperus procumbens (Endl.) Miq. CREEPING JUNIPER. JAPGARDEN JUNIPER. Some plants called *J. chinensis* 'Japonica', *J. chinensis* 'Japonica Procumbens', and *J. japonica* may belong here.
Juniperus procumbens 'Albo-variegata'. Listed name. Plants so called may be **J. davurica 'Expansa Variegata'**.
Juniperus procumbens 'Compacta Nana'. See **J. procumbens 'Nana'**.
Juniperus procumbens 'Nana'. DWARF JAPGARDEN JUNIPER.
Juniperus procumbens 'Variegata'. Listed name. Plants so called may be **J. davurica 'Expansa Aureo-spicata'** or **J. davurica 'Expansa Variegata'**.
Juniperus prostrata Pers. See **J. horizontalis**.
Juniperus prostrata 'Variegata'. See **J. horizontalis 'Variegata'**.
Juniperus reevesiana. Listed name, probably for **J. chinensis 'Foemina'**.
Juniperus 'Robust Green'. See **J. chinensis 'Robust Green'**.
Juniperus sabina L. SAVIN JUNIPER.
Juniperus sabina 'Admiral'. Listed name. Plants so called may be **J. horizontalis 'Admirabilis'** or **J. scopulorum 'Admiral'**.
Juniperus sabina 'Arcadia'.
Juniperus sabina 'Blue Danube'.
Juniperus sabina 'Broadmoor'.
Juniperus sabina 'Buffalo'.
Juniperus sabina 'Cupressifolia'. CYPRESS JUNIPER.
Juniperus sabina 'Fastigiata'.
Juniperus sabina 'Hicksii'. HICKS JUNIPER.
Juniperus sabina horizontalis 'Bar Harbor'. See **J. horizontalis 'Bar Harbor'**.
Juniperus sabina 'Scandia'.
Juniperus sabina 'Tamariscifolia'. SPANISH JUNIPER. TAM JUNIPER. TAMARIX JUNIPER.
Juniperus sabina 'Tamariscifolia No Blight'.
Juniperus sabina 'Tamariscifolia Variegata'. See **J. sabina 'Variegata'**.
Juniperus sabina 'Variegata'.
Juniperus sabina 'Von Ehren'. VON EHREN JUNIPER.
Juniperus 'San Jose'. See **J. chinensis 'San Jose'**.
Juniperus sargentii (A. Henry) Nakai. See **J. chinensis var. sargentii**.
Juniperus scopulorum Sarg. ROCKY MOUNTAIN JUNIPER.
Juniperus scopulorum 'Admiral'.
Juniperus scopulorum var. *argentea* D. Hill ex Rehd. See **J. scopulorum 'Hill's Silver'**.
Juniperus scopulorum 'Blue Haven'. See **J. scopulorum 'Blue Heaven'**.
Juniperus scopulorum 'Blue Heaven'.
Juniperus scopulorum 'Blue Point'. See **J. chinensis 'Blue Point'**.
Juniperus scopulorum 'Chandler's Silver'.
Juniperus scopulorum 'Cologreen'.
Juniperus scopulorum 'Colorado Green'. See **J. scopulorum 'Cologreen'**.
Juniperus scopulorum 'Columnar Sneed'.
Juniperus scopulorum 'Columnaris'.
Juniperus scopulorum 'Columnaris Sneedii'. See **J. scopulorum 'Columnar Sneed'**.
Juniperus scopulorum 'Cupressifolia Erecta'.
Juniperus scopulorum 'Cupressifolia Glauca'.
Juniperus scopulorum 'Dew Drop'. See **J. scopulorum 'Kenyonii.'**
Juniperus scopulorum 'Emerald Green'?
Juniperus scopulorum 'Erecta Glauca'.
Juniperus scopulorum 'Gareei'.
Juniperus scopulorum 'Globe'.
Juniperus scopulorum 'Globosa'. See **J. scopulorum 'Globe'**.
Juniperus scopulorum 'Gray Gleam'.
Juniperus scopulorum 'Hill's Silver'. SILVER COLUMN JUNIPER.
Juniperus scopulorum 'Horizontalis'.
Juniperus scopulorum 'Hughes'.
Juniperus scopulorum 'Kenyonii'.
Juniperus scopulorum 'Lakewood Globe'.
Juniperus scopulorum 'Marshall'.
Juniperus scopulorum 'Moffetii'.
Juniperus scopulorum 'Moonlight'.
Juniperus scopulorum 'North Star'.
Juniperus scopulorum 'Pathfinder'.
Juniperus scopulorum 'Pendula'.
Juniperus scopulorum 'Platinum'.
Juniperus scopulorum 'Schottii'. See **J. virginiana 'Schottii'**.
Juniperus scopulorum 'Silver Glow'.
Juniperus scopulorum 'Silver King'. WHITE'S SILVER KING JUNIPER.
Juniperus scopulorum 'Sutherland'.
Juniperus scopulorum 'Tabletop'.
Juniperus scopulorum 'Tabletop Blue'. See **J. scopulorum 'Tabletop'**.
Juniperus scopulorum 'Tolleson's Green Weeping'.
Juniperus scopulorum 'Tolleson's Weeping'.
Juniperus scopulorum 'Welchii'. WELCH ROCKY MOUNTAIN JUNIPER.
Juniperus scopulorum 'Wichita Blue'.
Juniperus sibirica Burgsd. See **J. communis var. montana**.
Juniperus squamata Buch.-Ham. ex Lamb. SCALY-LEAVED NEPAL JUNIPER. SINGLE-SEED JUNIPER.
Juniperus squamata 'Albo-variegata'. See **J. davurica 'Variegata'**.
Juniperus squamata 'Expansa Parsonii'. See **J. davurica 'Expansa'**.
Juniperus squamata 'Forrestii'.
Juniperus squamata 'Loderi'.
Juniperus squamata 'Meyeri'. MEYER JUNIPER. FISHTAIL JUNIPER. MEYER SINGLE-SEED JUNIPER.
Juniperus squamata 'Parsonii'. See **J. davurica 'Expansa'**.
Juniperus squamata 'Parsonii Aureo-spicata'. See **J. davurica 'Expansa Aureo-spicata'**.
Juniperus squamata 'Parsonii Variegata'. See **J. davurica 'Expansa Variegata'**.
Juniperus squamata 'Prostrata'.
Juniperus squamata 'Wilsonii'.
Juniperus 'Tamariscifolia'. See **J. sabina 'Tamariscifolia'**.
Juniperus 'Torulosa'. See **J. chinensis 'Kaizuka'**.
Juniperus 'Torulosa Glauca'. See **J. chinensis 'Kaizuka Blue'**.
Juniperus 'Torulosa Variegata'. See **J. chinensis 'Kaizuka Variegated'**.
Juniperus 'Tripartita'. See **J. virginiana 'Tripartita'**.
Juniperus utahensis (Engelm.) Lemmon. See **J. osteosperma**.

Juniperus 'Vase'. See **J. communis 'Vase'**.
Juniperus virginiana L. EASTERN RED CEDAR. RED CEDAR.
Juniperus virginiana 'Burkii'. BURK EASTERN RED CEDAR.
Juniperus virginiana 'Canaertii'. CANAERT UPRIGHT JUNIPER.
Juniperus virginiana 'Cupressifolia'. Some plants so called may be **J. virginiana 'Hillspire'**.
Juniperus virginiana 'DeForest Green'.
Juniperus virginiana 'Glauca'. SILVER RED CEDAR.
Juniperus virginiana 'Globosa'.
Juniperus virginiana 'Hillii'. HILL DUNDEE JUNIPER.
Juniperus virginiana 'Hillspire'. Some plants called **J. virginiana 'Cupressifolia'** belong here.
Juniperus virginiana 'Kosteri'. KOSTER JUNIPER.
Juniperus virginiana 'Manhattan Blue'.
Juniperus virginiana var. *prostrata* (Pers.) Torr. See **J. horizontalis**.
Juniperus virginiana 'Prostrata' (of Pacific Coast)?
Juniperus virginiana 'Prostrata Silver Spreader'. See **J. virginiana 'Silver Spreader'**.
Juniperus virginiana 'Pyramidiformis'.
Juniperus virginiana 'Pyramidiformis Hillii'. See **J. virginiana 'Hillii'**.
Juniperus virginiana 'Schottii'.
Juniperus virginiana 'Silver Spreader'.
Juniperus virginiana 'Skyrocket'.
Juniperus virginiana 'Sneedii'. See **J. scopulorum 'Columnar Sneed'**.
Juniperus virginiana 'Tripartita' TRIPARTITE JUNIPER.
Juniperus 'Wilsonii'. See **J. squamata 'Wilsonii'**.
Juniperus 'Wiltonii'. See **J. horizontalis 'Wiltonii'**.
Justicia adhatoda L. Syn.: *Adhatoda vasica*.
Justicia aurea Schlechtend. Syn.: *Justicia umbrosa*. *Jacobinia aurea*.
Justicia brandegeana Wasshausen & L. B. Sm. Syn.: *Beloperone guttata*. SHRIMP PLANT. SHRIMP BUSH.
Justicia brandegeana 'Yellow Queen'.
Justicia californica (Benth.) D. Gibson. Syn.: *Beloperone californica*. CHUPEROSA.
Justicia carnea Lindl. Syn.: *Jacobinia carnea*. Plants called *Jacobinia rosea* may belong here. BRAZILIAN PLUME FLOWER. FLAMINGO PLANT.
Justicia coccinea Hort. See **Odontonema stricta**.
Justicia floribunda Hort. See **Justicia rizzinii**.
Justicia fulvicoma Schlechtend. & Cham.
Justicia ghiesbreghtiana Lem. Syn.: *Jacobinia ghiesbreghtiana*. May not be cult. Plants so called are probably **Justicia spicigera**.
Justicia leonardii Wasshausen. Syn.: *Jacobinia incana*. Some plants called **Anisacanthus thurberi** belong here.
Justicia pauciflora (Nees) Griseb. See **J. rizzinii**.
Justicia rizzinii Wasshausen. Syn.: *Jacobinia pauciflora*. *Justicia floribunda*. *Justicia pauciflora*. *Libonia floribunda*.
Justicia spicigera Schlechtend. Syn.: *Jacobinia spicigera*. *Jacobinia mohintlii*. Plants called **Justicia ghiesbreghtiana** may belong here.
Justicia umbrosa Benth. See **J. aurea**.

K

KAFFIR BOOM: **Erythrina caffra**.
KAHILI: **Hedychium gardneranum**.
KAKI: **Diospyros kaki**.

Kalanchoe beharensis Drake. FELT-BUSH. FELT-PLANT. VELVET LEAF.
Kalanchoe orgyalis Baker.
Kalanchoe pumila Baker.
KALMIA, BOG: **Kalmia poliifolia**.
Kalmia angustifolia L. SHEEP LAUREL. LAMBKILL.
Kalmia angustifolia 'Rubra'.
Kalmia latifolia L. CALICO BUSH. IVYBUSH. MOUNTAIN LAUREL.
Kalmia latifolia 'Dexter Pink'.
Kalmia latifolia 'Ostbo Red'.
Kalmia microphylla (Hook.) Heller. Syn.: *K. poliifolia* var. *microphylla*. WESTERN LEDUM. ALPINE LEDUM.
Kalmia poliifolia Wangenh. PALE LAUREL. BOG KALMIA. BOG LAUREL.
Kalmia poliifolia var. *microphylla* (Hook.) Rehd. See **K. microphylla**.
Kalmiopsis leachiana (L. F. Henders.) Rehd.
Kalopanax pictus (Thunb.) Nakai. See **K. septemlobus**.
Kalopanax septemlobus (Thunb.) G. Koidz. Syn.: *K. pictus*. *Acanthopanax ricinifolius*.
KANGAROO-PAWS, ALBANY: **Anigosanthos flavidus**.
  GREEN: **Anigosanthos manglesii**.
KANGAROO THORN: **Acacia armata**.
KANGAROO VINE: **Cissus antarctica**.
KARANDA: **Carissa carandas**.
KARO: **Pittosporum crassifolium**.
KARRI: **Eucalyptus diversicolor**.
KASHMIR BOUQUET: **Clerodendrum bungei**.
KASHMIR BOUQUET: **Clerodendrum philippinum**.
KATSURA TREE: **Cercidiphyllum japonicum**.
KAURI: **Agathis australis**.
  QUEENSLAND: **Agathis robusta**.
KEAKI, JAPANESE: **Zelkova serrata**.
Kennedia comptoniana (Andr.) Link. See **Hardenbergia comptoniana**.
Kennedia prostrata R. Br. ex Ait. f.
Kennedia rubicunda (Schneev.) Vent.
Kensitia pillansii (Kensit) Fedde.
KENTIA, RED: **Lepidorrhachis mooreana**.
Kentia belmoreana C. Moore & F. J. Muell. See **Howea belmoreana**.
Kentia forsterana C. Moore & F. J. Muell. See **Howea forsterana**.
KERRIA, WHITE: **Rhodotypos scandens**.
Kerria japonica (L.) DC. JAPANESE ROSE.
Kerria japonica 'Pleniflora'.
Keteleeria davidiana (C. Bertrand) Beissner.
Keteleeria fortunei (A. Murr.) Carrière.
KHAT: **Catha edulis**.
Kigelia africana (Lam.) Benth. Syn.: *K. pinnata*. SAUSAGE TREE.
Kigelia pinnata (Jacq.) DC. See **K. africana**.
KING'S MANTLE: **Thunbergia erecta**.
KINNIKINNICK: **Arctostaphyllos uva-ursi**.
  COMPACT: **Arctostaphyllos uva-ursi 'Compacta'**.
KIWI: **Actinidia chinensis**.
KIWI VINE: **Actinidia chinensis**.
Kleinia mandraliscae Tineo. See **Senecio mandraliscae**.
Kleinia tomentosa (Haw.) Haw. See **Senecio haworthii**.
Kniphofia uvaria (L.) Oken. Syn.: *Tritoma uvaria*. RED-HOT-POKER. POKER PLANT. TORCH FLOWER. TORCH LILY.
KNOTWEED, JAPANESE: **Polygonum cuspidatum**.
  ROSE-CARPET: **Polygonum vacciniifolium**.
Koelreuteria bipinnata Franch. Syn.: *K. integrifoliola*. CHINESE FLAME TREE. EVERGREEN GOLDEN RAIN TREE.
Koelreuteria elegans (Seem.) A. C. Sm. Syn.: *K. formosana*. *K. henryi*. FLAMEGOLD. CHINESE FLAME TREE.

Koelreuteria formosana Hayata. See **K. elegans**.
Koelreuteria henryi Dümmer. See **K. elegans**.
Koelreuteria integrifoliola Merrill. See **K. bipinnata**.
Koelreuteria paniculata Laxm. GOLDEN RAIN TREE. VARNISH TREE.
KOHUHU: **Pittosporum tenuifolium**.
Kolkwitzia amabilis Graebn. BEAUTYBUSH.
KUMQUAT, NAGAMI: **Fortunella margarita**.
   MARUMI: **Fortunella japonica**.
   OVAL: **Fortunella margarita**.
   ROUND: **Fortunella japonica**.
Kunzea baxteri (Klotzsch) Schauer.
Kunzea pulchella (Lindl.) George.
Kunzea sericea (Labill.) Turcz.
KURRAJONG: **Brachychiton populneus**.
   DESERT: **Brachychiton gregori**.
   WHITE: **Brachychiton discolor**.

L

LA BELLA SOMBRA: **Phytolacca dioica**.
LABURNUM, EAST AFRICAN: **Calpurnia aurea**.
   SCOTCH: **Laburnum alpinum**.
+*Laburnocytisus adamii* (Poit.) Schneid. [*Laburnum anagyroides* + *Cytisus purpureus*]. Syn.: *Laburnum adamii*.
Laburnum adamii (Poit.) Lavall. See +**Laburnocytisus adamii**.
Laburnum alpinum (Mill.) Bercht. & J. Presl. SCOTCH LABURNUM. GOLDEN-CHAIN.
Laburnum alpinum 'Pendulum'. WEEPING GOLDEN-CHAIN.
Laburnum anagyroides Medic. Syn.: *L. vulgare*. GOLDEN-CHAIN TREE.
Laburnum anagyroides 'Pendulum'.
Laburnum 'Vossii'. See **L. × watereri 'Vossii'**.
Laburnum 'Vossii Pendula'. Listed name. Plants so called may be **L. alpinum 'Pendulum'** or **L. anagyroides 'Pendulum'**.
Laburnum vulgare Bercht. & J. Presl. See **L. anagyroides**.
Laburnum × *watereri* (Kirchn.) Dipp. [*L. anagyroides* × *L. alpinum*]. LONG-CLUSTER GOLDEN-CHAIN TREE.
Laburnum × *watereri* 'Vossii'.
LACEBARK: **Hoheria glabrata**.
   NEW ZEALAND: **Hoheria populnea**.
   QUEENSLAND: **Brachychiton discolor**.
LACE VINE, SILVER: **Polygonum aubertii**.
Lagerstroemia 'Alba'. See **L. indica 'Alba'**.
Lagerstroemia 'Grayi'?
Lagerstroemia hirsuta (Lam.) Willd. Syn.: *L. reginae*. CRAPE MYRTLE.
Lagerstroemia indica L. CRAPE MYRTLE.
Lagerstroemia indica 'Alba'. WHITE CRAPE MYRTLE.
Lagerstroemia indica 'Catawba'.
Lagerstroemia indica 'Conestoga'.
Lagerstroemia indica 'Country Red'.
Lagerstroemia indica 'Durant Red'. See **L. indica 'Country Red'**.
Lagerstroemia indica 'Dwarf Purple'.
Lagerstroemia indica 'Glendora White'. WHITE CRAPE MYRTLE.
Lagerstroemia indica 'Gray's Red'.
Lagerstroemia indica 'Kellogg's Purple'. KELLOGG'S PURPLE CRAPE MYRTLE.
Lagerstroemia indica 'Near East'.
Lagerstroemia indica 'New Snow'.
Lagerstroemia indica 'Parade Purple'.
Lagerstroemia indica 'Peppermint Lace'.
Lagerstroemia indica 'Petite Embers'.
Lagerstroemia indica 'Petite Orchid'.
Lagerstroemia indica 'Petite Pinkie'.
Lagerstroemia indica 'Petite Red Imp'.
Lagerstroemia indica 'Petite Ruby'.
Lagerstroemia indica 'Petite Snow'.
Lagerstroemia indica 'Pink Ruffle'.
Lagerstroemia indica 'Pomona Red'.
Lagerstroemia indica 'Potomac'.
Lagerstroemia indica 'Powhatan'.
Lagerstroemia indica 'Purpurea'. PURPLE CRAPE MYRTLE.
Lagerstroemia indica 'Rosea'. PINK CRAPE MYRTLE.
Lagerstroemia indica 'Rubra'. RED CRAPE MYRTLE.
Lagerstroemia indica 'Rubra Grayi'?
Lagerstroemia indica 'Shell Pink'.
Lagerstroemia indica 'Snow White'. See **L. indica 'New Snow'**.
Lagerstroemia indica 'Tiny Fire'.
Lagerstroemia indica 'Watermelon Red'.
Lagerstroemia indica 'William Toovey'.
Lagerstroemia 'Majestic Orchid'.
Lagerstroemia reginae Roxb. See **L. hirsuta**.
Lagerstroemia speciosa (L.) Pers. QUEEN'S CRAPE MYRTLE. PRIDE-OF-INDIA.
Lagerstroemia speciosa 'Royal Pink'.
Lagerstroemia subcostata Koehne.
Lagunaria patersonii (Andr.) G. Don. COW-ITCH. ORCHID TREE. PRIMROSE TREE.
LAMBKILL: **Kalmia angustifolia**.
LAMB'S EARS: **Stachys byzantina**.
LANCEWOOD: **Pseudopanax crassifolium**.
LANTANA, TRAILING: **Lantana montevidensis**.
   WEEPING: **Lantana montevidensis**.
Lantana camara L. YELLOW SAGE.
Lantana 'Carnival'.
Lantana 'Christine'.
Lantana 'Ciceron'.
Lantana 'Confetti'.
Lantana 'Cream Carpet'.
Lantana 'Dwarf Bronze'.
Lantana 'Dwarf Cream'.
Lantana 'Dwarf Orange'.
Lantana 'Dwarf Orange Red'.
Lantana 'Dwarf Pink'.
Lantana 'Dwarf White'.
Lantana 'Dwarf Yellow'.
Lantana 'Gold Coin'.
Lantana 'Gold Mound'.
Lantana 'Golden Glow'.
Lantana 'Golden Pillar'.
Lantana 'Golden Plume'.
Lantana 'Gold Rush'.
Lantana 'Hybrida'?
Lantana 'Irene'.
Lantana 'Janina'.
Lantana 'Kathleen'.
Lantana montevidensis (K. Spreng.) Briq. Syn.: *L. sellowiana*. WEEPING LANTANA. TRAILING LANTANA.
Lantana 'Moonglow'.
Lantana 'Pink Frolic'.
Lantana 'Radiation'.
Lantana sellowiana Link & Otto. See **L. montevidensis**.
Lantana 'Sensation'.
Lantana 'Seraphin'.
Lantana 'Snow Queen'.
Lantana 'Sparkler'.
Lantana 'Spreading Gold Mound'.

Lantana 'Spreading Sunset'.
Lantana 'Spreading Sunshine'.
Lantana 'Sunburst'.
Lantana 'Tangerine'.
Lantana 'Tethys'.
Lapageria rosea Ruiz & Pav. CHILEAN BELLFLOWER.
LARCH, ALPINE: **Larix laricina**.
   AMERICAN: **Larix laricina**.
   EUROPEAN: **Larix decidua**.
   GOLDEN: **Pseudolarix kaempferi**.
   JAPANESE: **Larix kaempferi**.
   WESTERN: **Larix occidentalis**.
Larix decidua Mill. Syn.: *L. europaea*. EUROPEAN LARCH.
Larix europaea DC. See **L. decidua**.
Larix kaempferi (Lamb.) Carrière. Syn.: *L. leptolepis*. JAPANESE LARCH.
Larix laricina (Du Roi) K. Koch. ALPINE LARCH. AMERICAN LARCH. HACKMATACK. TAMARACK.
Larix leptolepis (Siebold & Zucc.) Gord. See **L. kaempferi**.
Larix occidentalis Nutt. WESTERN LARCH. WESTERN TAMARACK. HACKMATACK.
Larrea divaricata Cav. CREOSOTE BUSH.
Latania borbonica Lam. See **L. lontaroides**. Plants cult. in Calif. as *Latania borbonica* are **Livistona chinensis**.
Latania lontaroides (Gaertn.) H. E. Moore. Syn.: *L. borbonica*.
LAUDANUM: **Cistus ladanifer**.
LAUREL, ALEXANDRIAN: **Danae racemosa**.
   AMERICAN CHERRY: **Prunus caroliniana**.
   ASIATIC GROUND: **Epigaea asiatica**.
   AZORES CHERRY: **Prunus lusitanica** subsp. **azorica**.
   BLACK: **Gordonia lasianthus**.
   BOG: **Kalmia poliifolia**.
   CALIFORNIA: **Umbellularia californica**.
   CANARY ISLAND: **Laurus azorica**.
   CAROLINA CHERRY: **Prunus caroliniana**.
   CHERRY: **Prunus laurocerasus**.
   CHILEAN: **Beilschmiedia miersii**.
   ENGLISH: **Prunus laurocerasus**.
   GRECIAN: **Laurus nobilis**.
   HIMALAYAN: **Aucuba himalaica**.
   INDIAN: **Ficus microcarpa**.
   JAPANESE: **Aucuba japonica**.
   MOUNTAIN: **Kalmia latifolia**.
   NEW ZEALAND: **Corynocarpus laevigata**.
   PALE: **Kalmia poliifolia**.
   PORTUGAL: **Prunus lusitanica**.
   SHEEP: **Kalmia angustifolia**.
   SIERRA: **Leucothoe davisiae**.
   SPURGE: **Daphne laureola**.
   TEXAS MOUNTAIN: **Sophora secundiflora**.
   ZABEL CHERRY: **Prunus laurocerasus** 'Zabeliana'.
Laurocerasus lusitanica (L.) M. J. Roemer. See **Prunus lusitanica**.
Laurocerasus officinalis M. J. Roemer. See **Prunus laurocerasus**.
Laurocerasus officinalis 'Nana'. See **Prunus laurocerasus** 'Nana'.
Laurocerasus 'Otto Luyken'. See **Prunus laurocerasus** 'Otto Luyken'.
Laurocerasus 'Zabeliana'. See **Prunus laurocerasus** 'Zabeliana'.
Laurus azorica (Seubert) Franco. Syn.: *L. canariensis*. CANARY ISLAND LAUREL.
Laurus canariensis Webb & Berth. See **L. azorica**.
Laurus nobilis L. SWEET BAY. GRECIAN BAY TREE. GRECIAN LAUREL.
LAURUSTINUS: **Viburnum tinus**.
   SHINING: **Viburnum tinus** 'Lucidum'.
   VARIEGATED: **Viburnum tinus** 'Variegatum'.

Lavandula angustifolia Mill. Syn.: *L. officinalis*. *L. spica*. *L. vera*. ENGLISH LAVENDER.
Lavandula angustifolia 'Compacta'.
Lavandula angustifolia 'Hidcote'.
Lavandula angustifolia 'Munstead'.
Lavandula angustifolia 'Twickel Purple'.
Lavandula dentata L. FRENCH LAVENDER.
Lavandula 'Hidcote'. See **L. angustifolia** 'Hidcote'.
Lavandula latifolia (L. f.) Medic. SPIKE LAVENDER.
Lavandula officinalis Chaix. See **L. angustifolia**.
Lavandula officinalis 'Munstead'. See **L. angustifolia** 'Munstead'.
Lavandula pedunculata (Mill.) Cav. See **L. stoechas** subsp. **pedunculata**.
Lavandula spica L. and all listed cvs. See **L. angustifolia** and cvs.
Lavandula stoechas L. FRENCH LAVENDER. SPANISH LAVENDER.
Lavandula stoechas subsp. **pedunculata** (Mill.) Cav. Syn.: *L. pedunculata*.
Lavandula vera DC. See **L. angustifolia**.
LAVATERA, TREE: **Lavatera olbia**.
Lavatera arborea L. TREE MALLOW.
Lavatera assurgentiflora Kell. CALIFORNIA TREE MALLOW.
Lavatera olbia L. TREE LAVATERA.
LAVENDER, DESERT: **Hyptis emoryi**.
   ENGLISH: **Lavandula angustifolia**.
   FRENCH: **Lavandula dentata**.
   FRENCH: **Lavandula stoechas**.
   SEA: **Limonium perezii**.
   SPANISH: **Lavandula stoechas**.
   SPIKE: **Lavandula latifolia**.
LEAD TREE, LITTLE-LEAF: **Leucaena retusa**.
LEATHERLEAF: **Chamaedaphne calyculata**.
LEATHERLEAF: **Viburnum wrightii**.
LEATHERWOOD, WESTERN: **Dirca occidentalis**.
× *Ledodendron* 'Brilliant'. [*Ledum glandulosum* × *Rhododendron* 'Elizabeth'].
LEDUM, ALPINE: **Kalmia microphylla**.
   WESTERN: **Kalmia microphylla**.
Ledum columbianum Piper. See **L. glandulosum** var. **columbianum**.
Ledum glandulosum Nutt.
Ledum glandulosum var. **columbianum** (Piper) C. L. Hitchc. Syn.: *L. columbianum*.
Ledum groenlandicum Oeder. LABRADOR TEA.
Leea coccinea Planch. WEST INDIAN HOLLY.
LEECHEE: **Litchi chinensis**.
Leiophyllum buxifolium (Bergius) Elliott. BOX SAND MYRTLE.
LEMANDARIN: **Citrus** × **limonia**.
LEMON: **Citrus limon**.
   CHINESE DWARF: **Citrus limon** 'Meyer'.
   DWARF: **Citrus limon** 'Meyer'.
   EUREKA: **Citrus limon** 'Eureka'.
   GIANT: **Citrus limon** 'Ponderosa'.
   LISBON: **Citrus limon** 'Lisbon'.
   MEYER: **Citrus limon** 'Meyer'.
   PONDEROSA: **Citrus limon** 'Ponderosa'.
   VILLAFRANCA: **Citrus limon** 'Villafranca'.
LEMONADE BERRY: **Rhus integrifolia**.
LEMONLEAF: **Gaultheria shallon**.
Leonotis leonurus (L.) R. Br. LION'S TAIL. LION'S EAR.
LEOPARD-PLANT: **Ligularia tussilaginea** 'Aureo-maculata'.
Lepechinia calycina (Benth.) Epling. PITCHER SAGE.
Lepidorrhachis mooreana (F. J. Muell.) O. F. Cook. Syn.: *Clinostigma mooreanum*. RED KENTIA. MOORE SENTRY PALM.
Lepidospartum squamatum (Gray) Gray.

Lepidozamia peroffskyana Regel. Syn.: *Macrozamia denisonii*.
Leptodactylon californicum Hook. & Arn. PRICKLY PHLOX.
Leptospermum citratum (J. F. Bailey & C. T. White) Chall., Cheel & Penf. See **L. petersonii**.
Leptospermum citrinum. Listed name, perhaps for **L. petersonii**.
Leptospermum 'Dwarf Double Rose'. See **L. scoparium 'Dwarf Double Rose'**.
Leptospermum 'Fairy Rose'. See **L. scoparium 'Fairy Rose'**.
Leptospermum flavescens Sm.
Leptospermum flavescens var. *citratum* J. F. Bailey & C. T. White. See **L. petersonii**.
Leptospermum flavescens var. *citriodorum* F. M. Bailey. Plants so called may be **L. liversidgei**.
Leptospermum flavescens var. *grandiflorum* (Lodd.) Benth. Syn.: *L. grandiflorum*.
Leptospermum 'Flore Plenum'. See **L. scoparium 'Flore Plenum'**.
Leptospermum grandiflorum Lodd. See **L. flavescens** var. **grandiflorum**.
Leptospermum 'Grandiflorum Roseum'. See **L. scoparium 'Grandiflorum Roseum'**.
Leptospermum 'Grandiflorum Rubrum'. See **L. scoparium 'Grandiflorum Rubrum'**.
Leptospermum humifusum A. Cunn. ex Schauer.
Leptospermum 'Keatleyi'. See **L. scoparium 'Keatleyi'**.
Leptospermum laevigatum (Soland. ex Gaertn.) F. J. Muell. AUSTRALIAN TEA TREE.
Leptospermum laevigatum 'Compactum'.
Leptospermum laevigatum 'Reevesii'. REEVES' AUSTRALIAN TEA TREE. COMPACT AUSTRALIAN TEA TREE. DWARF TEA TREE.
Leptospermum lanigerum (Ait.) Sm. Syn.: *L. pubescens*. WOOLLY TEA TREE.
Leptospermum lanigerum 'Macrocarpum'?
Leptospermum liversidgei R. T. Baker & H. G. Sm. Plants called *L. flavescens* var. *citriodorum* may belong here.
Leptospermum macrocarpum A. Cunn. ex Schauer.
Leptospermum nitidum Hook. f.
Leptospermum persiciflorum Rchb. See **L. squarrosum**.
Leptospermum petersonii F. M. Bailey. Syn.: *L. citratum*. *L. flavescens* var. *citratum*. Plants called *L. citrinum* belong here.
Leptospermum pictorgillii?
Leptospermum 'Pompon'. See **L. scoparium 'Pompon'**.
Leptospermum pubescens Lam. See **L. lanigerum**.
Leptospermum 'Red Damask'. See **L. scoparium 'Red Damask'**.
Leptospermum 'Red Sparkler'. See **L. scoparium 'Red Sparkler'**.
Leptospermum 'Reevesii'. See **L. laevigatum 'Reevesii'**.
Leptospermum 'Reevesii Nanum'. See **L. laevigatum 'Reevesii'**.
Leptospermum 'Rose Double'. See **L. scoparium 'Rose Double'**.
Leptospermum 'Roseum'. See **L. scoparium 'Roseum'**.
Leptospermum rotundifolium (Maiden & Betche) Domin. ROUND-LEAVED TEA TREE.
Leptospermum 'Ruby Glow'. See **L. scoparium 'Ruby Glow'**.
Leptospermum 'Sanders'. See **L. scoparium 'Sanders'**.
Leptospermum scoparium J. R. Forst. & G. Forst. NEW ZEALAND TEA TREE. MANUKA TEA TREE. MANUKA.
Leptospermum scoparium 'Boscawenii'.
Leptospermum scoparium 'Dwarf Double Rose'.
Leptospermum scoparium 'Fairy Rose'.
Leptospermum scoparium 'Flore Plenum'.
Leptospermum scoparium var. *grandiflorum* Hook. See **L. squarrosum**.
Leptospermum scoparium 'Grandiflorum Roseum'.
Leptospermum scoparium 'Grandiflorum Rubrum'.
Leptospermum scoparium 'Helene Strybing'.
Leptospermum scoparium 'Horizontalis'.
Leptospermum scoparium 'Keatleyi'.
Leptospermum scoparium 'Nanum'.
Leptospermum scoparium 'Nichollsii'.
Leptospermum scoparium 'Pompon'.
Leptospermum scoparium 'Red Damask'.
Leptospermum scoparium 'Red Sparkler'.
Leptospermum scoparium 'Rose Double'.
Leptospermum scoparium 'Roseum'.
Leptospermum scoparium 'Rubrum'. See **L. scoparium 'Ruby Glow'**.
Leptospermum scoparium 'Ruby Glow'.
Leptospermum scoparium 'Sandersii'. SANDERS' TEA TREE.
Leptospermum scoparium 'Scarlet Carnival'.
Leptospermum scoparium 'Snow Flurry'.
Leptospermum scoparium 'Snow White'.
Leptospermum scoparium 'Waerengii'?
Leptospermum 'Snow Flurry'. See **L. scoparium 'Snow Flurry'**.
Leptospermum 'Snow White'. See **L. scoparium 'Snow White'**.
Leptospermum squarrosum Soland. ex Gaertn. Syn.: *L. persiciflorum*. *L. scoparium* var. *grandiflorum*.
Leucaena glauca (L.) Benth.
Leucaena retusa Benth. LITTLE-LEAF LEAD TREE.
Leucodendron argenteum (L.) R. Br. SILVER TREE.
Leucodendron decorum R. Br. See **L. laureolum**.
Leucodendron discolor Phillips & Hutchinson.
Leucodendron eucalyptifolium Buek ex Meissn.
Leucodendron galpinii Phillips & Hutchinson.
Leucodendron laureolum (Lam.) Fourcade. Syn. *L. decorum*.
Leucodendron salicifolium (Salisb.) I. Williams.
Leucodendron salignum Bergius.
Leucodendron spissifolium (Salisb. ex J. Knight) I. Williams.
Leucodendron tinctum I. Williams.
Leucophyllum frutescens (Berl.) I. M. Johnst. Syn.: *L. texanum*. CENIZA. TEXAS RANGER. SERRISA. TEXAS SAGE. TEXAS SILVER-LEAF.
Leucophyllum frutescens 'Compactum'.
Leucophyllum texanum Benth. See **L. frutescens**.
Leucophyllum texanum 'Compactum'. See **L. frutescens 'Compactum'**.
Leucospermum attenuatum R. Br. See **L. cuneiforme**.
Leucospermum catherinae Compton.
Leucospermum cordifolium (Salisb. ex J. Knight) Fourcade. Syn.: *L. nutans*. NODDING PINCUSHION.
Leucospermum cuneiforme (Burm. f.) Rourke. Syn.: *L. attenuatum*.
Leucospermum incisum Phillips. See **L. vestitum**.
Leucospermum lineare R. Br.
Leucospermum nutans R. Br. See **L. cordifolium**.
Leucospermum reflexum Buek ex Meissn. ROCKET PINCUSHION.
Leucospermum tottum (L.) R. Br.
Leucospermum vestitum (Lam.) Rourke. Syn.: *L. incisum*.
LEUCOTHOE, DROOPING: **Leucothoe fontanesiana**.
Leucothoe axillaris (Lam.) D. Don. Syn.: *L. catesbaei*.
Leucothoe catesbaei (Walter) Gray. See **L. axillaris**. Most plants called *L. catesbaei* are **L. fontanesiana**.
Leucothoe catesbaei 'Rainbow'. See **L. fontanesiana 'Rainbow'**.
Leucothoe davisiae Torr. ex Gray. SIERRA LAUREL.
Leucothoe fontanesiana (Steud.) Sleumer. DROOPING LEUCOTHOE. Most plants called *L. catesbaei* belong here.
Leucothoe fontanesiana 'Rainbow'.
Leucothoe keiskei Miq.
Leucothoe populifolia (Lam.) Dipp.
Libocedrus chilensis (D. Don) Endl. See **Austrocedrus chilensis**.
Libocedrus decurrens Torr. See **Calocedrus decurrens**.
Libocedrus decurrens 'Compacta'. See **Calocedrus decurrens 'Compacta'**.
Libonia floribunda K. Koch. See **Justicia rizzinii**.

LICORICE PLANT: **Helichrysum petiolatum.**
Licuala grandis H. Wendl. LICUALA FAN PALM.
Ligularia kaempferi Siebold & Zucc. See **L. tussilaginea.**
Ligularia tussilaginea (Burm. f.) Makino. Syn.: *L. kaempferi.*
Ligularia tussilaginea 'Aureo-maculata'. LEOPARD-PLANT.
Ligustrum californicum. Listed name for **L. ovalifolium.**
Ligustrum coriaceum (Carrière) Lavall. See **L. japonicum 'Rotundifolium'.**
Ligustrum delavayanum Hariot. Syn.: *L. ionandrum.*
Ligustrum henryi Hemsl. HENRY'S PRIVET.
Ligustrum howardii. Listed name for **L. japonicum.**
Ligustrum × ibolium E. F. Coe. [**L. obtusifolium × L. ovalifolium**]. IBOLIUM PRIVET.
Ligustrum indicum (Lour.) Merrill. Syn.: *L. nepalense.*
Ligustrum ionandrum Diels. See **L. delavayanum.**
Ligustrum japonicum Thunb. JAPANESE PRIVET. WAX-LEAF PRIVET. LUSTER-LEAF PRIVET. TEXAS PRIVET. Some plants called **L. lucidum** belong here.
Ligustrum japonicum 'Compactum'?
Ligustrum japonicum var. *coriaceum* Carrière. See **L. japonicum 'Rotundifolium'.**
Ligustrum japonicum 'Lusterleaf'. See **L. japonicum.**
Ligustrum japonicum var. *rotundifolium* Blume. See **L. japonicum 'Rotundifolium'.**
Ligustrum japonicum 'Rotundifolium'. Syn.: *L. coriaceum. L. japonicum* var. *coriaceum. L. japonicum* var. *rotundifolium.*
Ligustrum japonicum 'Silver Star'.
Ligustrum japonicum texanum. Listed name for **L. japonicum.**
Ligustrum japonicum 'Texanum'. Listed name for **L. japonicum.**
Ligustrum japonicum texanum compactum?
Ligustrum japonicum 'Variegatum'.
Ligustrum 'Lodense'. See **L. vulgare 'Lodense'.**
Ligustrum lucidum Ait. f. GLOSSY PRIVET. Some plants called **L. japonicum** belong here.
Ligustrum nepalense Wallich. See **L. indicum.**
Ligustrum obtusifolium Siebold & Zucc.
Ligustrum ovalifolium Hassk. CALIFORNIA PRIVET.
Ligustrum ovalifolium 'Aureum'. GOLDEN CALIFORNIA PRIVET.
Ligustrum ovalifolium 'Aureo-marginatum'. See **L. ovalifolium 'Aureum'.**
Ligustrum ovalifolium 'Variegatum'. See **L. ovalifolium 'Aureum'.**
Ligustrum quihoui Carrière.
Ligustrum 'Silver Star'. See **L. japonicum 'Silver Star'.**
Ligustrum sinense Lour. CHINESE PRIVET.
Ligustrum 'Suwannee River'. SUWANEE RIVER PRIVET.
Ligustrum texanum. Listed name for **L. japonicum.**
Ligustrum texanum 'Silver Star'. See **L. japonicum 'Silver Star'.**
Ligustrum × vicaryi Rehd. [**L. ovalifolium 'Aureum' × L. ?vulgare**]. VICARY PRIVET. VICARY GOLDEN PRIVET.
Ligustrum vulgare L. PRIVET. ENGLISH PRIVET.
Ligustrum vulgare 'Argenteo-variegatum'.
Ligustrum vulgare 'Aureo-variegatum'.
Ligustrum vulgare 'Buxifolium'.
Ligustrum vulgare 'Lodense'.
LILAC, CATALINA MOUNTAIN: **Ceanothus arboreus.**
   CHINESE: **Syringa × chinensis.**
   COMMON: **Syringa vulgaris.**
   CUT-LEAF: **Syringa laciniata.**
   CUT-LEAF PERSIAN: **Syringa laciniata.**
   HARDY VIOLET VINE: **Hardenbergia violacea.**
   HUNGARIAN: **Syringa josikaea.**
   JAPANESE TREE: **Syringa reticulata.**
   LATE: **Syringa villosa.**
   PERSIAN: **Syringa × persica.**
   ROUEN: **Syringa × chinensis.**
   SUMMER: **Buddleia davidii.**
   VINE: **Hardenbergia comptoniana.**
LILY, FIRE: **Xerophyllum tenax.**
   GLOBE SPEAR: **Doryanthes excelsa.**
   PALM: **Cordyline stricta.**
   PALMER SPEAR: **Doryanthes palmeri.**
   PINK PORCELAIN: **Alpinia zerumbet.**
   TORCH: **Kniphofia uvaria.**
LILY-OF-THE-VALLEY SHRUB: **Pieris japonica.**
LILY-OF-THE-VALLEY TREE: **Clethra arborea.**
LILY-OF-THE-VALLEY TREE: **Crinodendron patagua.**
LILLY-PILLY TREE: **Acmena smithii.**
LIME: **Citrus aurantiifolia.**
   AUSTRALIAN FINGER: **Microcitrus australasica.**
   AUSTRALIAN WILD: **Microcitrus australasica.**
   BEARSS SEEDLESS: **Citrus aurantiifolia 'Bearss'.**
   KEY: **Citrus aurantiifolia.**
   MEXICAN: **Citrus aurantiifolia.**
   RANGPUR: **Citrus × limonia.**
   WEST INDIAN: **Citrus aurantiifolia.**
LIMEQUAT: **× Citrofortunella floridana.**
   EUSTIS: **× Citrofortunella floridana 'Eustis'.**
Limonium pectinatum (Ait.) Kuntze. Syn.: *Statice pectinata.*
Limonium perezii (Stapf) F. T. Hubb. Syn.: *Statice perezii.* SEA LAVENDER.
Limonium vulgare Mill.
LINDEN: **Tilia × vulgaris.**
   AFRICAN: **Sparmannia africana.**
   AMERICAN: **Tilia americana.**
   CRIMEAN: **Tilia × euchlora.**
   EUROPEAN: **Tilia × vulgaris.**
   GERMAN: **Sparmannia africana.**
   LARGE-LEAVED: **Tilia platyphyllos.**
   LITTLE-LEAF: **Tilia cordata.**
   SILVER: **Tilia tomentosa.**
Lindera benzoin (L.) Blume. SPICE-BUSH.
LINGBERRY: **Vaccinium vitis-idaea** var. **minus.**
LINGONBERRY: **Vaccinium vitis-idaea.**
LINGONBERRY: **Vaccinium vitis-idaea** var. **minus.**
Linnaea borealis L. TWINFLOWER.
Linospadix monostachya (Mart.) H. Wendl.) Syn.: *Bacularia monostachya.* WALKING-STICK PALM.
Linum flavum L. GOLDEN FLAX. Some plants so called may be **Reinwardtia indica.**
LION'S EAR: **Leonotis leonurus.**
LION'S TAIL: **Leonotis leonurus.**
Lippia citriodora (Ort.) HBK. See **Aloysia triphylla.**
LIPSTICK PLANT: **Aeschynanthus radicans.**
LIQUIDAMBAR: **Liquidambar styraciflua.**
   FORMOSAN: **Liquidambar formosana.**
Liquidambar formosana Hance. FORMOSAN LIQUIDAMBAR. CHINESE SWEET GUM. FORMOSAN SWEET GUM.
Liquidambar orientalis Mill. ORIENTAL SWEET GUM.
Liquidambar styraciflua L. AMERICAN SWEET GUM. LIQUIDAMBAR. RED GUM. SWEET GUM.
Liquidambar styraciflua 'Afterglow'.
Liquidambar styraciflua 'Burgundy'.
Liquidambar styraciflua 'Festival'.
Liquidambar styraciflua 'Golden Glow'.
Liquidambar styraciflua 'Kent'.
Liquidambar styraciflua 'Palo Alto'.
Liriodendron chinense (Hemsl.) Sarg. CHINESE TULIP TREE.
Liriodendron tulipifera L. TULIP TREE. YELLOW POPLAR.
Liriodendron tulipifera 'Arnold'.
Liriodendron tulipifera 'Fastigiatum'.

LITCHI: **Litchi chinensis.**
LITCHI NUT: **Litchi chinensis.**
Litchi chinensis Sonn. LEECHEE. LITCHI. LITCHI NUT. LYCHEE.
Lithocarpus densiflorus (Hook. & Arn.) Rehd. TANBARK OAK.
Lithodora diffusa (Lag.) I. M. Johnst. Syn.: *Lithospermum diffusum.*
Lithodora diffusa 'Grace Ward'.
Lithospermum diffusum Lag. See **Lithodora diffusa.**
Litsea glauca Siebold. See **Neolitsea sericea.**
Livistona australis (R. Br.) Mart. AUSTRALIAN FOUNTAIN PALM. SOUTHERN FAN PALM.
Livistona chinensis (Jacq.) R. Br. ex Mart. Syn.: *L. oliviformis.* CHINESE FAN PALM. CHINESE FOUNTAIN PALM. FOUNTAIN PALM. Plants in Calif. called *Latania borbonica* belong here.
Livistona decipiens Becc.
Livistona mariae F. J. Muell.
Livistona oliviformis (Hassk.) Mart. See **L. chinensis.**
LIZARD PLANT: **Tetrastigma voinieranum.**
LOCUST, BLACK: **Robinia pseudoacacia.**
  BRISTLY: **Robinia hispida.**
  CLAMMY: **Robinia viscosa.**
  HONEY: **Gleditsia triacanthos.**
  HYBRID: **Robinia × ambigua.**
  IDAHO: **Robinia × ambigua 'Idahoensis'.**
  IMPERIAL HONEY: **Gleditsia triacanthos** forma **inermis 'Imperial'.**
  MONUMENT: **Robinia fertilis 'Monument'.**
  MORAINE HONEY: **Gleditsia triacanthos** forma **inermis 'Moraine'.**
  MOSSY: **Robinia hispida.**
  PINK: **Robinia × ambigua 'Decaisneana'.**
  PINK-FLOWERING: **Robinia × ambigua 'Decaisneana'.**
  PINK IDAHO: **Robinia × ambigua 'Idahoensis'.**
  RUBY LACE HONEY: **Gleditsia triacanthos** forma **inermis 'Ruby Lace'.**
  SHADEMASTER HONEY: **Gleditsia triacanthos** forma **inermis 'Shademaster'.**
  SKYLINE HONEY: **Gleditsia triacanthos** forma **inermis 'Skyline'.**
  SUNBURST HONEY: **Gleditsia triacanthos** forma **inermis 'Sunburst'.**
  THORNLESS HONEY: **Gleditsia triacanthos** forma **inermis.**
Lonchocarpus nicou DC. Not cult. See **Millettia reticulata.**
Lonchocarpus speciosus Bolus. See **Bolusanthus speciosus.**
Lonicera ciliosa (Pursh) Poir.
Lonicera 'Clavey's Dwarf'. See **L. × xylosteoides 'Clavey's Dwarf'.**
Lonicera confusa DC.
Lonicera fragrantissima Lindl. & Paxt. WINTER HONEYSUCKLE.
Lonicera halliana. Listed name for **L. japonica 'Halliana'.**
Lonicera × heckrottii Rehd. [?**L. americana × L. sempervirens**]. CORAL HONEYSUCKLE. GOLD-FLAME HONEYSUCKLE.
Lonicera henryi Hemsl.
Lonicera hildebrandiana Coll. & Hemsl. GIANT BURMESE HONEYSUCKLE.
Lonicera hispidula Dougl. ex Torr. & Gray. CALIFORNIA HONEYSUCKLE.
Lonicera hispidula var. **vacillans** Gray. HAIRY HONEYSUCKLE.
Lonicera involucrata (Richardson) Banks ex K. Spreng. TWINBERRY.
Lonicera japonica Thunb. JAPANESE HONEYSUCKLE.
Lonicera japonica 'Aureo-reticulata'. GOLD-NET HONEYSUCKLE.
Lonicera japonica var. **chinensis** (P. W. Wats.) Baker. CHINESE HONEYSUCKLE.
Lonicera japonica 'Halliana'. HALL'S JAPANESE HONEYSUCKLE.

Lonicera japonica 'Purpurea'. PURPLE-LEAVED HONEYSUCKLE.
Lonicera korolkowii Stapf.
Lonicera korolkowii var. *zabelii* (Rehd.) Rehd. Syn.: *L. zabelii.*
Lonicera maximowiczii (Rupr.) Maxim.
Lonicera maximowiczii var. *sachalinensis* Fr. Schmidt.
Lonicera nitida Wils. Syn.: *L. pileata* forma *yunnanensis.* BOX HONEYSUCKLE.
Lonicera nitida 'Elegant'.
Lonicera periclymenum L. WOODBINE.
Lonicera periclymenum var. *belgica* Ait. DUTCH HONEYSUCKLE. DUTCH WOODBINE.
Lonicera pileata D. Oliver. PRIVET HONEYSUCKLE.
Lonicera pileata forma *yunnanensis* (Franch.) Rehd. See **L. nitida.**
Lonicera 'Pink Gold Flame'?
Lonicera sempervirens L. TRUMPET HONEYSUCKLE. CORAL HONEYSUCKLE.
Lonicera subspicata Hook. & Arn. WILD HONEYSUCKLE.
Lonicera subspicata var. *johnstonii* Keck.
Lonicera syringantha Maxim.
Lonicera tatarica L. TATARIAN HONEYSUCKLE.
Lonicera tatarica 'Alba'. WHITE TATARIAN HONEYSUCKLE.
Lonicera tatarica 'Rosea'. PINK TATARIAN HONEYSUCKLE.
Lonicera tatarica 'Sibirica'. RED TATARIAN HONEYSUCKLE.
Lonicera × xylosteoides Tausch. [*L. tatarica × L. xylosteum* L.].
Lonicera × xylosteoides 'Clavey's Dwarf'.
Lonicera zabelii Rehd. See **L. korolkowii** var. **zabelii.**
LOOKING-GLASS PLANT: **Coprosma repens.**
LOOSESTRIFE, PURPLE: **Lythrum salicaria.**
Lophomyrtus obcordata (Raoul) Burret. Syn.: *Myrtus obcordata.* RED MYRTLE.
LOQUAT: **Eriobotrya japonica.**
  BRONZE: **Eriobotrya deflexa.**
Loropetalum chinense (R. Br.) D. Oliver.
LOTUS VINE, RED: **Lotus berthelotii.**
  YELLOW: **Lotus corniculatus.**
Lotus berthelotii Masf. RED LOTUS VINE. PARROT'S BEAK. WINGED PEA.
Lotus corniculatus L. BIRD'S FOOT TREFOIL. YELLOW LOTUS VINE.
Lotus mascaensis Burchard.
Lotus scoparius (Nutt. ex Torr. & Gray) Ottley. DEERWEED.
LUCKY NUT: **Thevetia peruviana.**
Luculia gratissima Sweet.
Luehea divaricata Mart.
LULO: **Solanum quitoense.**
Luma apiculata (DC.) Burret. Syn.: *Eugenia apiculata. Myrceugenella apiculata. Myrceugenia apiculata.* TEMU.
Luma chequen (Molina) Gray. Syn.: *Myrceugenella chequen.*
LUPINE, BUSH: **Lupinus arboreus.**
  TREE: **Lupinus arboreus.**
Lupinus albifrons Benth.
Lupinus arboreus Sims. BUSH LUPINE. TREE LUPINE.
Lupinus longifolius (S. Wats.) Abrams.
LYCHEE: **Litchi chinensis.**
Lycianthes rantonnei (Carrière) Bitter. Syn.: *Solanum rantonnetii.* BLUE POTATO BUSH. BLUE SOLANUM.
Lyonothamnus asplenifolius Greene. See **L. floribundus** subsp. **asplenifolius.**
Lyonothamnus floribundus Gray. CATALINA IRONWOOD.
Lyonothamnus floribundus subsp. *asplenifolius* (Greene) Raven. Syn.: *L. asplenifolius.* FERN-LEAF CATALINA IRONWOOD. ISLAND IRONWOOD.
Lysiloma thornberi Britt. & Rose. FEATHER BUSH.
Lythrum 'Morden Dream'.
Lythrum salicaria L. PURPLE LOOSESTRIFE.

Lythrum salicaria 'Morden Pink'. See **L. virgatum 'Morden Pink'**.
Lythrum salicaria 'Roseum Superbum'.
Lythrum virgatum L.
Lythrum virgatum 'Morden Pink'.

M

Maackia amurensis Rupr. & Maxim.
MACADAMIA, BITTER: **Macademia ternifolia**.
   ROUGH-SHELL: **Macadamia tetraphylla**.
   SMOOTH-SHELL: **Macadamia integrifolia**.
MACADAMIA NUT: **Macadamia tetraphylla**.
Macadamia integrifolia Maiden & Betche. QUEENSLAND NUT. SMOOTH-SHELL MACADAMIA. Some plants called **M. ternifolia** belong here.
Macadamia ternifolia F. J. Muell. BITTER MACADAMIA. SMALL-FRUITED QUEENSLAND NUT. Probably not cult. Plants so called are **M. integrifolia** or **M. tetraphylla**.
Macadamia tetraphylla L. A. S. Johnson. AUSTRALIAN NUT. MACADAMIA NUT. ROUGH-SHELL MACADAMIA. Some plants called **M. ternifolia** belong here.
Macfadyena cynanchoides (Cham.) Morong. See **Dolichandra cynanchoides**.
Macfadyena unguis-cati (L.) A. Gentry. Syn.: *Bignonia unguis-cati*. *B. tweediana*. *Doxantha unguis-cati*. CAT'S-CLAW. YELLOW TRUMPET VINE.
Mackaya bella Harv. Some plants called **Ruellia tuberosa** or **Asystasia bella** belong here.
Maclura pomifera (Raf.) Schneid. OSAGE ORANGE. BOW-WOOD.
Macrozamia communis L. A. S. Johnson. Some plants called **M. spiralis** belong here.
Macrozamia denisonii C. Moore & F. J. Muell. See **Lepidozamia peroffskyana**.
Macrozamia diplomera (F. J. Muell.) L. A. S. Johnson.
Macrozamia fawcettii C. Moore.
Macrozamia heteromera C. Moore.
Macrozamia moorei F. J. Muell.
Macrozamia riedlei (Fisch. ex Gaud.-Beaup.) C. A. Gardner.
Macrozamia secunda C. Moore.
Macrozamia spiralis (Salisb.) Miq. Some plants so called are **M. communis**.
MADAKE: **Phyllostachys bambusoides**.
MADEIRA VINE: **Anredera cordifolia**.
MADRONE: **Arbutus menziesii**.
   PACIFIC: **Arbutus menziesii**.
   STRAWBERRY: **Arbutus unedo**.
MAGIC FLOWER: **Cantua buxifolia**.
MAGNOLIA, ALEXANDRINA: **Magnolia × soulangiana 'Alexandrina'**.
   ANISE: **Magnolia salicifolia**.
   ASHE: **Magnolia macrophylla** subsp. **ashei**.
   BIG-LEAF: **Magnolia macrophylla**.
   CAMPBELL: **Magnolia campbellii**.
   CHINESE: **Magnolia sieboldii** subsp. **sinensis**.
   CHINESE: **Magnolia × soulangiana**.
   DAWSON: **Magnolia dawsoniana**.
   DELAVAY: **Magnolia delavayi**.
   JAPANESE WHITEBARK: **Magnolia hypoleuca**.
   KOBUS: **Magnolia kobus**.
   LENNE: **Magnolia × soulangiana 'Lennei'**.
   LILY: **Magnolia quinquepeta**.
   LOEBNER: **Magnolia kobus** var. **loebneri**.
   OYAMA: **Magnolia sieboldii**.
   PURPLE LILY: **Magnolia quinquepeta 'Nigra'**.
   SAINT MARY: **Magnolia grandiflora 'Saint Mary'**.
   SAUCER: **Magnolia × soulangiana**.
   SOUTHERN: **Magnolia grandiflora**.
   STAR: **Magnolia kobus** var. **stellata**.
   THOMPSON: **Magnolia × thompsoniana**.
   UMBRELLA: **Magnolia tripetala**.
   VEITCH: **Magnolia × veitchii**.
   WATSON: **Magnolia × wieseneri**.
   WIESENER: **Magnolia × wieseneri**.
   WILSON: **Magnolia wilsonii**.
Magnolia acuminata (L.) L. CUCUMBER TREE.
Magnolia acuminata var. *cordata* (Michx.) Sarg. Syn.: *M. cordata*. YELLOW CUCUMBER TREE.
Magnolia acuminata 'Golden Glow'.
Magnolia 'Alexandrina'. See **M. × soulangiana 'Alexandrina'**.
Magnolia ashei Weatherby. See **M. macrophylla** subsp. **ashei**.
Magnolia 'Ballerina'. See **M. kobus** var. **loebneri 'Ballerina'**.
Magnolia × brooklynensis Kalmb. [**M. acuminata × M. quinquepeta**].
Magnolia × brooklynensis 'Evamaria'.
Magnolia × brooklynensis 'Woodsman'.
Magnolia campbellii Hook. f. & Thoms. CAMPBELL MAGNOLIA.
Magnolia campbellii 'Alba'.
Magnolia campbellii 'Charles Raffill'.
Magnolia campbellii 'Chyverton'.
Magnolia campbellii 'Hendrick's Park'.
Magnolia campbellii 'Kew's Surprise'.
Magnolia campbellii 'Lanarth'.
Magnolia campbellii 'Late Pink'.
Magnolia campbellii 'Maharaja'.
Magnolia campbellii 'Maharanee'.
Magnolia campbellii subsp. *mollicomata* (W. W. Sm.) Johnstone. Syn.: *M. mollicomata*.
Magnolia campbellii 'Strybing White'.
Magnolia 'Chyverton'. See **M. campbellii 'Chyverton'**.
Magnolia 'Chyverton Red'. See **M. dawsoniana 'Chyverton Red'**.
Magnolia conspicua Salisb. See **M. heptapeta**.
Magnolia cordata Michx. See **M. acuminata** var. **cordata**.
Magnolia cylindrica Wils.
Magnolia dawsoniana Rehd. & Wils. DAWSON MAGNOLIA.
Magnolia dawsoniana 'Chyverton Red'.
Magnolia delavayi Franch. DELAVAY MAGNOLIA.
Magnolia denudata Desr. See **M. heptapeta**.
Magnolia fraseri Walter.
Magnolia fraseri var. *pyramidata* (Bartr.) Pamp. Syn.: *M. pyramidata*.
Magnolia 'Freeman'. [**M. virginiana × M. grandiflora**].
Magnolia glauca L. See **M. virginiana**.
Magnolia globosa Hook. f. & Thoms.
Magnolia grandiflora L. BULL BAY. SOUTHERN MAGNOLIA.
Magnolia grandiflora 'Exmouth'.
Magnolia grandiflora 'Majestic Beauty'.
Magnolia grandiflora 'Russet'.
Magnolia grandiflora 'Saint Mary'. SAINT MARY MAGNOLIA.
Magnolia grandiflora 'Samuel Sommer'.
Magnolia grandiflora 'San Marino'.
Magnolia grandiflora 'Victoria'.
Magnolia heptapeta (Buc'hoz) Dandy. Syn.: *M. conspicua*. *M. denudata*. *M. yulan*. YULAN.
Magnolia × highdownensis Dandy. See **M. wilsonii**.
Magnolia hypoleuca Siebold & Zucc. Syn.: *M. obovata*. JAPANESE WHITEBARK MAGNOLIA.
Magnolia × kewensis Pearce. See **M. salicifolia**.

Magnolia × *kewensis* 'Wada's Memory'. See **M. salicifolia 'Wada's Memory'**.
Magnolia kobus DC. Syn.: *M. kobus* var. *borealis*. KOBUS MAGNOLIA.
Magnolia kobus var. *borealis* Sarg. See **M. kobus**.
Magnolia kobus var. **loebneri** (Kache) Spongberg. Syn.: *M.* × *loebneri*. LOEBNER MAGNOLIA.
Magnolia kobus var. **loebneri 'Ballerina'**.
Magnolia kobus var. **loebneri 'Leonard Messel'**.
Magnolia kobus var. **loebneri 'Merrill'**.
Magnolia kobus var. **stellata** (Siebold & Zucc.) Blackburn. Syn.: *M. stellata*. STAR MAGNOLIA.
Magnolia kobus var. **stellata 'Centennial'**.
Magnolia kobus var. **stellata 'Rosea'**.
Magnolia kobus var. **stellata 'Royal Star'**.
Magnolia kobus var. **stellata 'Rubra'**.
Magnolia kobus var. **stellata 'Waterlily'**.
Magnolia 'Lennei'. See **M.** × **soulangiana 'Lennei'**.
Magnolia liliiflora Desr. See **M. quinquepeta**.
Magnolia liliiflora 'Gracilis'. See **M. quinquepeta 'Gracilis'**.
Magnolia liliiflora 'Nigra'. See **M. quinquepeta 'Nigra'**.
Magnolia 'Lilliputian'. See **M.** × **soulangiana 'Lilliputian'**.
Magnolia × *loebneri* Kache. [*M. kobus* var. *kobus* × *M. kobus* var. *stellata*]. See **M. kobus** var. **loebneri**.
Magnolia × *loebneri* 'Ballerina'. See **M. kobus** var. **loebneri 'Ballerina'**.
Magnolia × *loebneri* 'Leonard Messel'. See **M. kobus** var. **loebneri 'Leonard Messel'**.
Magnolia × *loebneri* 'Merrill'. See **M. kobus** var. **loebneri 'Merrill'**.
Magnolia 'Lombardy Rose'. See **M.** × **soulangiana 'Lombardy Rose'**.
Magnolia macrophylla Michx. BIG-LEAF MAGNOLIA. LARGE-LEAVED CUCUMBER TREE.
Magnolia macrophylla subsp. **ashei** (Weatherby) Spongberg. Syn.: *M. ashei*. ASHE MAGNOLIA.
Magnolia 'Majestic Beauty'. See **M. grandiflora 'Majestic Beauty'**.
Magnolia 'Maharanee'. See **M. campbellii 'Maharanee'**.
Magnolia mollicomata W. W. Sm. See **M. campbellii** subsp. **mollicomata**.
Magnolia nigra. Listed name for **M. quinquepeta 'Nigra'**.
Magnolia nigricans. Listed name for **M. quinquepeta 'Nigra'**.
Magnolia nitida W. W. Sm.
Magnolia obovata Thunb. See **M. hypoleuca**.
Magnolia officinalis Rehd. & Wils.
Magnolia officinalis var. **biloba** Rehd. & Wils.
Magnolia 'Picture'. See **M.** × **soulangiana 'Picture'**.
Magnolia 'Pink Superba'. See **M.** × **soulangiana 'Pink Superba'**.
Magnolia × *proctoriana* Rehd. See **M. salicifolia**.
Magnolia pyramidata Bartr. See **M. fraseri** var. **pyramidata**.
Magnolia quinquepeta (Buc'hoz) Dandy. Syn.: *M. liliiflora*. LILY MAGNOLIA. Many plants called **M. quinquepeta 'Nigra'** belong here.
Magnolia quinquepeta 'Gracilis'.
Magnolia quinquepeta 'Nigra'. PURPLE LILY MAGNOLIA. Many plants so called are **M. quinquepeta**.
Magnolia quinquepeta 'Royal Crown'.
Magnolia rostrata W. W. Sm.
Magnolia 'Rouged Alabaster'. See **M.** × **soulangiana 'Rouged Alabaster'**.
Magnolia 'Royal Crown'. See **M. quinquepeta 'Royal Crown'**.
Magnolia 'Russet'. See **M. grandiflora 'Russet'**.
Magnolia 'Rustica Rubra'. See **M.** × **soulangiana 'Rustica Rubra'**.
Magnolia 'Saint Mary'. See **M. grandiflora 'Saint Mary'**.
Magnolia salicifolia (Siebold & Zucc.) Maxim. Syn.: *M.* × *kewensis*. *M.* × *proctoriana*. ANISE MAGNOLIA.
Magnolia salicifolia 'Wada's Memory'.
Magnolia salicifolia 'W. B. Clarke'.
Magnolia sargentiana Rehd. & Wils. var. **robusta** Rehd. & Wils.
Magnolia sieboldii K. Koch. OYAMA MAGNOLIA.
Magnolia sieboldii subsp. **sinensis** (Rehd. & Wils.) Spongberg. Syn.: *M. sinensis*. CHINESE MAGNOLIA.
Magnolia sinensis (Rehd. & Wils.) Stapf. See **M. sieboldii** subsp. **sinensis**.
Magnolia × *soulangiana* Soul.-Bod. [*M. heptapeta* × *M. quinquepeta*]. SAUCER MAGNOLIA. CHINESE MAGNOLIA. TULIP TREE.
Magnolia × *soulangiana* 'Alba'.
Magnolia × *soulangiana* 'Alba Superba'.
Magnolia × *soulangiana* 'Alexandrina'. ALEXANDRINA MAGNOLIA.
Magnolia × *soulangiana* 'Amabilis'.
Magnolia × *soulangiana* 'Brozzonii'.
Magnolia × *soulangiana* 'Burgundy'.
Magnolia × *soulangiana* 'Grace McDade'.
Magnolia × *soulangiana* 'Lennei'. LENNE MAGNOLIA.
Magnolia × *soulangiana* 'Lennei Alba'.
Magnolia × *soulangiana* 'Lilliputian'.
Magnolia × *soulangiana* 'Lombardy Rose'.
Magnolia × *soulangiana* 'Nigra'. See **M. quinquepeta 'Nigra'**.
Magnolia × *soulangiana* 'Picture'.
Magnolia × *soulangiana* 'Pink Superba'.
Magnolia × *soulangiana* 'Rouged Alabaster'.
Magnolia × *soulangiana* 'Rubra'. See **M.** × **soulangiana 'Rustica Rubra'**.
Magnolia × *soulangiana* 'Rustica'. See **M.** × **soulangiana 'Rustica Rubra'**.
Magnolia × *soulangiana* 'Rustica Rubra'.
Magnolia × *soulangiana* 'San Jose'.
Magnolia × *soulangiana* 'Speciosa'.
Magnolia × *soulangiana* 'Verbanica'.
Magnolia 'Speciosa'. See **M.** × **soulangiana 'Speciosa'**.
Magnolia sprengeri Pamp.
Magnolia sprengeri 'Diva'.
Magnolia stellata (Siebold & Zucc.) Maxim. and listed cvs. See **M. kobus** var. **stellata** and cvs.
Magnolia × *thompsoniana* (Loud.) Vos. [*M. tripetala* × *M. virginiana*]. THOMPSON MAGNOLIA.
Magnolia × *thompsoniana* 'Urbana'.
Magnolia tripetala L. UMBRELLA MAGNOLIA. UMBRELLA TREE.
Magnolia × *veitchii* Bean. [*M. heptapeta* × *M. campbellii*]. VEITCH MAGNOLIA.
Magnolia virginiana L. Syn.: *M. virginiana* var. *australis*. *M. glauca*. SWEET BAY.
Magnolia virginiana var. *australis* Sarg. See **M. virginiana**.
Magnolia 'Wada's Memory'. See **M. salicifolia 'Wada's Memory'**.
Magnolia × *watsonii* Hook. f. See **M.** × **wieseneri**.
Magnolia 'W. B. Clarke'. See **M. salicifolia 'W. B. Clarke'**.
Magnolia × *wieseneri* Carrière. [?*M. hypoleuca* × *M. sieboldii*]. Syn.: *M.* × *watsonii*. WATSON MAGNOLIA. WIESENER MAGNOLIA.
Magnolia wilsonii (Finet & Gagnep.) Rehd. Syn.: *M.* × *highdownensis*. WILSON MAGNOLIA.
Magnolia 'Woodsman'. See **M.** × **brooklynensis 'Woodsman'**.
Magnolia yulan Desf. See **M. heptapeta**.
MAGUEY: **Agave americana**.
MAHALA MATS: **Ceanothus prostratus**.
Mahernia verticillata L. See **Hermannia verticillata**.

× **Mahoberberis aquisargentii** Krüssm. [*Berberis* ?*sargentiana* × **Mahonia aquifolium**].
× **Mahoberberis miethkeana** Melander & Eade. [*Berberis* ?*manipurana* × **Mahonia aquifolium**].
× **Mahoberberis neubertii** (Lem.) Schneid. [*Berberis vulgaris* × **Mahonia aquifolium**]. Some plants called **Berberis ilicifolia** belong here.
MAHOE: **Hibiscus tiliaceus**.
MAHOGANY, ALDER-LEAF MOUNTAIN: **Cercocarpus betuloides** var. **blancheae**.
  CATALINA MOUNTAIN: **Cercocarpus betuloides** var. **traskiae**.
  CURL-LEAF MOUNTAIN: **Cercocarpus ledifolius**.
  DESERT MOUNTAIN: **Cercocarpus ledifolius**.
  MOUNTAIN: **Cercocarpus betuloides**.
  RED: **Eucalyptus resinifera**.
  SAN DIEGO MOUNTAIN: **Cercocarpus minutiflorus**.
  SWAMP: **Eucalyptus robusta**.
  WHITE: **Eucalyptus acmenioides**.
MAHONIA, CASCADE: **Mahonia nervosa**.
  CREEPING: **Mahonia repens**.
  DESERT: **Mahonia fremontii**.
  HOLLY: **Mahonia aquifolium**.
  LEATHERLEAF: **Mahonia bealei**.
  LONGLEAF: **Mahonia nervosa**.
  NEVIN: **Mahonia nevinii**.
Mahonia aquifolium (Pursh) Nutt. Syn.: *Berberis aquifolium*. OREGON GRAPE. HOLLY MAHONIA. OREGON HOLLY-GRAPE.
Mahonia aquifolium 'Compacta'. COMPACT OREGON GRAPE.
Mahonia aquifolium 'Golden Abundance'.
Mahonia aquifolium 'John Muir'.
Mahonia aquifolium 'Mayhan Dwarf'.
Mahonia aquifolium 'Nana'.
Mahonia aquifolium 'Orangee Flame'.
Mahonia 'Arthur Menzies'. [*M. lomariifolia* × *M.* ?*bealei*].
Mahonia bealei (Fort.) Carrière. LEATHERLEAF MAHONIA.
Mahonia 'Compacta'. See *M. aquifolium* 'Compacta'.
Mahonia fortunei (Lindl.) Fedde.
Mahonia fremontii (Torr.) Fedde. Syn.: *Berberis fremontii*. DESERT BARBERRY. DESERT MAHONIA.
Mahonia higginsiae (Munz) Ahrendt. Syn.: *Berberis higginsiae*. HIGGINS BARBERRY.
Mahonia japonica (Thunb.) DC.
Mahonia lomariifolia Takeda. CHINESE HOLLY-GRAPE.
Mahonia nervosa (Pursh) Nutt. Syn.: *Berberis nervosa*. LONGLEAF MAHONIA. CASCADE MAHONIA.
Mahonia nevinii (Gray) Fedde. Syn.: *Berberis nevinii*. NEVIN MAHONIA.
Mahonia pinnata (Lag.) Fedde. Syn.: *Berberis pinnata*. CALIFORNIA HOLLY-GRAPE. CALIFORNIA BARBERRY.
Mahonia pinnata 'Ken Hartman'.
Mahonia piperana Abrams. Syn.: *Berberis piperana*.
Mahonia pumila (Greene) Fedde.
Mahonia repens (Lindl.) G. Don. Syn.: *Berberis repens*. CREEPING BARBERRY. CREEPING MAHONIA.
Mahonia trifoliolata (Moricand) Fedde.
MAIDENHAIR TREE: **Ginkgo biloba**.
Malacothamnus davidsonii (B. L. Robinson) Greene. Syn.: *Malvastrum davidsonii*.
Malacothamnus fasciculatus (Nutt. ex Torr. & Gray) Greene. COMMON BUSH MALLOW.
MALLEE, BELL-FRUITED: **Eucalyptus preissiana**.
  BURRACOPPIN: **Eucalyptus burracoppinensis**.
  COARSE-FLOWERED: **Eucalyptus grossa**.
  COARSE-LEAVED: **Eucalyptus grossa**.
  DESMOND'S: **Eucalyptus desmondensis**.
  GOOSEBERRY: **Eucalyptus calycogona**.
  GREEN: **Eucalyptus viridis**.
  HORNED: **Eucalyptus eremophila**.
  KRUSE'S: **Eucalyptus kruseana**.
  LERP: **Eucalyptus incrassata**.
  OLDFIELD'S: **Eucalyptus oldfieldii**.
  PEAR-FRUITED: **Eucalyptus pyriformis**.
  RED-FLOWERED: **Eucalyptus erythronema**.
  RIDGE-FRUITED: **Eucalyptus incrassata**.
  ROSE: **Eucalyptus rhodantha**.
  ROUND-LEAVED: **Eucalyptus orbifolia**.
  SHARP-CAPPED: **Eucalyptus oxymitra**.
  SOUTHERN CROSS SILVER: **Eucalyptus crucis**.
  SQUARE-FRUITED: **Eucalyptus tetraptera**.
  SWAMP: **Eucalyptus spathulata**.
  TALL SAND: **Eucalyptus eremophila**.
MALLET, GARDNER'S: **Eucalyptus gardneri**.
MALLOW, APRICOT: **Sphaeralcea ambigua**.
  CALIFORNIA TREE: **Lavatera assurgentiflora**.
  COMMON BUSH: **Malacothamnus fasciculatus**.
  TREE: **Lavatera arborea**.
  WAX: **Malvaviscus arboreus**.
Malpighia coccigera L. MINIATURE HOLLY. SINGAPORE HOLLY.
Malpighia glabra L. BARBADOS CHERRY.
Malus 'Aldenhamensis'. ALDENHAM CRAB APPLE.
Malus 'Almey'.
Malus × arnoldiana (Rehd.) Rehd. [*M. floribunda* × *M. baccata*]. ARNOLD CRAB APPLE.
Malus × atrosanguinea (F. L. Spaeth) Schneid. [*M. halliana* × *M. sieboldii*]. CARMEN CRAB APPLE.
Malus baccata (L.) Borkh. Syn.: *Pyrus baccata*. SIBERIAN CRAB APPLE.
Malus baccata 'Columnaris'.
Malus baccata var. **mandshurica** (Maxim.) Schneid. MANCHURIAN CRAB APPLE.
Malus 'Bechtel'. See *M. ioensis* 'Plena'.
Malus coronaria (L.) Mill. Syn.: *Pyrus coronaria*. AMERICAN CRAB APPLE. WILD SWEET CRAB APPLE.
Malus coronaria 'Charlottae'.
Malus 'Dolgo'.
Malus domestica Borkh. Syn.: *Pyrus malus*. COMMON APPLE.
Malus 'Dorothea'.
Malus 'Echtermeyer'. See *M.* 'Oekonomierat Echtermeyer'.
Malus × eleyi (Bean) Hesse. [*M. niedzwetzkyana* × *M. spectabilis*]. Syn.: *M.* × *purpurea* forma *eleyi*. *M. sylvestris* var. *eleyi*. ELEY CRAB APPLE.
Malus 'Ferrill's Crimson'.
Malus floribunda Siebold ex Van Houtte. JAPANESE FLOWERING CRAB APPLE.
Malus halliana Koehne. **'Parkmanii'**. PARKMAN CRAB APPLE.
Malus 'Hopa'. [*M. niedzwetzkyana* × *M.* ?*baccata*].
Malus hupehensis (Pamp.) Rehd. Syn.: *M. theifera*. TEA CRAB APPLE.
Malus ioensis (Wood) Britt. Syn.: *Pyrus ioensis*. PRAIRIE CRAB APPLE.
Malus ioensis 'Klemi'. Listed name of uncertain application. Sometimes also spelled 'Klems' or 'Klehm'.
Malus ioensis 'Plena'. BECHTEL CRAB APPLE.
Malus × *kaido* (Siebold ex Wenz.) Pardé. See *M.* × *micromalus*.
Malus kansuensis (Batal.) Schneid.
Malus 'Katherine'.
Malus × micromalus Makino. [*M. baccata* × *M. spectabilis*]. Syn.: *M. kaido*. MIDGET CRAB APPLE. KAIDO CRAB APPLE.
Malus niedzwetzkyana Dieck. Syn.: *M. pumila* 'Niedzwetzkyana'. RED-VEIN CRAB APPLE.

Malus 'Oekonomierat Echtermeyer'. ECHTERMEYER CRAB APPLE. WEEPING CRAB APPLE.
Malus prunifolia (Willd.) Borkh. PLUM-LEAVED APPLE. CHINESE APPLE.
Malus pumila Mill. PARADISE APPLE.
Malus pumila 'Niedzwetzkyana'. See *M. niedzwetzkyana*.
Malus × *purpurea* (Barbier) Rehd. [*M.* × *atrosanguinea* × *M. niedzwetzkyana*].
Malus × *purpurea* forma *eleyi* (Bean) Rehd. See *M.* × *eleyi*.
Malus 'Red Jade'.
Malus 'Red Silver'.
Malus sargentii Rehd. SARGENT CRAB APPLE.
Malus × *scheideckeri* F. L. Spaeth ex Zabel. [*M. floribunda* × *M. prunifolia*]. SCHEIDECKER CRAB APPLE.
Malus sieboldii (Regel) Rehd. Syn.: *M. toringo*. TORINGO CRAB APPLE.
Malus sikkimensis (Wenz.) Koehne ex Schneid.
Malus spectabilis (Ait.) Borkh. CHINESE FLOWERING APPLE.
Malus sylvestris Mill. APPLE. CRAB APPLE.
Malus sylvestris var. *eleyi* (Bean) L. H. Bailey. See *M.* × *eleyi*.
Malus theifera Rehd. See *M. hupehensis*.
Malus toringo Carrière. See *M. sieboldii*.
Malus yunnanensis (Franch.) Schneid.
Malus yunnanensis 'Angel's Wing's'?
Malus × *zumi* (Matsum.) Rehd. [*M. baccata* var. *mandshurica* × *M. sieboldii*].
Malus × *zumi* 'Calocarpa'.
Malvastrum davidsonii B. L. Robinson. See *Malacothamnus davidsonii*.
Malvaviscus arboreus Cav. WAX MALLOW. TURK'S CAP.
Malvaviscus arboreus var. *mexicanus* Schlechtend. TURK'S CAP.
MANDARIN: *Citrus reticulata*.
  ALGERIAN: *Citrus reticulata* 'Clementine'.
  DANCY: *Citrus reticulata* 'Dancy'.
  KING: *Citrus* × *nobilis* 'King'.
  KINNOW: *Citrus reticulata* 'Kinnow'.
  OWARI: *Citrus reticulata* 'Owari'.
Mandevilla 'Alice du Pont'. See *M.* × *amabilis* 'Alice du Pont'.
Mandevilla × *amabilis* (Backh.) Dress. Syn.: *Dipladenia* × *amoena*. Plants called *Dipladenia rosea* may belong here.
Mandevilla × *amabilis* 'Alice du Pont'.
Mandevilla laxa (Ruiz & Pav.) Woodson. Syn.: *M. suaveolens*. CHILEAN JASMINE.
Mandevilla splendens (Hook. f.) Woodson. Syn.: *Dipladenia splendens*.
Mandevilla suaveolens Lindl. See *M. laxa*.
Manettia bicolor Hook. f. See *M. inflata*.
Manettia inflata Sprague. Syn.: *M. bicolor*. BRAZILIAN FIRE-CRACKER. FIRE-CRACKER VINE.
Mangifera indica L. MANGO.
MANGO: *Mangifera indica*.
MANNA PLANT: *Tamarix gallica*.
MANUKA: *Leptospermum scoparium*.
MANZANITA, BAKER: *Arctostaphylos stanfordiana* subsp. *bakeri*.
  BIG-BERRY: *Arctostaphylos glauca*.
  CREEPING: *Arctostaphylos uva-ursi*.
  DUNE: *Arctostaphylos pumila*.
  EASTWOOD: *Arctostaphylos glandulosa*.
  FORT BRAGG: *Arctostaphylos nummularia*.
  HAIRY: *Arctostaphylos columbiana*.
  HEART-LEAF: *Arctostaphylos andersonii*.
  HOARY: *Arctostaphylos canescens*.
  HOOKER: *Arctostaphylos hookeri*.
  HOWELL: *Arctostaphylos stanfordiana* subsp. *hispidula*.
  IONE: *Arctostaphylos myrtifolia*.
  ISLAND: *Arctostaphylos insularis*.
  JAMES WEST: *Arctostaphylos densiflora* 'James West'.
  LAUREL HILL: *Arctostaphylos hookeri* subsp. *franciscana*.
  LITTLE SUR: *Arctostaphylos edmundsii*.
  LOMPOC: *Arctostaphylos pechoensis* var. *viridissima*.
  MEXICAN: *Arctostaphylos pungens*.
  MONTEREY: *Arctostaphylos hookeri*.
  MORRO BAY: *Arctostaphylos morroensis*.
  MT. DIABLO: *Arctostaphylos auriculata*.
  OTAY: *Arctostaphylos otayensis*.
  PAJARO: *Arctostaphylos pajaroensis*.
  PARRY: *Arctostaphylos manzanita*.
  PINE-MAT: *Arctostaphylos nevadensis*.
  POINT-LEAF: *Arctostaphylos pungens*.
  PROSTRATE: *Arctostaphylos uva-ursi*.
  SAND-MAT: *Arctostaphylos pumila*.
  SERPENTINE: *Arctostaphylos obispoensis*.
  SHAGBARK: *Arctostaphylos rudis*.
  SHAGGY-BARK: *Arctostaphylos tomentosa*.
  STANFORD: *Arctostaphylos stanfordiana*.
  STATE HYBRID: *Arctostaphylos* × *media*.
  VINE HILL: *Arctostaphylos densiflora*.
  WHITE-LEAF: *Arctostaphylos viscida*.
  WOOLLY: *Arctostaphylos tomentosa*.
MANZANOTE: *Olmediella betschlerana*.
MAPLE, AMUR: *Acer ginnala*.
  ASH-LEAVED: *Acer negundo*.
  BIG-LEAF: *Acer macrophyllum*.
  BLACK: *Acer saccharum* subsp. *nigrum*.
  BLOODLEAF: *Acer palmatum* 'Atropurpureum'.
  COLISEUM: *Acer cappadocicum*.
  CRIMSON KING: *Acer platanoides* 'Crimson King'.
  CUT-LEAF GREEN: *Acer palmatum* 'Dissectum'.
  DAVID: *Acer davidii*.
  DWARF JAPANESE: *Acer palmatum*.
  ENGLISH: *Acer campestre*.
  EVERGREEN: *Acer oblongum*.
  EVERGREEN: *Acer paxii*.
  FAASSEN REDLEAF: *Acer platanoides* 'Faassen's Black'.
  FERN-LEAF FULL-MOON: *Acer japonicum* 'Aconitifolium'.
  FLOWERING: *Abutilon hybridum*.
  FLOWERING: *Abutilon megapotamicum*.
  FLOWERING: *Abutilon* 'Red Monarch'.
  FULL-MOON: *Acer japonicum*.
  GINNALA: *Acer ginnala*.
  GLOBE: *Acer platanoides* 'Globosum'.
  GLOBE SUGAR: *Acer saccharum* 'Globosum'.
  GOLDEN FULL-MOON: *Acer japonicum* 'Aureum'.
  HARD: *Acer saccharum*.
  HAWTHORN: *Acer crataegifolium*.
  HEDGE: *Acer campestre*.
  JAPANESE: *Acer japonicum*.
  JAPANESE: *Acer palmatum*.
  LACE-LEAF JAPANESE: *Acer palmatum* 'Dissectum'.
  MONTPELIER: *Acer monspessulanum*.
  MOUNTAIN: *Acer spicatum*.
  NIKKO: *Acer maximowiczianum*.
  NORWAY: *Acer platanoides*.
  OBLONG: *Acer oblongum*.
  OREGON: *Acer macrophyllum*.
  PAPERBARK: *Acer griseum*.
  PENNSYLVANIA: *Acer pensylvanicum*.
  PURPLEBLOW: *Acer truncatum*.
  PYRAMIDAL NORWAY: *Acer platanoides* 'Columnare'.
  RED: *Acer rubrum*.
  RED COLISEUM: *Acer cappadocicum* 'Rubrum'.
  RED JAPANESE: *Acer palmatum* 'Atropurpureum'.

RED JAPANESE: **Acer palmatum 'Rubrum'**.
RED LACE-LEAF JAPANESE: **Acer palmatum 'Ornatum'**.
RED-VEIN: **Acer rufinerve**.
RIBBONLEAF PURPLE: **Acer palmatum 'Atropurpureum'**.
ROCK: **Acer saccharum**.
ROCKY MOUNTAIN: **Acer glabrum**.
SCARLET: **Acer rubrum**.
SCHWEDLER NORWAY: **Acer platanoides 'Schwedleri'**.
SCHWEDLER PURPLE: **Acer platanoides 'Schwedleri'**.
SHANTUNG: **Acer truncatum**.
SILVER: **Acer saccharinum**.
SILVER VARIEGATED NORWAY: **Acer platanoides 'Drummondii'**.
SOFT: **Acer saccharinum**.
SPAETH: **Acer pseudo-platanus 'Atropurpureum'**.
STRIPED: **Acer pensylvanicum**.
SUGAR: **Acer saccharum**.
SUMMERSHADE NORWAY: **Acer platanoides 'Summershade'**.
SYCAMORE: **Acer pseudo-platanus**.
SYRIAN: **Acer syriacum**.
TATARIAN: **Acer tataricum**.
THREAD-LEAF: **Acer palmatum 'Dissectum'**.
TRIDENT: **Acer buergeranum**.
VARIEGATED FLOWERING: **Abutilon pictum 'Thompsonii'**.
VARIEGATED NORWAY: **Acer platanoides 'Variegatum'**.
VINE: **Acer circinatum**.
WHITE: **Acer saccharinum**.
WIER: **Acer saccharinum 'Wieri'**.
WIER CUTLEAF: **Acer saccharinum 'Wieri'**.
Marcetella moquiniana (Webb & Berth.) Sventenius.
MARGUERITE: **Chrysanthemum frutescens**.
  BLUE: **Felicia amelloides**.
Marianthus ringens (J. Drummond & Harv.) F. J. Muell.
Markhamia hildebrandtii (Baker) Sprague.
MARLOCK, FORREST'S: **Eucalyptus forrestiana**.
  LONG-FLOWERED: **Eucalyptus macrandra**.
  WHITE-LEAVED: **Eucalyptus tetragona**.
MARMALADE BUSH: **Streptosolen jamesonii**.
MARRI: **Eucalyptus calophylla**.
Mascarena verschaffeltii (H. Wendl.) L. H. Bailey.
  See **Hyophorbe verschaffeltii**.
MASTIC TREE: **Pistacia lentiscus**.
  CHIOS: **Pistacia lentiscus**.
  MT. ATLAS: **Pistacia atlantica**.
  PERUVIAN: **Schinus molle**.
MATTRESS-VINE: **Muehlenbeckia complexa**.
MAY-POP: **Passiflora incarnata**.
MAYTEN: **Maytenus boaria**.
Maytenus boaria Molina. MAYTEN.
Maytenus phyllanthoides Benth.
MAZZARD: **Prunus avium**.
MEDALLION TREE, GOLD: **Cassia leptophylla**.
Medicago arborea L. TREE MEDICK. TREE ALFALFA.
MEDAKE: **Arundinaria simonii**.
MEDICK, TREE: **Medicago arborea**.
MEDLAR: **Mespilus germanica**.
MELALEUCA, DOTTED: **Melaleuca hypericifolia**.
  DROOPING: **Melaleuca armillaris**.
  HEATH: **Melaleuca ericifolia**.
  LILAC: **Melaleuca decussata**.
  PINK: **Melaleuca nesophila**.
Melaleuca acuminata F. J. Muell.
Melaleuca armillaris (Soland. ex Gaertn.) Sm. DROOPING MELALEUCA. BRACELET BOTTLEBRUSH.
Melaleuca decora (Salisb.) Britten. Syn.: *M. genistifolia*.
Melaleuca decussata R. Br. ex Ait. f. LILAC MELALEUCA.
Melaleuca elliptica Labill.

Melaleuca ericifolia Sm. HEATH MELALEUCA. SWAMP PAPERBARK.
Melaleuca erubescens (Benth.) Otto.
Melaleuca fulgens R. Br.
Melaleuca genistifolia Sm. See **M. decora**.
Melaleuca glaberrima F. J. Muell.
Melaleuca huegelii Endl. HONEY MYRTLE.
Melaleuca hypericifolia Sm. DOTTED MELALEUCA.
Melaleuca lanceolata Otto. Syn.: *M. pubescens*.
Melaleuca lateritia Otto. ROBIN-REDBREAST BUSH.
Melaleuca leucadendron (L.) L. Not cult. in U.S.A. Plants so called are **M. quinquenervia**.
Melaleuca linariifolia Sm. FLAX-LEAF PAPERBARK.
Melaleuca longicoma Benth. See **M. macronycha**.
Melaleuca macronycha Turcz. Syn.: *M. longicoma*.
Melaleuca nesophila F. J. Muell. ROSE BOTTLEBRUSH. PINK MELALEUCA. WESTERN TEA MYRTLE.
Melaleuca nodosa Sm.
Melaleuca preissiana Schauer.
Melaleuca pubescens Schauer. See **M. lanceolata**.
Melaleuca quinquenervia (Cav.) S. T. Blake. CAJEPUT TREE. PAPERBARK TREE. PUNK TREE. SWAMP TEA TREE. Plants called **M. leucadendron** in U.S.A. belong here.
Melaleuca radula Lindl.
Melaleuca squarrosa Sm.
Melaleuca steedmanii C. A. Gardner.
Melaleuca styphelioides Sm. PRICKLY PAPERBARK.
Melaleuca thymifolia Sm.
Melaleuca wilsonii F. J. Muell.
Melastoma candidum D. Don.
Melastoma malabathricum L.
Melia azedarach L. CHINABERRY. BEAD TREE. PRIDE-OF-INDIA. PRIDE-OF-CHINA.
Melia azedarach 'Umbraculifera'. TEXAS UMBRELLA TREE.
Melia azedarach 'Umbraculiformis'. See **M. azedarach 'Umbraculifera'**.
Melianthus dregeana Sond.
Melianthus major L. HONEYBUSH.
Melianthus minor L.
Meratia praecox (L.) Rehd. & Wils. See **Chimonanthus praecox**.
Meryta sinclairii (Hook. f.) Seem. PUKA.
Mespilus germanica L. MEDLAR.
MESQUITE: **Prosopis glandulosa**.
MESQUITE, FALSE: **Calliandra eriophylla**.
  SCREW BEAN: **Prosopis pubescens**.
  WESTERN HONEY: **Prosopis glandulosa** var. **torreyana**.
MESSMATE, QUEENSLAND: **Eucalyptus cloeziana**.
  YELLOW: **Eucalyptus cloeziana**.
METAKE: **Sasa japonica**.
Metasequoia glyptostroboides Hu & Cheng. DAWN REDWOOD.
Metrosideros 'Aureus'. See **M. excelsus 'Aureus'**.
Metrosideros excelsus Soland. ex Gaertn. Syn.: *M. tomentosus*. NEW ZEALAND CHRISTMAS TREE. NEW ZEALAND IRON TREE.
Metrosideros excelsus 'Aureus'.
Metrosideros floridus Sm. See **M. fulgens**.
Metrosideros fulgens Soland. ex Gaertn. Syn.: *M. floridus*. *M. scandens*.
Metrosideros kermadecensis W. Oliver. Syn.: *M. villosus*. CRIMSON IRON TREE.
Metrosideros lucidus (G. Forst.) A. Rich. See **M. umbellatus**.
Metrosideros robustus A. Cunn. NORTH ISLAND RATA. NEW ZEALAND CHRISTMAS TREE.
Metrosideros scandens (J. R. Forst. & G. Forst.) Druce. See **M. fulgens**.
Metrosideros tomentosus A. Rich. See **M. excelsus**.

Metrosideros umbellatus Cav. Syn.: *M. lucidus*. SOUTHERN RATA.
Metrosideros villosus T. Kirk. See **M. kermadecensis**.
Michelia compressa (Maxim.) Sarg.
Michelia doltsopa Buch.-Ham. ex DC.
Michelia figo (Lour.) K. Spreng. Syn.: *M. fuscata*. BANANA SHRUB.
Michelia fuscata (Andr.) Blume. See **M. figo**.
MICKEY MOUSE PLANT: **Ochna serrulata**.
Microcitrus australasica (F. J. Muell.) Swingle. AUSTRALIAN WILD LIME. AUSTRALIAN FINGER LIME.
Microcoelum weddellianum (H. Wendl.) H. E. Moore. Syn.: *Syagrus weddelliana*. WEDDELL PALM.
Micromeria chamissonis (Benth.) Greene. See **Satureja douglasii**.
MIGNONETTE VINE: **Anredera cordifolia**.
MILK BUSH, AFRICAN: **Synadenium grantii**.
MILK TREE, AFRICAN: **Euphorbia trigona**.
Millettia reticulata Benth. Plants called **Lonchocarpus nicou** belong here.
MIMOSA: **Albizia julibrissin**.
    FLORIST'S: **Acacia dealbata**.
Mimosa pudica L. SENSITIVE PLANT.
Mimulus aurantiacus Curtis. See **Diplacus aurantiacus**.
Mimulus bifidus Penn. See **Diplacus grandiflorus**.
Mimulus flemingii Munz. See **Diplacus parviflorus**.
Mimulus longiflorus (Nutt.) A. L. Grant. See **Diplacus longiflorus**.
Mimulus longiflorus subsp. *calycinus* (Eastw.) Munz. See **Diplacus longiflorus** var. **calycinus**.
Mimulus longiflorus var. *rutilus* A. L. Grant. See **Diplacus longiflorus** var. **rutilus**.
Mimulus puniceus (Nutt.) Steud. See **Diplacus puniceus**.
MINTBUSH, ROUND-LEAVED: **Prostanthera rotundifolia**.
    SNOWY: **Prostanthera nivea**.
MIRROR-PLANT: **Coprosma repens**.
MIRROR-SHRUB: **Coprosma repens**.
Miscanthus sinensis Anderss. EULALIA.
Miscanthus sinensis 'Zebrina'. ZEBRA GRASS.
Mitchella repens L. PARTRIDGEBERRY. TWINBERRY.
Mitriostigma axillare Hochst. Syn.: *Gardenia citriodora*.
Modiola lateritia (Hook.) K. Schum. See **Modiolastrum lateritium**.
Modiolastrum lateritium (Hook.) Krapovickas. Syn.: *Modiola lateritia*.
MOITCH: **Eucalyptus rudis**.
MONKEY FLOWER: **Diplacus longiflorus**.
    BUSH: **Diplacus aurantiacus**.
    ISLAND: **Diplacus parviflorus**.
    RED: **Diplacus longiflorus** var. **rutilus**.
    PLUMAS: **Diplacus grandiflorus**.
    SOUTHERN: **Diplacus longiflorus**.
    WOOLLY: **Diplacus longiflorus** var. **calycinus**.
MONKEY-HAND TREE: **Chiranthodendron pentadactylon**.
MONKEY-PUZZLE TREE: **Araucaria araucana**.
Monochaetum humboldtianum (Kunth & Bouché) ex Walp. MOUNTAIN MEADOW BEAUTY.
Monstera deliciosa Liebm. Syn.: *Philodendron pertusum*. CERIMAN. SPLIT-LEAF PHILODENDRON.
Montanoa arborescens (DC.) Schultz-Bip. GIANT TREE DAISY.
Montanoa bipinnatifida (Kunth) K. Koch.
Montanoa grandiflora (DC.) Schultz-Bip. GIANT TREE DAISY.
Montanoa hibiscifolia (Benth.) K. Koch.
MOORT, RED-FLOWERED: **Eucalyptus nutans**.
    ROUND-LEAVED: **Eucalyptus platypus**.
MOOSEWOOD: **Acer pensylvanicum**.

Moraea bicolor (Sweet) Spae. See **Dietes bicolor**.
Moraea iridioides L. See **Dietes iridioides**.
Moraea polystachya (Thunb.) Ker-Gawl.
MORNING-GLORY: **Ipomoea fistulosa**.
    BUSH: **Convolvulus cneorum**.
    GROUND: **Convolvulus sabatius**.
MORNING-NOON-AND-NIGHT: **Brunfelsia australis**.
MORNING-NOON-AND-NIGHT: **Brunfelsia pauciflora** 'Eximia'.
Morus alba L. WHITE MULBERRY. SILKWORM MULBERRY.
Morus alba 'Chaparral'.
Morus alba 'Kingan'. KINGAN FRUITLESS MULBERRY.
Morus alba 'Pendula'. WEEPING MULBERRY.
Morus alba 'Stribling'. STRIBLING FRUITLESS MULBERRY.
Morus alba var. *tatarica* (Pall.) Ser. RUSSIAN MULBERRY.
Morus 'Kingan'. See **M. alba** 'Kingan'.
Morus nigra L. PERSIAN MULBERRY. BLACK MULBERRY.
Morus persica. Listed name, probably for **M. nigra**.
Morus rubra L. RED MULBERRY. AMERICAN MULBERRY.
MOTHER-IN-LAW PLANT: **Dieffenbachia seguine**.
MOTHER-IN-LAW'S TONGUE: **Sansevieria trifasciata**.
MOTHER-OF-THYME: **Thymus praecox** subsp. *arcticus*.
MOUNTAIN-LOVER: **Paxistima canbyi**.
MOUNTAIN MEADOW BEAUTY: **Monochaetum humboldtianum**.
MOUNTAIN-PRIDE: **Penstemon newberryi**.
Muehlenbeckia axillaris (Hook. f.) Walp. Syn.: *M. nana*. CREEPING WIRE VINE.
Muehlenbeckia complexa (A. Cunn.) Meissn. MATTRESS VINE. WIRE VINE.
Muehlenbeckia nana Thurst. See **M. axillaris**.
Muehlenbeckia platyclada (F. J. Muell.) Meissn. See **Homalocladium platycladum**.
MUGWORT, WESTERN: **Artemisia ludoviciana**.
    WHITE: **Artemisia lactiflora**.
MULBERRY, AMERICAN: **Morus rubra**.
    BLACK: **Morus nigra**.
    KINGAN FRUITLESS: **Morus alba** 'Kingan'.
    PAPER: **Broussonetia papyrifera**.
    PERSIAN: **Morus nigra**.
    RED: **Morus rubra**.
    RUSSIAN: **Morus alba** var. *tatarica*.
    SILKWORM: **Morus alba**.
    STRIBLING FRUITLESS: **Morus alba** 'Stribling'.
    WEEPING: **Morus alba** 'Pendula'.
    WHITE: **Morus alba**.
MULE-FAT: **Baccharis viminea**.
MULGA: **Acacia aneura**.
Murraya exotica L. See **M. paniculata**.
Murraya paniculata (L.) Jack. Syn.: *M. exotica*. COSMETIC-BARK TREE. ORANGE JESSAMINE.
Musa acuminata Colla. Syn.: *M. cavendishii*. *M. chinensis*. *M. zebrina*. PLANTAIN. EDIBLE BANANA. Some plants called *M. sumatrana* may belong here.
Musa acuminata subsp. *banksii* (F. J. Muell.) N. W. Simmonds. Syn.: *M. paradisiaca* subsp. *seminifera*.
Musa acuminata 'Dwarf Cavendish'. CHINESE DWARF BANANA. DWARF BANANA. LADYFINGER BANANA. CAVENDISH BANANA. Some plants called *M. cavendishii* and **M. nana** belong here.
Musa acuminata subsp. *microcarpa* (Becc.) N. W. Simmonds. Syn.: *M. microcarpa*.
Musa alba?
Musa balbisiana Colla.
Musa cavendishii Lamb. ex Paxt. See **M. acuminata**. Some plants called *M. cavendishii* are **M. acuminata** 'Dwarf Cavendish'.
Musa chinensis Sweet. See **M. acuminata**.
Musa coccinea Andr.

Musa ensete J. F. Gmelin. See **Ensete ventricosum**.
Musa mannii H. Wendl.
Musa martinii Van Geert. A confused name. Plants so called are sometimes **M. balbisiana**.
Musa maurelii. Listed name for an undetermined **Ensete**.
Musa microcarpa Becc. See **M. acuminata** subsp. **microcarpa**.
Musa nana Lour. DWARF BANANA. Some plants so called are **M. acuminata 'Dwarf Cavendish'**.
Musa ornata Roxb. INDIA BANANA. FLOWERING BANANA.
Musa × *paradisiaca* L. [**M. acuminata** × **M. balbisiana**]. Syn.: *M.* × *sapientum*. EDIBLE BANANA. PLANTAIN.
Musa × *paradisiaca* subsp. *seminifera* (Lour.) Baker. See **M. acuminata** subsp. **banksii**.
Musa sanguinea Hook. f.
Musa × *sapientum* L. See **M.** × **paradisiaca**.
Musa sumatrana Becc. A confused name. Plants so called may be **M. acuminata**.
Musa velutina H. Wendl. & Drude.
Musa 'Velutina Hybrid'?
Musa zebrina Van Houtte ex Planch. See **M. acuminata**.
MYALL, WEEPING: **Acacia pendula**.
Myoporum carsonii. Listed name for **M. laetum 'Carsonii'**.
Myoporum carsonii compactum. Listed name, probably for **M. laetum 'Carsonii'**.
Myoporum compactum. Listed name, probably for **M. laetum 'Carsonii'**.
Myoporum debile R. Br.
Myoporum insulare R. Br. BOOBYALLA.
Myoporum laetum G. Forst.
Myoporum laetum 'Carsonii'. Some plants called *M. laetum* 'Compactum' may belong here.
Myoporum laetum 'Compactum'. A confused name. Plants so called may be **M. laetum 'Carsonii'**.
Myoporum parvifolium R. Br.
Myoporum parvifolium 'Horshum'. Listed name, perhaps not distinct from **M. parvifolium**.
Myoporum parvifolium 'Prostratum'. Listed name for **M. parvifolium**.
Myoporum platycarpum R. Br.
Myoporum sandwicense (A. DC.) Gray. BASTARD SANDALWOOD.
Myoporum sinclairii?
Myrceugenella apiculata (DC.) Kausel. See **Luma apiculata**.
Myrceugenella chequen (Molina) Kausel. See **Luma chequen**.
Myrceugenia apiculata (DC.) Niedenzu. See **Luma apiculata**.
Myrciaria edulis (Vell.) Skeels. Syn.: *Eugenia edulis*. Some plants called **Myrciaria edulis** are **Eugenia aggregata**. CHERRY-OF-THE-RIO GRANDE.
Myrica azoricum. Listed name, perhaps for **M. faya**.
Myrica californica Cham. & Schlechtend. WAX MYRTLE. PACIFIC WAX MYRTLE.
Myrica caroliniensis Hort. See **M. pensylvanica**.
Myrica cerifera L. WAX MYRTLE. WAXBERRY. CANDLEBERRY.
Myrica faya Ait. CANDLEBERRY MYRTLE.
Myrica pensylvanica Loisel. Syn.: *M. caroliniensis*. BAYBERRY.
Myroxylon senticosum (Hance) Warb. See **Xylosma congestum**.
Myrrhinium atropurpureum Schott. Syn.: *M. rubriflorum*. *M. salicinum*.
Myrrhinium rubriflorum (Camb.) Berg. See **M. atropurpureum**.
Myrrhinium salicinum Gand. See **M. atropurpureum**.
Myrsine africana L. AFRICAN BOXWOOD.
Myrsine nummularia Hook. f.
MYRTLE: **Myrtus communis**.
MYRTLE: **Vinca major**.
MYRTLE: **Vinca minor**.
AUSTRALIAN WILLOW: **Agonis flexuosa**.
BOXLEAF: **Myrtus communis 'Buxifolia'**.
BOX SAND: **Leiophyllum buxifolium**.
CANDLEBERRY: **Myrica faya**.
COMPACT: **Myrtus communis 'Compacta'**.
COMPACT VARIEGATED: **Myrtus communis 'Compacta Variegata'**.
CRAPE: **Lagerstroemia hirsuta**.
CRAPE: **Lagerstroemia indica**.
GUM: **Angophora costata**.
GUM: **Angophora floribunda**.
HONEY: **Melaleuca huegelii**.
JUNIPER WILLOW: **Agonis juniperina**.
KELLOGG'S PURPLE CRAPE: **Lagerstroemia indica 'Kellogg's Purple'**.
OREGON: **Umbellularia californica**.
PACIFIC WAX: **Myrica californica**.
PINK CRAPE: **Lagerstroemia indica 'Rosea'**.
PURPLE CRAPE: **Lagerstroemia indica 'Purpurea'**.
QUEEN'S CRAPE: **Lagerstroemia speciosa**.
RED: **Lophomyrtus obcordata**.
RED CRAPE: **Lagerstroemia indica 'Rubra'**.
ROSEMARY: **Myrtus communis 'Microphylla'**.
SMALL-LEAF: **Myrtus communis 'Microphylla'**.
TARENTINE: **Myrtus communis** subsp. **tarentina**.
VARIEGATED: **Myrtus communis 'Variegata'**.
VARIEGATED SMALL-LEAF: **Myrtus communis 'Microphylla Variegata'**.
WAX: **Myrica californica**.
WAX: **Myrica cerifera**.
WESTERN TEA: **Melaleuca nesophila**.
WHITE CRAPE: **Lagerstroemia indica 'Alba'**.
WHITE CRAPE: **Lagerstroemia indica 'Glendora White'**.
Myrtus communis L. MYRTLE.
Myrtus communis 'Boetica'.
Myrtus communis 'Buxifolia'. BOXLEAF MYRTLE.
Myrtus communis 'Compacta'. COMPACT MYRTLE.
Myrtus communis 'Compacta Variegata'. COMPACT VARIEGATED MYRTLE.
Myrtus communis 'Microphylla'. ROSEMARY MYRTLE. SMALL-LEAF MYRTLE.
Myrtus communis 'Microphylla Variegata'. VARIEGATED SMALL-LEAF MYRTLE.
Myrtus communis 'Minima'.
Myrtus communis subsp. **tarentina** (L.) Arcang. TARENTINE MYRTLE.
Myrtus communis 'Variegata'. VARIEGATED MYRTLE.
Myrtus lechlerana (Miq.) Sealy. See **Amomyrtus luma**.
Myrtus 'Microphylla'. See **M. communis 'Microphylla'**.
Myrtus 'Microphylla Variegata'. See **M. communis 'Microphylla Variegata'**.
Myrtus molinae Barnéoud. See **Ugni molinae**.
Myrtus nummularia Poir.
Myrtus obcordata (Raoul) Hook. f. See **Lophomyrtus obcordata**.
Myrtus ugni Molina. See **Ugni molinae**.

N

Nandina domestica Thunb. SACRED BAMBOO. HEAVENLY BAMBOO.
Nandina domestica 'Compacta'.
Nandina domestica 'Compacta Nana'. See **N. domestica 'Purpurea'**.
Nandina domestica 'Moyers Red'.

Nandina domestica 'Nana Purpurea'. See **N. domestica 'Purpurea'**.
Nandina domestica **'Purpurea'**.
Nandina domestica **'Royal Princess'**.
Nandina domestica sinensis. Listed name for **N. domestica 'Royal Princess'**.
Nandina sinensis. Listed name for **N. domestica 'Royal Princess'**.
Nannorrhops ritchiana (Griffith) Aitchison. MAZARI PALM.
  NANNY-BERRY: **Viburnum prunifolium**.
    RUSTY: **Viburnum rufidulum**.
  NARANJILLA: **Solanum quitoense**.
  NASHI: **Pyrus pyrifolia** var. **culta**.
Neanthe bella. Listed name for **Chamaedorea elegans**.
  NECTARINE: **Prunus persica** var. **nucipersica**.
    AZALEA-FLOWERED: **Prunus persica 'Alma Stultz'**.
  NEEDLE BUSH: **Hakea suaveolens**.
Neolitsea sericea (Blume) G. Koidz. Syn.: *Litsea glauca*.
Neopanax arboreus (J. Murr.) Allan. Syn.: *Nothopanax arboreus*.
Neopanax laetus (T. Kirk) Allan. Syn.: *Nothopanax laetus*.
Nerium oleander L. OLEANDER.
Nerium oleander **'Betty'**.
Nerium oleander **'Blood Red'**.
Nerium oleander **'Cardinal'**.
Nerium oleander **'Cherry Red'**.
Nerium oleander **'Cherry Ripe'**.
Nerium oleander **'Compte Barthelemy'**.
Nerium oleander **'Dawn'**.
Nerium oleander **'Doctor Golfin'**.
Nerium oleander 'Double Salmon'. See **N. oleander 'Mrs. George Roeding'**.
Nerium oleander **'Ethel'**.
Nerium oleander **'Hawaii'**.
Nerium oleander **'Isle of Capri'**.
Nerium oleander **'Jannoch Red'**.
Nerium oleander **'Madonna'**.
Nerium oleander **'Madonna Grandiflora'**.
Nerium oleander **'Mme. Jannoch'**.
Nerium oleander **'Mrs. George Roeding'**.
Nerium oleander **'Mrs. Swanson'**.
Nerium oleander 'Namken Yellow'?
Nerium oleander **'Palm Springs'**.
Nerium oleander **'Petite Pink'**.
Nerium oleander **'Petite Salmon'**.
Nerium oleander **'Pink Beauty'**.
Nerium oleander **'Pink Berry'**.
Nerium oleander **'Pomona Dwarf'**.
Nerium oleander **'Professor Bodkin'**.
Nerium oleander **'Rosea'**.
Nerium oleander **'Rubra'**.
Nerium oleander **'Ruby Glow'**.
Nerium oleander **'Sarah Bernhardt'**.
Nerium oleander **'Sealy Pink'**.
Nerium oleander **'Sister Agnes'**.
Nerium oleander **'Sorrento'**.
Nerium oleander **'Variegatum'**.
Nerium oleander **'Willow Street'**.
  NETBUSH, GILES: **Calothamnus gilesii**.
    HAIRY: **Calothamnus villosus**.
Nicotiana glauca R. Graham. TREE TOBACCO.
Nierembergia frutescens Durieu. See **N. scoparia**.
Nierembergia repens Ruiz & Pav. Syn.: *N. rivularis*. WHITE-CUP.
Nierembergia rivularis Miers. See **N. repens**.
Nierembergia scoparia Sendtn. Syn.: *N. frutescens*. CUPFLOWER.

NIGHTSHADE, CATALINA: **Solanum wallacei**.
  COSTA RICAN: **Solanum wendlandii**.
  JASMINE: **Solanum jasminoides**.
NIHON NASHI: **Pyrus 'Okusankichi'**.
NINEBARK: **Physocarpus capitatus**.
NOGAL: **Juglans major**.
NOLINA, WOLF'S: **Nolina parryi** subsp. **wolfii**.
Nolina longifolia (Schult. & Schult. f.) Hemsl. Syn.: *Dasylirion longifolium*. GRASS PALM.
Nolina parryi S. Wats.
Nolina parryi subsp. **wolfii** Munz. WOLF'S NOLINA.
Nothofagus moorei (F. J. Muell.) Krasser. SOUTHERN HEMISPHERE BEECH.
Nothopanax arboreus (G. Forst.) Seem. See **Neopanax arboreus**.
Nothopanax guilfoylei (Bull) Merrill. See **Polyscias guilfoylei**.
Nothopanax laetus (T. Kirk) Cheesem. See **Neopanax laetus**.
NUTGALL TREE: **Rhus chinensis**.
NUTMEG, CALIFORNIA: **Torreya californica**.
Nymania capensis (Thunb.) S. Lindb.
Nyssa aquatica L. COTTON GUM. TUPELO GUM.
Nyssa sinensis D. Oliver.
Nyssa sylvatica Marsh. TUPELO. SOUR GUM. BLACK GUM. PEPPERIDGE.

OAK, AMERICAN WHITE: **Quercus alba**.
  BLUE: **Quercus douglasii**.
  BURR: **Quercus macrocarpa**.
  CALIFORNIA BLACK: **Quercus kelloggii**.
  CALIFORNIA LIVE: **Quercus agrifolia**.
  CALIFORNIA SCRUB: **Quercus dumosa**.
  CALIFORNIA SCRUB: **Quercus turbinella** subsp. **californica**.
  CALIFORNIA WHITE: **Quercus lobata**.
  CANYON: **Quercus chrysolepis**.
  CHESTNUT-LEAVED: **Quercus acutissima**.
  COAST LIVE: **Quercus agrifolia**.
  CORK: **Quercus suber**.
  DROOPING SHE: **Casuarina stricta**.
  DWARF INTERIOR LIVE: **Quercus wislizeni** var. **frutescens**.
  EASTERN BLACK: **Quercus velutina**.
  EASTERN RED: **Quercus rubra**.
  ENGELMANN: **Quercus engelmannii**.
  ENGLISH: **Quercus robur**.
  GARRY: **Quercus garryana**.
  GOLDEN CUP: **Quercus chrysolepis**.
  HOLLY: **Quercus ilex**.
  HOLM: **Quercus ilex**.
  INTERIOR LIVE: **Quercus wislizeni**.
  KELLOGG: **Quercus kelloggii**.
  LARGE-LEAVED: **Quercus macrophylla**.
  LEATHER: **Quercus durata**.
  MAUL: **Quercus chrysolepis**.
  MESA: **Quercus engelmannii**.
  MOSSY-CUP: **Quercus macrocarpa**.
  MOUNTAIN SHE: **Casuarina stricta**.
  OREGON WHITE: **Quercus garryana**.
  OVERCUP: **Quercus lyrata**.
  PIN: **Quercus palustris**.
  POSSUM: **Quercus nigra**.
  PYRAMIDAL ENGLISH: **Quercus robur 'Fastigiata'**.

OAK, continued
- RED: **Quercus rubra.**
- RIVER SHE: **Casuarina cunninghamiana.**
- SAND LIVE: **Quercus virginiana** var. **maritima.**
- SAW-TOOTH: **Quercus acutissima.**
- SCARLET: **Quercus coccinea.**
- SHRUB LIVE: **Quercus turbinella.**
- SILK: **Grevillea robusta.**
- SOUTHERN LIVE: **Quercus virginiana.**
- STUNTED SHE: **Casuarina nana.**
- SWAMP POST: **Quercus lyrata.**
- SWAMP WHITE: **Quercus bicolor.**
- TANBARK: **Lithocarpus densiflora.**
- TRUFFLE: **Quercus ilex.**
- TRUFFLE: **Quercus robur.**
- TURKEY: **Quercus cerris.**
- VALLEY: **Quercus lobata.**
- WATER: **Quercus nigra.**
- WILLOW: **Quercus phellos.**
- YELLOW-BARKED: **Quercus velutina.**

OCEAN-SPRAY: **Holodiscus discolor.**

Ochna atropurpurea DC. Plants cult. under this name are **O. serrulata.**

Ochna multiflora DC. Probably not in cult. Plants so called are mostly **O. serrulata.**

Ochna serrulata (Hochst.) Walp. BIRD'S-EYE BUSH. MICKEY MOUSE PLANT. Some plants called **O. atropurpurea** and **O. multiflora** belong here.

OCOTILLO: **Fouquieria splendens.**

OCTOPUS TREE: **Brassaia actinophylla.**

Odontonema stricta (Nees) Kuntze. Plants called *Justicia coccinea* belong here.

Odontospermum maritimum (L.) Schultz-Bip. See **Asteriscus maritimus.**

Odontospermum sericeum (L. f.) Schultz-Bip. See **Asteriscus sericeus.**

Oemleria cerasiformis (Torr. & Gray ex Hook. & Arn.) Landon. Syn.: *Osmaronia cerasiformis.* OSOBERRY. INDIAN PLUM.

- OLD-MAN: **Artemisia abrotanum.**
- OLD-MAN'S BEARD: **Chionanthus virginicus.**
- OLD-MAN'S BEARD: **Clematis vitalba.**
- OLD-WOMAN: **Artemisia stellerana.**

Olea africana Mill. Syn.: *O. chrysophylla. O. verrucosa.*
Olea chrysophylla Lam. See **O. africana.**
Olea europaea L. OLIVE.
Olea europaea 'Ascolano'.
Olea europaea 'Barouni'.
Olea europaea 'Bonita'.
Olea europaea 'Fruitless'?
Olea europaea 'Manzanillo'.
Olea europaea 'Mission'.
Olea europaea 'San Fernando'.
Olea europaea 'Sevillano'. SPANISH QUEEN OLIVE.
Olea europaea 'Swan Hill'. SWAN HILL FRUITLESS OLIVE.
Olea europaea 'Tetley's Special'.
Olea ferruginea Royle.
Olea verrucosa (Willd.) Link. See **O. africana.**

OLEANDER: **Nerium oleander.**
- YELLOW: **Thevetia peruviana.**

Olearia arborescens (G. Forst.) Cockayne & Laing.
Olearia × *haastii* Hook. f. [**O. avicenniifolia** (Raoul) Hook. f. × **O. moschata** Hook. f.]. DAISYBUSH.
Olearia phlogopappa (Labill.) DC. See **O. stellulata.**
Olearia stellulata (Labill.) DC. Syn.: *O. phlogopappa.*
Olearia traversii (F. J. Muell.) Hook. f.

OLEASTER: **Elaeagnus angustifolia.**
OLIVE: **Olea europaea.**
- AMERICAN: **Osmanthus americanus.**
- DESERT: **Forestiera neomexicana.**
- FALSE: **Cassine orientalis.**
- HOLLY: **Osmanthus heterophyllus.**
- RUSSIAN: **Elaeagnus angustifolia.**
- SPANISH QUEEN: **Olea europaea** 'Sevillano'.
- SPURGE: **Cneorum tricoccon.**
- SWAN HILL FRUITLESS: **Olea europaea** 'Swan Hill'.
- SWEET: **Osmanthus fragrans.**
- WILD: **Halesia carolina.**
- WILD: **Osmanthus americanus.**

Olmediella betschlerana (Goepp.) Loes. COSTA RICAN HOLLY. PUERTO RICAN HOLLY. MANZANOTE. GUATEMALAN HOLLY.

Olneya tesota Gray. DESERT IRONWOOD.

OMBU: **Phytolacca dioica.**

Opsiandra maya O. F. Cook. OPSIANDRA PALM.

Opuntia ficus-indica (L.) Mill. INDIAN FIG. TUNA. SPINELESS CACTUS.

ORANGE, BERGAMOT: **Citrus aurantium** subsp. **bergamia.**
- BITTER: **Citrus aurantium.**
- BOUQUET: **Citrus aurantium** 'Bouquet'.
- DILLER: **Citrus sinensis** 'Diller'.
- DOUBLE MOCK: **Philadelphus** × **virginalis.**
- EVERGREEN MOCK: **Philadelphus mexicanus.**
- FLAME MOCK: **Chorizema varium.**
- FLYING DRAGON TRIFOLIATE: **Poncirus trifoliata** 'Monstrosa'.
- HAMLIN: **Citrus sinensis** 'Hamlin'.
- MANDARIN: **Citrus reticulata.**
- MEXICAN: **Choisya ternata.**
- MOCK: **Choisya ternata.**
- MOCK: **Pittosporum tobira.**
- OSAGE: **Maclura pomifera.**
- OTAHEITE: **Citrus otaitensis.**
- ROBERTSON NAVEL: **Citrus sinensis** 'Robertson'.
- SATSUMA: **Citrus reticulata.**
- SEEDLESS VALENCIA: **Citrus sinensis** 'Seedless Valencia'.
- SEVILLE: **Citrus aurantium.**
- SHAMOUTI: **Citrus sinensis** 'Shamouti'.
- SOUR: **Citrus aurantium.**
- SUMMERNAVEL: **Citrus sinensis** 'Summernavel'.
- SWEET: **Citrus sinensis.**
- SWEET MOCK: **Philadelphus coronarius.**
- TAROCCO: **Citrus sinensis** 'Tarocco'.
- TEMPLE: **Citrus** × **nobilis** 'Temple'.
- TRIFOLIATE: **Poncirus trifoliata.**
- TROVITA: **Citrus sinensis** 'Trovita'.
- VALENCIA: **Citrus sinensis** 'Valencia'.
- WASHINGTON NAVEL: **Citrus sinensis** 'Washington'.
- WILD MOCK: **Philadelphus lewisii.**

ORANGE-RED FLOWER: **Hamelia patens.**
ORCHID BELL: **Erica chamissonis.**
ORCHID TREE: **Bauhinia purpurea.**
ORCHID TREE: **Bauhinia variegata.**
ORCHID TREE: **Lagunaria patersonii.**
- HONG KONG: **Bauhinia blakeana.**
- PURPLE: **Bauhinia variegata.**
- WHITE: **Bauhinia variegata** 'Candida'.

ORCHID VINE: **Stigmaphyllon ciliatum.**

Oreodoxa regia HBK. See **Roystonia regia.**
Oreopanax capitatus (Jacq.) Decne. & Planch.
Oreopanax peltatus Linden ex Regel. Syn.: *O. salvinii.*
Oreopanax salvinii Hemsl. See **O. peltatus.**
Origanum dictamnus L. Syn.: *Amaracus dictamnus.* CRETE DITTANY.

Oroxylum indicum (L.) Vent.
Orphium frutescens (L.) E. H. Mey.
OSIER, PURPLE: **Salix purpurea**.
OSMANTHUS, CHINESE: **Osmanthus armatus**.
   DELAVAY: **Osmanthus delavayi**.
   FORREST'S: **Osmanthus yunnanensis**.
   HOLLY-LEAF: **Osmanthus heterophyllus**.
   SAN JOSE: **Osmanthus × fortunei 'San Jose'**.
   SWEET: **Osmanthus fragrans**.
   YELLOW SWEET: **Osmanthus fragrans** forma **aurantiacus**.
Osmanthus americanus (L.) Gray. AMERICAN OLIVE. WILD OLIVE. DEVILWOOD.
Osmanthus aquifolium Siebold & Zucc. See **O. heterophyllus**.
Osmanthus armatus Diels. CHINESE OSMANTHUS.
Osmanthus aurantiacus (Makino) Nakai. See **O. fragrans** forma **aurantiacus**.
Osmanthus × burkwoodii (Burkw. & Skipw.) P. S. Green. [**O. delavayi × O. decorus**]. Syn.: × *Osmarea burkwoodii*.
Osmanthus decorus (Boiss. & Bal.) Kasapligil. Syn.: *Phillyrea decora*. *P. vilmoriniana*.
Osmanthus delavayi Franch. Syn.: *Siphonosmanthus delavayi*.
   DELAVAY OSMANTHUS.
Osmanthus forrestii Rehd. See **O. yunnanensis**.
Osmanthus × fortunei Carrière. [?**O. fragrans × O. heterophyllus**].
Osmanthus × fortunei 'San Jose'. SAN JOSE OSMANTHUS.
Osmanthus fragrans (Thunb.) Lour. SWEET OSMANTHUS. SWEET OLIVE.
Osmanthus fragrans forma **aurantiacus** (Makino) P. S. Green. Syn.: *O. aurantiacus*. YELLOW SWEET OSMANTHUS.
Osmanthus heterophyllus (G. Don) P. S. Green. Syn.: *O. aquifolium*. *O. heterophyllus* 'Ilicifolius'. *O. ilicifolius*. HOLLY OLIVE. CHINESE HOLLY. FALSE HOLLY. HOLLY-LEAF OSMANTHUS. Plants which have juvenile leaves with spiny margins and are called *O. heterophyllus* 'Ilicifolius', *O. ilicifolius*, and *O. aquifolium* belong here.
Osmanthus heterophyllus 'Aureus'.
Osmanthus heterophyllus 'Gulftide'.
Osmanthus heterophyllus 'Ilicifolius'. See **O. heterophyllus**. Plants called *O. heterophyllus* 'Ilicifolius' are **O. heterophyllus** having juvenile leaves with spiny margins.
Osmanthus heterophyllus 'Ilicifolius Variegatus'. See **O. heterophyllus** 'Variegatus'.
Osmanthus heterophyllus 'Purpurascens'. See **O. heterophyllus** 'Purpureus'.
Osmanthus heterophyllus 'Purpureus'.
Osmanthus heterophyllus 'Rotundifolius'.
Osmanthus heterophyllus 'Variegatus'.
Osmanthus ilicifolius (Hassk.) Carrière. See **O. heterophyllus**.
Osmanthus ilicifolius 'Albo-marginatus'. See **O. heterophyllus** 'Variegatus'.
Osmanthus ilicifolius 'Argenteo-marginatus'. See **O. heterophyllus** 'Variegatus'.
Osmanthus ilicifolius 'Aureo-marginatus'. See **O. heterophyllus** 'Aureus'.
Osmanthus ilicifolius 'Gulftide'. See **O. heterophyllus** 'Gulftide'.
Osmanthus ilicifolius 'Purpurascens'. See **O. heterophyllus** 'Purpureus'.
Osmanthus ilicifolius 'Purpureus'. See **O. heterophyllus** 'Purpureus'.
Osmanthus ilicifolius 'Variegatus'. See **O. heterophyllus** 'Variegatus'.
Osmanthus 'San Jose'. See **O. × fortunei 'San Jose'**.

Osmanthus yunnanensis (Franch.) P. S. Green. Syn.: *O. forrestii*. FORREST'S OSMANTHUS.
× *Osmarea burkwoodii* Burkw. & Skipw. See **Osmanthus × burkwoodii**.
Osmaronia cerasiformis (Torr. & Gray ex Hook. & Arn.) Greene. See **Oemleria cerasiformis**.
OSOBERRY: **Oemleria cerasiformis**.
Osteomeles schweriniae Schneid. CHINESE BONE-BERRY.
Osteospermum 'Burgundy Mound'. See **O. fruticosum** 'Burgundy Mound'.
Osteospermum fruticosum (L.) Norlindh. TRAILING AFRICAN DAISY.
Osteospermum fruticosum 'Burgundy Mound'.
Ostrya carpinifolia Scop. EUROPEAN HOP HORNBEAM.
Ostrya virginiana (Mill.) K. Koch. AMERICAN HOP HORNBEAM.
OTATE: **Yushania aztecorum**.
OUR LORD'S CANDLE: **Yucca whipplei**.
Oxera pulchella Labill. ROYAL CLIMBER.
Oxydendrum arboreum (L.) DC. SOURWOOD. SORREL TREE.
Oxypetalum caeruleum (D. Don) Decne.

P

Pachistima. See **Paxistima**.
Pachypodium densiflorum Baker.
Pachypodium geayi Costantin & Bois.
Pachypodium lamieri Drake.
Pachypodium saundersii N. E. Br.
PACHYSANDRA, JAPANESE: **Pachysandra terminalis**.
   VARIEGATED JAPANESE: **Pachysandra terminalis** 'Variegata'.
Pachysandra terminalis Siebold & Zucc. JAPANESE PACHYSANDRA. JAPANESE SPURGE.
Pachysandra terminalis 'Variegata'. VARIEGATED JAPANESE PACHYSANDRA. VARIEGATED JAPANESE SPURGE.
Pachystachys lutea Nees.
Pachystima. See **Paxistima**.
Paeonia suffruticosa Andr. TREE PEONY.
PAGODA-TREE, JAPANESE: **Sophora japonica**.
   WEEPING JAPANESE: **Sophora japonica** 'Pendula'.
PALM, ALEXANDER: **Ptychosperma elegans**.
   ALEXANDRA: **Archontophoenix alexandrae**.
   ASSAI: **Euterpe edulis**.
   AUSTRALIAN FOUNTAIN: **Livistona australis**.
   BAMBOO: **Chamaedorea erumpens**.
   BELMORE SENTRY: **Howea belmoreana**.
   BIG-FRUIT FAN: **Chamaerops humilis** var. **elatior**.
   BLUE FAN: **Brahea armata**.
   BLUE HESPER: **Brahea armata**.
   BOTTLE: **Beaucarnea recurvata**.
   BUTTERFLY: **Chrysalidocarpus lutescens**.
   BROOM: **Coccothrinax argentea**.
   CABADA: **Chrysalidocarpus cabadae**.
   CALIFORNIA FAN: **Washingtonia filifera**.
   CANARY ISLAND DATE: **Phoenix canariensis**.
   CANE: **Chrysalidocarpus lutescens**.
   CANTON FISHTAIL: **Caryota ochlandra**.
   CARNAUBA: **Copernicia prunifera**.
   CAUDATE FEATHER: **Polyandrococos caudescens**.
   CHILEAN: **Jubaea chilensis**.
   CHINESE FAN: **Livistona chinensis**.
   CHINESE FOUNTAIN: **Livistona chinensis**.
   CHINESE WINDMILL: **Trachycarpus fortunei**.

PALM, continued
  CHRISTMAS: **Veitchia merrillii.**
  CLIFF DATE: **Phoenix rupicola.**
  CLUMP: **Phoenix reclinata.**
  CLUSTERED FISHTAIL: **Caryota mitis.**
  COCOS: **Arecastrum romanzoffianum.**
  COSTA RICAN PARLOR: **Chamaedorea costaricana.**
  COYOLI: **Acrocomia mexicana.**
  CUBAN ROYAL: **Roystonea regia.**
  DATE: **Phoenix dactylifera.**
  DRACAENA: **Cordyline australis.**
  DWARF DATE: **Phoenix roebelenii.**
  EAST INDIAN WINE: **Phoenix rupicola.**
  EUROPEAN FAN: **Chamaerops humilis.**
  FEATHER DUSTER: **Rhopalostylis sapida.**
  FERN: **Cycas circinalis.**
  FISHTAIL WINE: **Caryota urens.**
  FLORIDA ROYAL: **Roystonea elata.**
  FORSTER SENTRY: **Howea forsterana.**
  FOUNTAIN: **Livistona chinensis.**
  FRANCESCHI: **Brahea elegans.**
  GRASS: **Nolina longifolia.**
  GRUGRU: **Acrocomia totai.**
  GUADALUPE: **Brahea edulis.**
  HAIR: **Chamaerops humilis.**
  HARDY BLUE COCOS: **Butia capitata.**
  HEMP: **Trachycarpus fortunei.**
  HURRICANE: **Ptychosperma macarthurii.**
  INDIA DATE: **Phoenix rupicola.**
  INDIA DATE: **Phoenix sylvestris.**
  KING: **Archontophoenix alexandrae.**
  KING: **Archontophoenix cunninghamiana.**
  LADY: **Rhapis excelsa.**
  LICUALA FAN: **Licuala grandis.**
  LICURI: **Syagrus coronata.**
  LICURY: **Syagrus coronata.**
  MACARTHUR: **Ptychosperma macarthurii.**
  MACARTHUR FEATHER: **Ptychosperma macarthurii.**
  MANILLA: **Veitchia merrillii.**
  MAZARI: **Nannorrhops ritchiana.**
  MEDITERRANEAN FAN: **Chamaerops humilis.**
  MEXICAN BLUE: **Brahea armata.**
  MEXICAN FAN: **Washingtonia robusta.**
  MOORE SENTRY: **Lepidorrhachis mooreana.**
  NEEDLE: **Rhapidophyllum hystrix.**
  NIKAU: **Rhopalostylis baueri.**
  NIKAU: **Rhopalostylis cheesemanii.**
  NIKAU: **Rhopalostylis sapida.**
  OPSIANDRA: **Opsiandra maya.**
  OVERTOP: **Rhyticocos amara.**
  PARADISE: **Howea forsterana.**
  PARLOR: **Chamaedorea elegans.**
  PINDO: **Butia capitata.**
  PLUME: **Arecastrum romanzoffianum.**
  PYGMY DATE: **Phoenix loureirii.**
  PYGMY DATE: **Phoenix roebelenii.**
  QUEEN: **Arecastrum romanzoffianum.**
  RATTAN: **Rhapis humilis.**
  ROCK: **Brahea dulcis.**
  ROEBELEN: **Phoenix roebelenii.**
  ROEBELIN: **Phoenix roebelenii.**
  SAGO: **Cycas circinalis.**
  SAGO: **Cycas revoluta.**
  SAN JOSE HESPER: **Brahea brandegeei.**
  SAW CABBAGE: **Acoelorrhaphe wrightii.**
  SEAM-BERRY: **Coccothrinax crinita.**
  SENEGAL DATE: **Phoenix reclinata.**
  SENTRY: **Howea belmoreana.**
  SENTRY: **Howea forsterana.**
  SHINING BRUSH: **Rhopalostylis sapida.**
  SILVER: **Coccothrinax argentata.**
  SILVER DATE: **Phoenix sylvestris.**
  SILVER THATCH: **Coccothrinax argentea.**
  SLENDER LADY: **Rhapis humilis.**
  SOLITAIRE: **Ptychosperma elegans.**
  SONORAN: **Sabal uresana.**
  SONORAN FAN: **Washingtonia robusta.**
  SOUTH AMERICAN JELLY: **Butia capitata.**
  SOUTHERN FAN: **Livistona australis.**
  SPINDLE: **Hyophorbe verschaffeltii.**
  SYRUP: **Jubaea chilensis.**
  THATCH: **Coccothrinax crinita.**
  UMBRELLA: **Hedyscepe canterburyana.**
  WALKING-STICK: **Linospadix monostachya.**
  WEDDELL: **Microcoelum weddellianum.**
  WINDMILL: **Trachycarpus fortunei.**
  WINE: **Jubaea chilensis.**
  WOOLLY BUTIA: **Butia eriospatha.**
  YELLOW: **Chrysalidocarpus lutescens.**
PALMA CHRISTI: **Ricinus communis.**
PALMETTO: **Sabal palmetto.**
  BLACKBURN: **Sabal blackburniana.**
  BLUE: **Sabal palmetto.**
  BUSH: **Sabal minor.**
  CABBAGE: **Sabal palmetto.**
  DWARF SAW: **Rhapidophyllum hystrix.**
  HISPANIOLAN: **Sabal blackburniana.**
  MEXICAN: **Sabal mexicana.**
  OAXACANA: **Sabal mexicana.**
  RIO GRANDE: **Sabal mexicana.**
  SAVANNAH: **Sabal minor.**
  SAW: **Serenoa repens.**
  SCRUB: **Serenoa repens.**
  SILVER SAW: **Acoelorrhaphe wrightii.**
  TEXAS: **Sabal mexicana.**
  VICTORIA: **Sabal mexicana.**
PALO VERDE: **Cercidium floridum.**
  BLUE: **Cercidium floridum.**
  LITTLELEAF: **Cercidium microphyllum.**
  MEXICAN: **Parkinsonia aculeata.**
Pandanus utilis Bory. SCREW PINE.
PANDA PLANT: **Philodendron bipennifolium.**
Pandorea brycei (N. E. Br.) Rehd. See **Podranea brycei.**
Pandorea jasminoides (Lindl.) K. Schum. Syn.: *Bignonia jasminoides. Tecoma jasminoides.* BOWER PLANT. BOWER VINE.
Pandorea jasminoides 'Alba'.
Pandorea jasminoides 'Rosea'. PINK TRUMPET VINE.
Pandorea pandorana (Andr.) Steenis. Syn.: *Tecoma australis.* WONGA-WONGA VINE.
Pandorea ricasoliana (Tanfani) Baill. See **Podranea ricasoliana.**
PAPAYA: **Carica papaya.**
  MOUNTAIN: **Carica pubescens.**
PAPERBARK, FLAX-LEAF: **Melaleuca linariifolia.**
  PRICKLY: **Melaleuca styphelioides.**
  SWAMP: **Melaleuca ericifolia.**
PAPERBARK TREE: **Melaleuca quinquenervia.**
PAPER-FLOWER: **Bougainvillea glabra.**
PAPER PLANT: **Cyperus papyrus.**
PAPYRUS: **Cyperus papyrus.**
PARADISE FLOWER: **Solanum wendlandii.**

Parahebe × *bidwillii* (Hook.) W. Oliver. [*P. decora* × *P. lyallii*]. Syn.: *Hebe* × *bidwillii*. *Veronica* × *bidwillii* Hook.
Parahebe canescens (T. Kirk) W. Oliver. Syn.: *Hebe canescens*. *Veronica canescens*.
Parahebe catarractae (G. Forst.) W. Oliver. Syn.: *Hebe catarractae*. *Parahebe diffusa*. *Veronica catarractae*.
Parahebe decora Ashwin. Syn.: *Veronica bidwillii* Hook. f.
Parahebe diffusa (Hook. f.) W. Oliver. See *P. catarractae*.
Parahebe lyallii (Hook. f.) W. Oliver. Syn.: *Hebe lyallii*. *Veronica lyallii*.
Paranomus spicatus (Bergius) Kuntze.
PARA-PARA: **Pisonia umbellifera**.
PARASOL TREE, CHINESE: **Firmiana simplex**.
Parkinsonia aculeata L. JERUSALEM THORN. MEXICAN PALO VERDE. RATAMA.
PARROTIA, PERSIAN: **Parrotia persica**.
Parrotia persica (DC.) C. A. Mey. PERSIAN PARROTIA.
PARROT'S-BEAK: **Clianthus puniceus**.
PARROT'S BEAK: **Lotus berthelotii**.
Parthenocissus 'Engelmannii'. See *P. quinquefolia* 'Engelmannii'.
Parthenocissus henryana (Hemsl.) Graebn. ex Diels & Gilg. Syn.: *Ampelopsis henryana*. *Vitis henryana*.
Parthenocissus quinquefolia (L.) Planch. Syn.: *Ampelopsis quinquefolia*. *A. virginiana*. VIRGINIA CREEPER. AMERICAN IVY. FIVE-LEAVED IVY.
Parthenocissus quinquefolia 'Engelmannii'. Syn.: *Ampelopsis quinquefolia* var. *engelmannii*. ENGELMANN CREEPER.
Parthenocissus tricuspidata (Siebold & Zucc.) Planch. Syn.: *Ampelopsis tricuspidata*. BOSTON IVY. JAPANESE IVY.
Parthenocissus tricuspidata 'Beverly Brooks'.
Parthenocissus tricuspidata 'Lowii'.
Parthenocissus tricuspidata 'Veitchii'.
PARTRIDGEBERRY: **Mitchella repens**.
PARTRIDGE-BREAST: **Aloe variegata**.
Passiflora alata Dryander.
Passiflora × *alatocaerulea* Lindl. [*P. alata* × *P. caerulea*]. Syn.: *P. pfordtii*. PASSION VINE.
Passiflora caerulea L. BLUE PASSION VINE. BLUE-CROWN PASSION FLOWER.
Passiflora coccinea Aublet. RED PASSION FLOWER. RED GRANADILLA.
Passiflora edulis Sims. PASSION FRUIT. PURPLE GRANADILLA.
Passiflora incarnata L. MAY-POP. APRICOT VINE.
Passiflora jamesonii (M. T. Mast.) L. H. Bailey. RED PASSION FLOWER.
Passiflora laurifolia L. YELLOW GRANADILLA.
Passiflora manicata (Juss.) Pers. RED PASSION FLOWER.
Passiflora mixta L. f.
Passiflora mollissima (HBK) L. H. Bailey. PINK PASSION FLOWER. BANANA PASSION FRUIT.
Passiflora pfordtii M. T. Mast. See *P.* × *alatocaerulea*.
Passiflora princeps Lodd. See *P. racemosa*.
Passiflora racemosa Brot. Syn.: *P. princeps*. RED PASSION FLOWER.
PASSION FLOWER, BLUE-CROWN: **Passiflora caerulea**.
  PINK: **Passiflora mollissima**.
  RED: **Passiflora coccinea**.
  RED: **Passiflora jamesonii**.
  RED: **Passiflora manicata**.
  RED: **Passiflora racemosa**.
PASSION FRUIT: **Passiflora edulis**.
  BANANA: **Passiflora mollissima**.
PASSION VINE: **Passiflora** × **alatocaerulea**.
  BLUE: **Passiflora caerulea**.
PATTERN PLANT, GRECIAN: **Acanthus mollis**.
PAULOWNIA, ROYAL: **Paulownia tomentosa**.

Paulownia fargesii Franch.
Paulownia fortunei (Seem.) Hance.
Paulownia imperialis Siebold & Zucc. See *P. tomentosa*.
Paulownia lilacina Sprague.
Paulownia tomentosa (Thunb.) Steud. Syn.: *P. imperialis*. EMPRESS TREE. PRINCESS TREE. ROYAL PAULOWNIA.
Paurotis wrightii (Griseb. & H. Wendl.) Britt. See **Acoelorrhaphe wrightii**.
Pavonia multiflora Juss. See **Triplochlamys multiflora**.
PAWPAW: **Asimina triloba**.
Paxistima canbyi Gray. RAT-STRIPPER. CLIFF-GREEN. MOUNTAIN-LOVER.
Paxistima myrsinites (Pursh) Raf. OREGON BOXWOOD.
PEA, AUSTRALIAN FLAME: **Chorizema cordatum**.
  BUSH FLAME: **Chorizema varium**.
  CHAPARRAL: **Pickeringia montana**.
  DARLING: **Swainsonia galegifolia**.
  DESERT: **Clianthus formosus**.
  GLORY: **Clianthus formosus**.
  GLORY: **Clianthus puniceus**.
  GLORY: **Sesbania punicea**.
  HEART-LEAF FLAME: **Chorizema cordatum**.
  HOLLY FLAME: **Chorizema ilicifolium**.
  WINGED: **Lotus berthelotii**.
  WINTER SWEET: **Swainsonia galegifolia**.
PEACH: **Prunus persica**.
  BURBANK FLOWERING: **Prunus persica 'Burbank'**.
  DESERT: **Prunus andersonii**.
  DWARF FLOWERING: **Prunus persica 'Dwarf Mandarin'**.
  WILD: **Prunus fasciculata**.
PEAR: **Pyrus communis**.
  ALLIGATOR: **Persea americana**.
  BRADFORD: **Pyrus calleryana 'Bradford'**.
  CALLERY: **Pyrus calleryana**.
  EVERGREEN: **Pyrus kawakamii**.
  JAPANESE SAND: **Pyrus pyrifolia**.
  LATE JAPANESE: **Pyrus 'Okusankichi'**.
  OCTOBER JAPANESE: **Pyrus 'Okusankichi'**.
  SAND: **Pyrus pyrifolia**.
PEARLBUSH: **Exochorda racemosa**.
PEA SHRUB, SIBERIAN: **Caragana arborescens**.
  SWAN RIVER: **Brachysema lanceolatum**.
  SWEET: **Polygala** × **dalmaisiana**.
PEA VINE, AUSTRALIAN: **Dolichos lignosus**.
PECAN: **Carya illinoensis**.
PEEPUL: **Ficus religiosa**.
Pelargonium angulosum (Mill.) L'Hér. ex Ait.
Pelargonium capitatum (L.) L'Hér. ex Ait. ROSE-SCENTED GERANIUM.
Pelargonium crispum (L.) L'Hér. ex Ait. LEMON GERANIUM.
Pelargonium crispum 'Prince Rupert'.
Pelargonium cucullatum (L.) L'Hér. ex Ait.
Pelargonium × *domesticum* L. H. Bailey. [Hybrid origin involving *P. cucullatum*, *P. angulosum*, *P. grandiflorum* (Andr.) Willd, and others]. REGAL GERANIUM. FANCY GERANIUM. LADY WASHINGTON GERANIUM. MARTHA WASHINGTON GERANIUM. PANSY-FLOWERED GERANIUM. SHOW GERANIUM.
Pelargonium × *fragrans* Willd. [*P. exstipulatum* (Cav.) L'Hér. ex Ait. × *P. odoratissimum*]. NUTMEG GERANIUM. NUTMEG-SCENTED GERANIUM.
Pelargonium graveolens L'Hér. ex Ait. ROSE GERANIUM. SWEET-SCENTED GERANIUM.
Pelargonium grossularioides (L.) L'Hér. ex Ait. GOOSEBERRY GERANIUM.

Pelargonium × *hortorum* L. H. Bailey. [Of complex hybrid origin, involving **P. inquinans, P. zonale** and others].
  GERANIUM. GARDEN GERANIUM. BEDDING GERANIUM.
Pelargonium inquinans (L.) L'Hér. ex Ait.
Pelargonium × *nervosum* Sweet. [Parentage not known]. LIME GERANIUM.
Pelargonium odoratissimum (L.) L'Hér. ex Ait.
  APPLE GERANIUM. APPLE-SCENTED GERANIUM.
Pelargonium peltatum (L.) L'Hér. ex Ait. IVY GERANIUM.
Pelargonium quercifolium (L. f.) L'Hér. ex Ait. OAK-LEAVED GERANIUM. VILLAGE-OAK GERANIUM.
Pelargonium tomentosum Jacq. HERB-SCENTED GERANIUM. PEPPERMINT GERANIUM.
Pelargonium vitifolium (L.) L'Hér. ex Ait. GRAPE-LEAVED GERANIUM.
Pelargonium zonale (L.) L'Hér. ex Ait. HORSESHOE GERANIUM.
Peltophorum africanum Sond.
PENSTEMON, BRIDGES': **Penstemon bridgesii.**
  CLIMBING: **Penstemon cordifolius.**
  FOOTHILL BLUE: **Penstemon heterophyllus.**
  NEWBERRY: **Penstemon newberryi.**
  SHOWY: **Penstemon spectabilis.**
  YELLOW BUSH: **Penstemon antirrhinoides.**
Penstemon antirrhinoides Benth. YELLOW BUSH PENSTEMON.
Penstemon barbatus (Cav.) Roth.
Penstemon barbatus 'Rose Elf'.
Penstemon barrettiae Gray.
Penstemon breviflorus Lindl.
Penstemon bridgesii Gray. BRIDGES' PENSTEMON.
Penstemon centranthifolius Benth. SCARLET-BUGLER.
Penstemon cordifolius Benth. CLIMBING PENSTEMON.
Penstemon gloxinioides. Listed name for **P. hartwegii** or hybrids between **P. hartwegii** and **P. cobaea** Nutt.
Penstemon hartwegii Benth. Some plants called *P. gloxinioides* belong here.
Penstemon heterophyllus Lindl. FOOTHILL BLUE PENSTEMON.
Penstemon heterophyllus subsp. **purdyi** Keck.
Penstemon laetus Gray.
Penstemon laetus subsp. **roezlii** (Regel) Keck. Syn.: **P. roezlii**.
Penstemon newberryi Gray. MOUNTAIN-PRIDE. NEWBERRY PENSTEMON.
Penstemon newberryi 'Mt. Shasta'.
Penstemon roezlii Regel. See **P. laetus** subsp. **roezlii.**
Penstemon rupicola (Piper) T. J. Howell.
Penstemon spectabilis Thurber ex Gray. SHOWY PENSTEMON.
Pentapterygium 'Ludgvan Cross'. See **Agapetes 'Ludgvan Cross'.**
Pentapterygium rugosum (Hook. f. & Thoms. ex Hook.) Hook. f. See **Agapetes incurvata.**
Pentapterygium serpens (Wight) Klotzsch. See **Agapetes serpens.**
Pentas lanceolata (Forssk.) Deflers. STAR-CLUSTER. EGYPTIAN STAR-CLUSTER.
Pentas lanceolata 'Pioneer Pink'.
PEONY, TREE: **Paeonia suffruticosa.**
PEPPER, JAPAN: **Zanthoxylum piperitum.**
PEPPERBUSH, SWEET: **Clethra alnifolia.**
PEPPERIDGE: **Nyssa sylvatica.**
PEPPERMINT, BLACK: **Eucalyptus amygdalina.**
  NICHOL'S WILLOW-LEAVED: **Eucalyptus nicholii.**
  RIVER: **Eucalyptus elata.**
  SILVER: **Eucalyptus tenuiramis.**
  WHITE: **Eucalyptus pulchella.**
PEPPERMINT TREE: **Agonis flexuosa.**
PEPPER TREE: **Drimys lanceolata.**
  BRAZILIAN: **Schinus terebinthifolius.**
  CALIFORNIA: **Schinus molle.**
PEPPERWOOD: **Umbellularia californica.**
Peraphyllum ramosissimum Nutt. SQUAW APPLE.
PERIWINKLE: **Vinca major.**
PERIWINKLE: **Vinca minor.**
PERNETTYA, NEW ZEALAND: **Pernettya nana.**
Pernettya leucocarpa DC. See **P. pumila** var. **leucocarpa.**
Pernettya mucronata (L. f.) Gaud.-Beaup. ex K. Spreng.
Pernettya nana Colenso. NEW ZEALAND PERNETTYA.
Pernettya pumila (L. f.) Hook.
Pernettya pumila var. **leucocarpa** (DC.) Kausel. Syn.: *P. leucocarpa*.
Persea americana Mill. Syn.: *P. gratissima*. ALLIGATOR PEAR. AVOCADO.
Persea americana var. **drymifolia** (Schlechtend. & Cham.) S. F. Blake. Syn.: *P. drymifolia*. MEXICAN AVOCADO.
Persea borbonia (L.) K. Spreng. RED BAY. SWAMP RED BAY.
Persea gratissima C. F. Gaertner. See **P. americana.**
Persea indica (L.) K. Spreng.
Persea meyeniana Nees.
PERSIMMON, DATE PLUM: **Diospyros lotus.**
  FUYU: **Diospyros kaki** 'Fuyu'.
  HACHIYA: **Diospyros kaki** 'Hachiya'.
  JAPANESE: **Diospyros kaki.**
  VIRGINIA: **Diospyros virginiana.**
Phaedranthus buccinatorius (DC.) Miers. See **Distictis buccinatoria.**
Phaseolus caracalla L. See **Vigna caracalla.**
Phaseolus giganteus. Listed name, perhaps for **Vigna caracalla.**
Phellodendron amurense Rupr. AMUR CORK TREE.
Philadelphus 'Atlas'. See **P.** × **polyanthus 'Atlas'.**
Philadelphus 'Belle Etoile'. See **P.** × **purpureo-maculatus 'Belle Etoile'.**
Philadelphus coronarius L. SWEET MOCK ORANGE.
Philadelphus coronarius 'Aureus'.
Philadelphus coulteri S. Wats.
Philadelphus 'Frosty Morn'.
Philadelphus gordonianus Lindl. See **P. lewisii.**
Philadelphus grandiflorus Willd. See **P. inodorus** var. **grandiflorus.**
Philadelphus 'Innocence'. See **P.** × **lemoinei 'Innocence'.**
Philadelphus inodorus L.
Philadelphus inodorus var. **grandiflorus** (Willd.) Gray. Syn.: *P. grandiflorus*.
Philadelphus karwinskyanus Koehne. Some plants called *P. mexicanus* belong here.
Philadelphus × **lemoinei** Lemoine. [*P. coronarius* × *P. microphyllus*].
Philadelphus × **lemoinei** 'Innocence'.
Philadelphus × **lemoinei** 'Silver Showers'.
Philadelphus lewisii Pursh. Syn.: *P. gordonianus*. WILD MOCK ORANGE.
Philadelphus mexicanus Schlechtend. EVERGREEN SYRINGA. EVERGREEN MOCK ORANGE. Some plants so called may be *P. karwinskyanus*.
Philadelphus microphyllus Gray.
Philadelphus 'Minnesota Snowflake'. See **P.** × **virginalis 'Minnesota Snowflake'.**
Philadelphus × **polyanthus** Rehd. [**P.** × **lemoinei** × *P.* ?*insignis*].
Philadelphus × **polyanthus** 'Atlas'.
Philadelphus × **purpureo-maculatus** Lemoine. [*P. coulteri* × *P.* × *lemoinei*].
Philadelphus × **purpureo-maculatus** 'Belle Etoile'.
Philadelphus 'Silver Showers'. See **P.** × **lemoinei 'Silver Showers'.**

Philadelphus 'Virginal'. See **P. × *virginalis* 'Virginal'**.
Philadelphus × *virginalis* Rehd. [Parentage doubtful].
DOUBLE MOCK ORANGE.
Philadelphus × *virginalis* 'Minnesota Snowflake'.
Philadelphus × *virginalis* 'Virginal'.
Phillyrea angustifolia L.
Phillyrea decora Boiss. & Bal. See ***Osmanthus decorus***.
Phillyrea latifolia L.
Phillyrea latifolia var. ***media*** (L.) Schneid. Syn.: *P. media*.
Phillyrea media L. See ***P. latifolia*** var. ***media***.
Phillyrea vilmoriniana Boiss. & Bal. See ***Osmanthus decorus***.
PHILODENDRON, BLUSHING: **Philodendron erubescens**.
    FIDDLE-LEAF: **Philodendron bipennifolium**.
    HEART-LEAF: **Philodendron cordatum**.
    HEART-LEAF: **Philodendron scandens**.
    HORSEHEAD: **Philodendron bipennifolium**.
    LEATHER-LEAF: **Philodendron guttiferum**.
    RED-LEAF: **Philodendron erubescens**.
    SPADE-LEAF: **Philodendron domesticum**.
    SPLIT-LEAF: **Monstera deliciosa**.
Philodendron andersonii?
Philodendron 'Barryi'. Listed name for hybrids of **P. selloum × P. bipinnatifidum**.
Philodendron bipennifolium Schott. FIDDLE-LEAF PHILODENDRON. PANDA PLANT. HORSEHEAD PHILODENDRON. Some plants called **P. panduriforme** belong here.
Philodendron bipinnatifidum Schott ex Endl.
Philodendron cordatum (Vell.) Kunth. HEART-LEAVED PHILODENDRON. Some plants so called are **P. scandens** subsp. **oxycardium**.
Philodendron domesticum Bunting. SPADE-LEAF PHILODENDRON. Some plants called **P. hastatum** belong here.
Philodendron dubium Chodat & Vischer. Not cult. Plants so called are **P. radiatum**.
Philodendron eichleri Engler.
Philodendron erubescens K. Koch & Augustin. RED-LEAF PHILODENDRON. BLUSHING PHILODENDRON.
***Philodendron* 'Evansii'**. [?**P. selloum × P. speciosum**].
***Philodendron* 'Fantasy'**.
***Philodendron* 'Florida'**. [**P. pedatum × P. squamiferum**].
***Philodendron* 'Florida Compacta'**. [**P. pedatum** var. **palmatisectum × P. squamiferum**].
Philodendron guttiferum Kunth. LEATHER-LEAF PHILODENDRON.
Philodendron hastatum K. Koch & H. Sello. ELEPHANT-EAR. Some plants so called are **P. domesticum**.
Philodendron imbe Schott ex Endl.
***Philodendron* 'Jungle Gardens'**.
***Philodendron* 'Jungle Queen'**.
Philodendron laciniatum (Vell.) Engler. See **P. pedatum**.
Philodendron lundii Warming.
***Philodendron* 'Lynette'**. [?**P. wendlandii × P. elaphoglossoides** Schott].
***Philodendron* 'Mandaianum'**. [?**P. domesticum × P. erubescens**].
Philodendron ornatum Schott. Syn.: *P. sodiroi*.
Philodendron oxycardium Schott. See **P. scandens** subsp. **oxycardium**.
Philodendron panduriforme (HBK) Kunth. May not be cult. Plants so called are **P. bipennifolium**.
Philodendron pedatum (Hook.) Kunth. Syn.: *P. laciniatum*.
Philodendron pedatum var. ***palmatisectum*** (Engler) A. & F. Jonker.
Philodendron pertusum Kunth & Bouché. See **Monstera deliciosa**.
Philodendron radiatum Schott. Some plants called **P. dubium** belong here.
***Philodendron* 'Ruby Queen'**.
***Philodendron* 'São Paulo'**.
Philodendron scandens K. Koch & H. Sello. HEART-LEAF PHILODENDRON.
Philodendron scandens subsp. ***oxycardium*** (Schott) Bunting. Syn.: *P. oxycardium*. PARLOR IVY. Some plants called **P. cordatum** belong here.
Philodendron selloum K. Koch.
Philodendron sodiroi Bellair & St.-Léger. See **P. ornatum**.
Philodendron speciosum Schott ex Endl.
Philodendron squamiferum Poepp.
***Philodendron* 'Tropicana'**.
Philodendron verrucosum Mathieu ex Schott.
Philodendron wendlandii Schott.
Phlomis fruticosa L. JERUSALEM SAGE.
PHLOX, PRICKLY: **Leptodactylon californicum**.
Phoenix canariensis Chabaud. CANARY ISLAND DATE PALM.
Phoenix dactylifera L. DATE PALM. EDIBLE DATE.
Phoenix loureirii Kunth. PYGMY DATE PALM.
Phoenix reclinata Jacq. SENEGAL DATE PALM. CLUMP PALM.
Phoenix roebelenii O'Brien. DWARF DATE PALM. PYGMY DATE PALM. ROEBELEN PALM. ROEBELIN PALM.
Phoenix rupicola T. Anderson. EAST INDIAN WINE PALM. INDIA DATE PALM. CLIFF DATE PALM.
Phoenix sylvestris (L.) Roxb. SILVER DATE PALM. INDIA DATE PALM.
Phormium 'Atropurpureum'. See **P. tenax** 'Atropurpureum'.
Phormium colensoi Hook. f. See **P. cookianum**.
Phormium cookianum Le Jolis. Syn.: *P. colensoi*.
Phormium tenax J. R. Forst. & G. Forst. NEW ZEALAND FLAX.
***Phormium tenax* 'Atropurpureum'**.
Phormium tenax 'Bronze'?
Phormium tenax 'Golden'?
***Phormium tenax* 'Purpureum'**.
***Phormium tenax* 'Red Dwarf'**.
Phormium tenax 'Red Pendula'?
***Phormium tenax* 'Rubrum'**. RED NEW ZEALAND FLAX.
***Phormium tenax* 'Variegatum'**. VARIEGATED NEW ZEALAND FLAX.
PHOTINIA, CHINESE: **Photinia serrulata**.
    JAPANESE: **Photinia glabra**.
Photinia arbutifolia (Ait.) Lindl. See **Heteromeles arbutifolia**.
Photinia arbutifolia var. *macrocarpa* Munz. See **Heteromeles arbutifolia** var. **macrocarpa**.
Photinia davidsoniae Rehd. & Wils.
Photinia deflexa Hemsl. See **Eriobotrya deflexa**.
Photinia × fraseri Dress. [**P. serrulata × P. glabra**].
***Photinia × fraseri* 'Birmingham'**.
***Photinia × fraseri* 'Red Robin'**.
***Photinia × fraseri* 'Robusta'**.
Photinia glabra (Thunb.) Maxim. JAPANESE PHOTINIA.
Photinia glabra fraseri. Listed name for **P. × fraseri**.
Photinia glabra 'Red Robin'. See **P. × fraseri 'Red Robin'**.
Photinia 'Lineata'. See **P. serrulata 'Aculeata'**.
Photinia serrulata Lindl. CHINESE PHOTINIA.
***Photinia serrulata* 'Aculeata'**. RED-TWIG CHINESE TOYON.
Photinia serrulata 'Lineata'. See **P. serrulata 'Aculeata'**.
Photinia serrulata 'Nova'. See **P. serrulata 'Aculeata'**.
Photinia serrulata 'Nova Lineata'. See **P. serrulata 'Aculeata'**.
Photinia villosa (Thunb.) DC.
Phygelius capensis E. H. Mey. ex Benth. CAPE FUCHSIA.
Phyllodoce empetriformis (Sm.) D. Don. MOUNTAIN HEATHER.
Phyllodoce glanduliflora (Hook.) Cov. MOUNTAIN HEATHER.
Phyllostachys angusta McClure. STONE BAMBOO.
Phyllostachys aurea Carrière ex A. & C. Rivière. Syn.: *Bambusa aurea*. GOLDEN BAMBOO. FISH-POLE BAMBOO.
Phyllostachys aureosulcata McClure. YELLOW-GROOVE BAMBOO.

Phyllostachys bambusoides Siebold & Zucc. Syn.: *P. reticulata*. JAPANESE TIMBER BAMBOO. GIANT TIMBER BAMBOO. MADAKE.
Phyllostachys bambusoides 'Castillon'.
Phyllostachys bambusoides var. *marliacea* (Mitf.) Makino. WRINKLED BAMBOO.
Phyllostachys bambusoides 'Slender Crookstem'.
Phyllostachys bissetii McClure. DAVID BISSET BAMBOO.
Phyllostachys congesta Rendle.
Phyllostachys dulcis McClure. SWEET-SHOOT BAMBOO.
Phyllostachys edulis Houzeau de Lehaie. See **P. pubescens**.
Phyllostachys flexuosa A. & C. Rivière. ZIG-ZAG BAMBOO.
Phyllostachys glauca McClure.
Phyllostachys makinoi Hayata. MAKINO BAMBOO.
Phyllostachys meyeri McClure. MEYER BAMBOO.
Phyllostachys mitis A. & C. Rivière. See **P. viridis**.
Phyllostachys nidularia Munro.
Phyllostachys nigra (Lodd. ex Lindl.) Munro. BLACK BAMBOO.
Phyllostachys nigra 'Bory'. BORY BAMBOO.
Phyllostachys nigra 'Henon'. HENON BAMBOO.
Phyllostachys nigra var. *muchisasa* (Houzeau de Lehaie) Nakai.
Phyllostachys nuda McClure.
Phyllostachys pubescens Mazel ex Houzeau de Lehaie. Syn.: *P. edulis*. MOSO BAMBOO.
Phyllostachys purpurata McClure.
Phyllostachys purpurata 'Solid Stem'.
Phyllostachys purpurata 'Straight Stem'.
Phyllostachys reticulata (Rupr.) K. Koch. See **P. bambusoides**.
Phyllostachys rubromarginata McClure.
Phyllostachys sulphurea A. & C. Rivière var. *viridis* R. A. Young. See **P. viridis**.
Phyllostachys viridis (R. A. Young) McClure. Syn.: *P. mitis*. *P. sulphurea* var. *viridis*.
Phyllostachys viridis 'Robert Young'.
Phyllostachys vivax McClure.
Phymatocarpus porphyrocephalus F. J. Muell.
Phymosia umbellata (Cav.) Kearney. Syn.: *Sphaeralcea umbellata*.
Physianthus albens Mart. See **Araujia sericofera**.
Physocarpus capitatus (Pursh) Kuntze. NINEBARK.
Physocarpus opulifolius (L.) Maxim. Syn.: *Spiraea opulifolia*.
Physocarpus opulifolius 'Nanus'.
Phytolacca dioica L. UMBU. OMBU. LA BELLA SOMBRA.
Picea abies (L.) Karst. Syn.: *P. excelsa*. NORWAY SPRUCE.
Picea abies 'Aurea'.
Picea abies 'Brevifolia'.
Picea abies 'Clanbrassiliana'.
Picea abies 'Crippsii'. See **P. abies 'Brevifolia'**.
Picea abies 'Densata'. See **P. glauca 'Densata'**.
Picea abies 'Echiniformis'. Some plants so called may be **P. glauca 'Echiniformis'**.
Picea abies 'Maxwellii'.
Picea abies 'Mucronata'.
Picea abies 'Nidiformis'. NEST SPRUCE.
Picea abies 'Pendula'.
Picea abies 'Procumbens'.
Picea abies 'Pumila'. DWARF SPRUCE.
Picea abies 'Pygmaea'.
Picea abies 'Remontii'.
Picea abies 'Repens'.
Picea abies 'Sherwood Gem'. SHERWOOD GEM SPRUCE.
Picea abies 'Sherwoodii'. SHERWOOD MULTONOMAH SPRUCE.
Picea albertiana S. Br. See **P. glauca** var. **albertiana**. Some plants called *P. albertiana* are **P. glauca 'Conica'**.
Picea brewerana S. Wats. BREWER'S WEEPING SPRUCE.
Picea engelmannii Parry ex Engelm. ENGELMANN SPRUCE.
Picea engelmannii 'Glauca'.
Picea excelsa (Lam.) Link, and all listed cvs. See **P. abies** and cvs.
Picea excelsa var. *brevifolia* Webster. See **P. abies 'Brevifolia'**.
Picea glauca (Moench) Voss. WHITE SPRUCE.
Picea glauca var. *albertiana* (S. Br.) Sarg. Syn.: *P. albertiana*. ALBERTA SPRUCE. Some plants so called are **P. glauca 'Conica'**.
Picea glauca 'Conica'. DWARF ALBERTA SPRUCE. DWARF WHITE SPRUCE. Some plants called *P. albertiana* and *P. glauca* var. *albertiana* belong here.
Picea glauca 'Densata'. BLACK HILLS SPRUCE.
Picea glauca 'Echiniformis'. Some plants called **P. abies 'Echiniformis'** belong here.
Picea jezoensis (Siebold & Zucc.) Carrière. YEDDO SPRUCE.
Picea jezoensis var. *hondoensis* (Mayr) Rehd. HONDO SPRUCE.
Picea 'Koster'. See **P. pungens 'Koster'**.
Picea 'Kosterana Prostrata'?
Picea omorika (Pančić) Purkyne. SERBIAN SPRUCE.
Picea orientalis (L.) Link. ORIENTAL SPRUCE. EASTERN SPRUCE.
Picea orientalis 'Nana'.
Picea polita (Siebold & Zucc.) Carrière. See **P. torano**.
Picea pungens Engelm. COLORADO SPRUCE. COLORADO BLUE SPRUCE.
Picea pungens 'Glauca'. COLORADO BLUE SPRUCE. A collective name for all glaucous-leaved cvs. not otherwise named.
Picea pungens 'Glauca Globosa'. See **P. pungens 'Globosa'**.
Picea pungens 'Glauca Koster'. See **P. pungens 'Koster'**.
Picea pungens 'Glauca Moerheimii'. See **P. pungens 'Moerheimii'**.
Picea pungens 'Glauca Montgomery'. See **P. pungens 'Montgomery'**.
Picea pungens 'Glauca Pendula'. KOSTER'S WEEPING BLUE SPRUCE.
Picea pungens 'Glauca Procumbens'.
Picea pungens 'Glauca Prostrata'.
Picea pungens 'Globosa'.
Picea pungens 'Hoopsii'.
Picea pungens 'Koster'. KOSTER BLUE SPRUCE.
Picea pungens 'Kosterana'. See **P. pungens 'Glauca Pendula'**.
Picea pungens 'Kosterana Prostrata'?
Picea pungens 'Moerheimii'. MOERHEIM BLUE SPRUCE.
Picea pungens 'Montgomery'. MONTGOMERY SPRUCE.
Picea pungens 'Pendens'. Listed name. May be **P. pungens 'Glauca Pendula'**.
Picea rubens Sarg. RED SPRUCE.
Picea sitchensis (Bong.) Carrière. SITKA SPRUCE. TIDELAND SPRUCE.
Picea smithiana (Wallich) Boiss. HIMALAYAN SPRUCE.
Picea torano (Siebold) Koehne. Syn.: *P. polita*. TIGER-TAIL SPRUCE.
Picea wilsonii M. T. Mast. WILSON SPRUCE.
Pickeringia montana Nutt. ex Torr. & Gray. CHAPARRAL PEA.
PIERIS, CHINESE: **Pieris formosa** var. **forrestii**.
   HIMALAYAN: **Pieris formosa**.
   JAPANESE: **Pieris japonica**.
   MOUNTAIN: **Pieris floribunda**.
Pieris floribunda (Pursh) Benth. & Hook. f. Syn.: *Andromeda floribunda*. ANDROMEDA. FETTER-BUSH. MOUNTAIN PIERIS.
Pieris formosa (Wallich) D. Don. HIMALAYAN PIERIS.
Pieris formosa var. *forrestii* (Harrow ex W. W. Sm.) Airy Shaw. Syn.: *P. forrestii*. CHINESE ANDROMEDA. CHINESE PIERIS.
Pieris formosa var. *forrestii* 'Wakehurst'.
Pieris 'Forrest Flame'. [*P. formosa* var. *forrestii* 'Wakehurst' × *P. japonica*].
Pieris forrestii Harrow ex W. W. Sm. See **P. formosa** var. **forrestii**.

Pieris japonica (Thunb.) D. Don ex G. Don. ANDROMEDA.
    JAPANESE PIERIS. LILY-OF-THE-VALLEY SHRUB.
Pieris japonica 'Bert Chandler'.
Pieris japonica 'Christmas Cheer'.
Pieris japonica 'Coleman'.
Pieris japonica 'Compacta'.
Pieris japonica 'Crispa'.
Pieris japonica 'Daisen'.
Pieris japonica 'Dorothy Wyckoff'.
Pieris japonica 'Flamingo'.
Pieris japonica 'Mountain Fire'.
Pieris japonica 'Pygmaea'.
Pieris japonica 'Valentine'. [*P. japonica* 'Flamingo' ×
    P. japonica 'Valley Rose'].
Pieris japonica 'Valley Rose'.
Pieris japonica 'Variegata'.
Pieris japonica 'White Cascade'.
Pieris taiwanensis Hayata.
Pieris taiwanensis 'Snow Drift'?
Pieris 'Valley Fire'. [*P. formosa* var. *forrestii* × *P. japonica*
    'White Caps'].
PIGEON BERRY: **Duranta repens.**
Pileostegia viburnoides Hook. f. & Thoms.
Pimelea coarctica. Listed name for **P. prostrata.**
Pimelea ferruginea Labill. ROSY RICEFLOWER.
Pimelea prostrata (J. R. Forst. & G. Forst.) Willd.
PINCUSHION, NODDING: **Leucospermum cordifolium.**
    ROCKET: **Leucospermum reflexum.**
PINCUSHION TREE: **Hakea laurina.**
PINE, AFRICAN FERN: **Podocarpus gracilior.**
    ALEPPO: **Pinus halepensis.**
    ARMAND DAVID'S: **Pinus armandii.**
    AUSTRIAN: **Pinus nigra.**
    AUSTRIAN BLACK: **Pinus nigra.**
    BEACH: **Pinus contorta.**
    BENGUET: **Pinus insularis.**
    BHUTAN: **Pinus wallichiana.**
    BIGCONE: **Pinus coulteri.**
    BISHOP: **Pinus muricata.**
    BLACK: **Pinus nigra.**
    BLUE: **Pinus wallichiana.**
    BRISTLECONE: **Pinus aristata.**
    BRISTLECONE: **Pinus longaeva.**
    BRUTIA: **Pinus brutia.**
    CALABRIAN: **Pinus brutia.**
    CANARY ISLAND: **Pinus canariensis.**
    CHINESE: **Pinus tabuliformis.**
    CHIR: **Pinus roxburghii.**
    CLUSTER: **Pinus pinaster.**
    COLORADO PINYON: **Pinus edulis.**
    CORSICAN: **Pinus nigra** subsp. **laricio.**
    COULTER: **Pinus coulteri.**
    CYPRESS: **Callitris preissii.**
    DIGGER: **Pinus sabiniana.**
    DWARF BEACH: **Pinus contorta** var. **bolanderi.**
    DWARF MUGO: **Pinus mugo.**
    DWARF SIBERIAN: **Pinus pumila.**
    DWARF STONE: **Pinus pumila.**
    EASTERN WHITE: **Pinus strobus.**
    ELDARICA: **Pinus eldarica.**
    FOUR-NEEDLED PINYON: **Pinus quadrifolia.**
    GREAT BASIN BRISTLECONE: **Pinus longaeva.**
    HICKORY: **Pinus aristata.**
    HIMALAYAN WHITE: **Pinus wallichiana.**
    HOOP: **Araucaria cunninghamii.**
    INTERMOUNTAIN BRISTLECONE: **Pinus longaeva.**

ITALIAN STONE: **Pinus pinea.**
JACK: **Pinus banksiana.**
JAPANESE BLACK: **Pinus thunbergiana.**
JAPANESE FIVE-NEEDLED: **Pinus parviflora.**
JAPANESE RED: **Pinus densiflora.**
JAPANESE TABLE: **Pinus densiflora** 'Umbraculifera'.
JAPANESE UMBRELLA: **Pinus densiflora** 'Umbraculifera'.
JAPANESE WHITE PINE: **Pinus parviflora.**
JEFFREY: **Pinus jeffreyi.**
JELECOTE: **Pinus patula.**
JERSEY: **Pinus virginiana.**
JERUSALEM: **Pinus halepensis.**
KAURI: **Agathis australis.**
KNOBCONE: **Pinus attenuata.**
LACEBARK: **Pinus bungeana.**
LIMBER: **Pinus flexilis.**
LODGEPOLE: **Pinus murrayana.**
LONGLEAF: **Pinus palustris.**
LUZON: **Pinus insularis.**
MARITIME: **Pinus pinaster.**
MEXICAN: **Pinus patula.**
MEXICAN STONE: **Pinus cembroides.**
MEXICAN WHITE: **Pinus ayacahuite.**
MEXICAN WHITE: **Pinus strobiformis.**
MONTEREY: **Pinus radiata.**
MONTEZUMA: **Pinus montezumae.**
MORETON BAY: **Araucaria cunninghamii.**
MOUNTAIN: **Pinus mugo.**
MUGO: **Pinus mugo.**
NEW CALEDONIAN: **Araucaria columnaris.**
NORFOLK ISLAND: **Araucaria heterophylla.**
NUT: **Pinus edulis.**
NUT: **Pinus monophylla.**
ONE-NEEDLED PINYON: **Pinus monophylla.**
OREGON: **Pseudotsuga menziesii.**
ORIENTAL YEW: **Podocarpus macrophyllus** var. **maki.**
PARRY PINYON: **Pinus quadrifolia.**
PINYON: **Pinus edulis.**
PINYON: **Pinus monophylla.**
PONDEROSA: **Pinus ponderosa.**
RED: **Pinus resinosa.**
RIGA: **Pinus sylvestris** var. **rigensis.**
ROCKY MOUNTAIN YELLOW: **Pinus ponderosa** var. **scopulorum.**
SANTA CRUZ ISLAND: **Pinus muricata.**
SCOTCH: **Pinus sylvestris.**
SCOTS: **Pinus sylvestris.**
SCREW: **Pandanus utilis.**
SHORE: **Pinus contorta.**
SHORTLEAF: **Pinus echinata.**
SHRUBBY SWISS MOUNTAIN: **Pinus mugo** var. **pumilio.**
SHRUBBY YEW: **Podocarpus macrophyllus** var. **maki.**
SLASH: **Pinus caribaea.**
SOUTHERN YELLOW: **Pinus palustris.**
SPRUCE: **Pinus virginiana.**
STAR: **Araucaria heterophylla.**
SUGAR: **Pinus lambertiana.**
SWISS MOUNTAIN: **Pinus mugo.**
SWISS STONE: **Pinus cembra.**
TABLE MOUNTAIN: **Pinus pungens.**
TAMARACK: **Pinus murrayana.**
TANYOSHO: **Pinus densiflora** 'Umbraculifera'.
THREE-NEEDLED PINYON: **Pinus cembroides.**
TORREY: **Pinus torreyana.**
TOTARA: **Podocarpus totara.**
TURKISH: **Pinus brutia.**
TWO-NEEDLED PINYON: **Pinus edulis.**

PINE, continued
    UMBRELLA: **Pinus pinea.**
    UMBRELLA: **Sciadopitys verticillata.**
    VIRGINIA: **Pinus virginiana.**
    WEEPING JAPANESE RED: **Pinus densiflora 'Pendula'.**
    WESTERN WHITE: **Pinus monticola.**
    WESTERN YELLOW: **Pinus ponderosa.**
    WHITEBARK: **Pinus albicaulis.**
    YEW: **Podocarpus macrophyllus.**
  PINK-TIPS: **Callistemon salignus.**
Pinus albicaulis Engelm. WHITEBARK PINE.
Pinus aristata Engelm. BRISTLECONE PINE. HICKORY PINE.
Pinus armandii Franch. ARMAND DAVID'S PINE.
Pinus attenuata Lemmon. Syn.: *P. tuberculata.* KNOBCONE PINE.
Pinus × **attenuradiata** Stockw. & Righter. [**P. attenuata** × **P. radiata**].
Pinus australis Michx. f. See **P. palustris.**
Pinus ayacahuite C. A. Ehrenb. MEXICAN WHITE PINE.
Pinus banksiana Lamb. JACK PINE.
Pinus bolanderi Parl. See **P. contorta** var. **bolanderi.**
Pinus brutia Ten. Syn.: *P. halepensis* var. *brutia.* BRUTIA PINE. CALABRIAN PINE. TURKISH PINE. Some trees so called are **P. halepensis.**
Pinus bungeana Zucc. ex Endl. LACEBARK PINE.
Pinus canariensis Sweet ex K. Spreng. CANARY ISLAND PINE.
Pinus caribaea Morelet. SLASH PINE.
Pinus cembra L. SWISS STONE PINE.
Pinus cembroides Zucc. MEXICAN STONE PINE. THREE-NEEDLED PINYON PINE.
Pinus cembroides var. *edulis* (Engelm.) Voss. See **P. edulis.**
Pinus cembroides var. *monophylla* (Torr. & Frem.) Voss. See **P. monophylla.**
Pinus cembroides var. *parryana* (Engelm.) Voss. See **P. quadrifolia.**
Pinus contorta Dougl. ex Loud. SHORE PINE. BEACH PINE.
Pinus contorta var. **bolanderi** (Parl.) Vasey. Syn.: *P. bolanderi.* DWARF BEACH PINE.
Pinus contorta var. *latifolia* Engelm. See **P. murrayana.**
Pinus contorta var. *murrayana* (Grev. & Balf.) Engelm. See **P. murrayana.**
Pinus coulteri D. Don. BIGCONE PINE. COULTER PINE.
Pinus densiflora Siebold & Zucc. JAPANESE RED PINE.
Pinus densiflora 'Pendula'. WEEPING JAPANESE RED PINE.
Pinus densiflora tanyosho. Listed name for **P. densiflora 'Umbraculifera'.**
Pinus densiflora 'Umbraculifera'. TANYOSHO. JAPANESE TABLE PINE. JAPANESE UMBRELLA PINE. TANYOSHO PINE.
Pinus echinata Mill. SHORTLEAF PINE.
Pinus edulis Engelm. Syn.: *P. cembroides* var. *edulis.* NUT PINE. PINYON PINE. COLORADO PINYON PINE. TWO-NEEDLED PINYON PINE.
Pinus eldarica Medvedev. ELDARICA PINE.
Pinus excelsa Wallich. See **P. wallichiana.**
Pinus flexilis James. LIMBER PINE.
Pinus flexilis var. *reflexa* Engelm. See **P. strobiformis.**
Pinus griffithii McClelland. See **P. wallichiana.**
Pinus halepensis Mill. JERUSALEM PINE. ALEPPO PINE. Some trees called **P. brutia** belong here.
Pinus halepensis var. *brutia* (Ten.) A. Henry. See **P. brutia.**
Pinus insignis Dougl. ex Loud. See **P. radiata.**
Pinus insularis Endl. BENGUET PINE. LUZON PINE.
Pinus jeffreyi Grev. & Balf. JEFFREY PINE.
Pinus lambertiana Dougl. SUGAR PINE.
Pinus longaeva D. K. Bailey. GREAT BASIN BRISTLECONE PINE. INTERMOUNTAIN BRISTLECONE PINE. BRISTLECONE PINE.

Pinus longifolia Roxb. See **P. roxburghii.**
Pinus maritima Mill. Some plants so called are **P. nigra** and **P. pinaster.**
Pinus monophylla Torr. & Frem. Syn.: *P. cembroides* var. *monophylla.* PINYON PINE. ONE-NEEDLED PINYON PINE. NUT PINE.
Pinus montana Mill. See **P. mugo.**
Pinus montezumae Lamb. MONTEZUMA PINE.
Pinus monticola Dougl. ex D. Don. WESTERN WHITE PINE.
Pinus mugho Poir. See **P. mugo.**
Pinus mughus Scop. See **P. mugo.**
Pinus mugo Turra. Syn.: *P. montana. P. mugho. P. mughus. P. mugo* var. *mughus.* MUGO PINE. DWARF MUGO PINE. SWISS MOUNTAIN PINE. MOUNTAIN PINE.
Pinus mugo var. *mughus* (Scop.) Zenari. See **P. mugo.**
Pinus mugo var. **pumilio** (Haenke) Zenari. SHRUBBY SWISS MOUNTAIN PINE.
Pinus muricata D. Don. Syn.: *P. remorata.* BISHOP PINE. SANTA CRUZ ISLAND PINE.
Pinus murrayana Grev. & Balf. Syn.: *P. contorta* var. *latifolia. P. contorta* var. *murrayana.* LODGEPOLE PINE. TAMARACK PINE.
Pinus nigra Arnold. BLACK PINE. AUSTRIAN PINE. AUSTRIAN BLACK PINE. Some plants called *P. maritima* belong here.
Pinus nigra subsp. **laricio** Maire. CORSICAN PINE.
Pinus nigra var. *poiretiana* (Loud.) Schneid. See **P. nigra** subsp. **laricio.**
Pinus palustris Mill. Syn.: *P. australis.* LONGLEAF PINE. GEORGIA PINE. SOUTHERN YELLOW PINE.
Pinus parryana Engelm. See **P. quadrifolia.**
Pinus parviflora Siebold & Zucc. JAPANESE FIVE-NEEDLED PINE. JAPANESE WHITE PINE.
Pinus patula Schlechtend. & Cham. MEXICAN PINE. JELECOTE PINE.
Pinus pinaster Ait. MARITIME PINE. CLUSTER PINE. Some plants called *P. maritima* belong here.
Pinus pinea L. ITALIAN STONE PINE. UMBRELLA PINE.
Pinus ponderosa Dougl. ex P. Laws. & C. Laws. WESTERN YELLOW PINE. PONDEROSA PINE.
Pinus ponderosa var. **scopulorum** Engelm. Syn.: *P. scopulorum.* ROCKY MOUNTAIN YELLOW PINE.
Pinus pseudostrobus Lindl.
Pinus pumila (Pall.) Regel. DWARF STONE PINE. DWARF SIBERIAN PINE.
Pinus pungens Lamb. TABLE MOUNTAIN PINE.
Pinus quadrifolia Parl. ex Sudw. Syn.: *P. cembroides* var. *parryana. P. parryana.* FOUR-NEEDLED PINYON PINE. PARRY PINYON PINE.
Pinus radiata D. Don. Syn.: *P. insignis.* MONTEREY PINE.
Pinus radiata 'Eldorado'.
Pinus remorata Mason. See **P. muricata.**
Pinus resinosa Ait. RED PINE.
Pinus roxburghii Sarg. Syn.: *P. longifolia.* CHIR PINE.
Pinus sabiniana Dougl. ex D. Don. DIGGER PINE.
Pinus scopulorum (Engelm.) Shaw. See **P. ponderosa** var. **scopulorum.**
Pinus strobiformis Engelm. Syn.: *P. flexilis* var. *reflexa.* MEXICAN WHITE PINE.
Pinus strobus L. EASTERN WHITE PINE.
Pinus strobus 'Brevifolia'. Some plants called *P. strobus* 'Nana' may belong here.
Pinus strobus 'Densa'. Some plants called *P. strobus* 'Nana' may belong here.
Pinus strobus 'Fastigiata'.
Pinus strobus 'Minima'. Some plants called *P. strobus* 'Nana' may belong here.

Pinus strobus 'Nana'. Plants so called may be **P. strobus 'Brevifolia', P. strobus 'Densa', P. strobus 'Minima', P. strobus 'Pumila', P. strobus 'Radiata', P. strobus 'Umbraculifera',** or other dwarf cvs.
Pinus strobus 'Prostrata'.
Pinus strobus 'Pumila'. Some plants called P. strobus 'Nana' may belong here.
Pinus strobus 'Radiata'. Some plants called P. strobus 'Nana' may belong here.
Pinus strobus 'Umbraculifera'. Some plants called P. strobus 'Nana' may belong here.
Pinus sylvestris L. SCOTS PINE. SCOTCH PINE.
Pinus sylvestris 'Argentea'.
Pinus sylvestris 'Argentea Compacta'.
Pinus sylvestris 'Beauvronensis'.
Pinus sylvestris 'Compressa'.
Pinus sylvestris 'Fastigiata'.
Pinus sylvestris 'Gauca'?
Pinus sylvestris 'Glauca Nana'?
Pinus sylvestris 'Pendula'.
Pinus sylvestris var. **rigensis** Loud. RIGA PINE.
Pinus sylvestris 'Watereri'.
Pinus tabuliformis Carrière. CHINESE PINE.
Pinus tanyosho. Listed name for **P. densiflora 'Umbraculifera'.**
Pinus thunbergiana Franco. Syn.: *P. thunbergii.* JAPANESE BLACK PINE.
Pinus thunbergii Parl. See **P. thunbergiana.**
Pinus torreyana Parry ex Carrière. TORREY PINE.
Pinus tuberculata Gord. See **P. attenuata.**
Pinus virginiana Mill. VIRGINIA PINE. JERSEY PINE. SPRUCE PINE.
Pinus wallichiana A. B. Jackson. Syn.: *P. excelsa. P. griffithii.* HIMALAYAN WHITE PINE. BLUE PINE. BHUTAN PINE.
PIPE VINE, CALIFORNIA: **Aristolochia californica.**
Pisonia brunoniana Endl. See **P. umbellifera.**
Pisonia umbellifera (J. R. Forst. & G. Forst.) Seem. Syn.: *Heimerliodendron brunonianum. Pisonia brunoniana.* BIRD-CATCHER TREE. PARA-PARA.
PISTACHE, CHINESE: **Pistacia chinensis.**
EVERGREEN: **Pistacia lentiscus.**
PISTACHIO: **Pistacia vera.**
CHINESE: **Pistacia chinensis.**
PISTACHIO NUT: **Pistacia vera.**
Pistacia atlantica Desf. MT. ATLAS MASTIC TREE.
Pistacia chinensis Bunge. CHINESE PISTACHE. CHINESE PISTACHIO.
Pistacia chinensis 'Keith Davey'.
Pistacia integerrima J. L. Stewart ex Brandis.
Pistacia lentiscus L. MASTIC TREE. CHIOS MASTIC TREE. EVERGREEN PISTACHE.
Pistacia terebinthus L. CYPRUS TURPENTINE.
Pistacia vera L. PISTACHIO NUT. PISTACHIO.
PITANGA: **Eugenia uniflora.**
Pithecoctenium echinatum (Jacq.) K. Schum.
PITTOSPORUM, CAPE: **Pittosporum viridiflorum.**
DIAMOND-LEAF: **Pittosporum rhombifolium.**
EVERGREEN: **Pittosporum crassifolium.**
NARROW-LEAVED: **Pittosporum phillyraeoides.**
ORANGE-BERRY: **Pittosporum undulatum.**
QUEENSLAND: **Pittosporum rhombifolium.**
WILLOW: **Pittosporum phillyraeoides.**
Pittosporum crassifolium Banks & Soland. ex A. Cunn. KARO. EVERGREEN PITTOSPORUM.
Pittosporum crassifolium 'Compactum'.
Pittosporum daphniphylloides Hayata.
Pittosporum eriocarpum Royle.
Pittosporum erioloma C. Moore & F. J. Muell.
Pittosporum eugenioides A. Cunn. TARATA.

Pittosporum floribundum Wight & Arn. Some plants cult. as **P. napaulense** may belong here.
Pittosporum glabratum Lindl.
Pittosporum heterophyllum Franch.
Pittosporum napaulense (DC.) Rehd. & Wils. GOLDEN FRAGRANCE. Some plants so called are **P. floribundum.**
Pittosporum nigricans. Listed name for **P. tenuifolium.**
Pittosporum odoratum Merrill.
Pittosporum phillyraeoides DC. DESERT WILLOW. WILLOW PITTOSPORUM. NARROW-LEAVED PITTOSPORUM.
Pittosporum ralphii T. Kirk.
Pittosporum rhombifolium A. Cunn. ex Hook. QUEENSLAND PITTOSPORUM. DIAMOND-LEAF PITTOSPORUM.
Pittosporum tenuifolium Banks & Soland. ex Gaertn. TAWHIWHI. Plants called P. *nigricans* belong here.
Pittosporum tobira Ait. MOCK ORANGE. TOBIRA.
Pittosporum tobira 'Compacta'.
Pittosporum tobira 'Variegata'. VARIEGATED TOBIRA.
Pittosporum tobira 'Wheeler's Dwarf'.
Pittosporum tobira 'White Spot'.
Pittosporum undulatum Vent. VICTORIAN BOX. ORANGE-BERRY PITTOSPORUM.
Pittosporum viridiflorum Sims. CAPE PITTOSPORUM.
PLANE, LONDON: **Platanus × acerifolia.**
ORIENTAL: **Platanus orientalis.**
PLANTAIN: **Musa acuminata.**
PLANTAIN: **Musa × paradisiaca.**
Platanus × acerifolia (Ait.) Willd. [*P. occidentalis × P. orientalis*]. LONDON PLANE. Some plants cult. as **P. orientalis** belong here.
Platanus occidentalis L. AMERICAN SYCAMORE. BUTTONWOOD.
Platanus orientalis L. EUROPEAN SYCAMORE. ORIENTAL PLANE. Rare in cult. Some plants so called are **P. × acerifolia.**
Platanus racemosa Nutt. WESTERN SYCAMORE. CALIFORNIA SYCAMORE. BUTTONWOOD.
Platanus racemosa var. *wrightii* (S. Wats.) L. Benson. See **P. wrightii.**
Platanus wrightii S. Wats. Syn.: *P. racemosa* var. *wrightii.* ARIZONA SYCAMORE.
Platycladus orientalis (L.) Franco. Syn.: *Thuja orientalis.* ORIENTAL ARBORVITAE.
Platycladus orientalis 'Aureus'.
Platycladus orientalis 'Aureus Nanus'. BERCKMANN ARBORVITAE. DWARF GOLDEN ARBORVITAE.
Platycladus orientalis 'Bakeri'. BAKER'S ARBORVITAE.
Platycladus orientalis 'Berckmannii'. See **P. orientalis 'Aureus Nanus'.**
Platycladus orientalis 'Beverleyensis'. BEVERLY HILLS ARBORVITAE. GOLDEN PYRAMID ARBORVITAE.
Platycladus orientalis 'Blue Cone'.
Platycladus orientalis 'Blue Spire'?
Platycladus orientalis 'Bonita'.
Platycladus orientalis 'Bonita Upright'. See **P. orientalis 'Bonita'.**
Platycladus orientalis 'Chase's Golden'?
Platycladus orientalis 'Compactus'. See **P. orientalis 'Sieboldii'.**
Platycladus orientalis 'Decussatus'. See **P. orientalis 'Juniperoides'.**
Platycladus orientalis 'Elegantissimus'.
Platycladus orientalis 'Excelsus'.
Platycladus orientalis 'Fruitlandii'. FRUITLAND ARBORVITAE.
Platycladus orientalis 'Hohman'. HOHMAN ARBORVITAE.
Platycladus orientalis 'Juniperoides'.
Platycladus orientalis 'Mordigan'?
Platycladus orientalis 'Raffles'. RAFFLES ARBORVITAE.
Platycladus orientalis 'Rosedale'. ROSEDALE ARBORVITAE.

Platycladus orientalis 'Sieboldii'.
Platycladus orientalis 'Stewart'?
Platycladus orientalis 'Strictus'.
Platycladus orientalis 'Westmont'.
Pleioblastus distichus (Mitf.) Nakai. See **Arundinaria disticha**.
Pleioblastus variegatus (Siebold ex Miq.) Makino.
  See **Arundinaria variegata**.
Pleioblastus viridi-striatus var. *vagans*. Listed name, probably for **Arundinaria pygmaea**.
Pleomele reflexa (Lam.) N. E. Br. See **Dracaena reflexa**.
Pleomele thalioides (E. Morr.) N. E. Br. See **Dracaena thalioides**.
PLEROMA: **Tibouchina urvilleana.**
Pleroma grandiflora. Listed name for **Tibouchina urvilleana**.
Pleroma splendens. Listed name for **Tibouchina urvilleana**.
PLUM, BURBANK: **Prunus salicina 'Burbank'**
  CHERRY: **Prunus cerasifera.**
  DATE: **Diospyros lotus.**
  DATE: **Diospyros virginiana.**
  DWARF RED-LEAF: **Prunus × cistena.**
  EUROPEAN: **Prunus domestica.**
  FLOWERING: **Prunus × blireiana.**
  HOLLYWOOD FLOWERING: **Prunus cerasifera 'Hollywood'.**
  INDIAN: **Oemleria cerasiformis.**
  JAPANESE: **Prunus salicina.**
  KAFFIR: **Harpephyllum caffrum.**
  MYROBALAN: **Prunus cerasifera.**
  NATAL: **Carissa macrocarpa.**
  PISSARD: **Prunus cerasifera 'Atropurpurea'.**
  PURPLE-LEAF: **Prunus cerasifera 'Atropurpurea'.**
  SIERRA: **Prunus subcordata.**
  THUNDERCLOUD FLOWERING: **Prunus cerasifera 'Thundercloud'.**
  TRAILBLAZER: **Prunus 'Trailblazer'.**
  VESUVIUS FLOWERING: **Prunus cerasifera 'Vesuvius'.**
  WEEPING FLOWERING: **Prunus cerasifera 'Pendula'.**
PLUMBAGO, BLUE CAPE: **Plumbago auriculata.**
  BURMESE: **Ceratostigma griffithii.**
  CHINESE: **Ceratostigma willmottianum.**
  DWARF: **Ceratostigma plumbaginoides.**
  WHITE CAPE: **Plumbago auriculata 'Alba'.**
Plumbago auriculata Lam. Syn.: *P. capensis. P. grandiflora.* BLUE CAPE PLUMBAGO.
Plumbago auriculata 'Alba'. WHITE CAPE PLUMBAGO.
Plumbago capensis Thunb. See **P. auriculata**.
Plumbago capensis 'Alba'. See **Plumbago auriculata 'Alba'**.
Plumbago grandiflora Ten. See **P. auriculata**.
Plumbago larpentiae Lindl. See **Ceratostigma plumbaginoides**.
PLUME FLOWER, BRAZILIAN: **Justicia carnea.**
PLUMERIA, SINGAPORE: **Plumeria obtusa.**
Plumeria obtusa L. SINGAPORE PLUMERIA.
Plumeria rubra L. FRANGIPANI.
Podalyria calyptrata Willd.
PODOCARPUS, BROAD-LEAF: **Podocarpus nagi.**
Podocarpus alpinus R. Br. ex Hook. f.
Podocarpus andinus Poepp. ex Endl. PLUM FIR.
Podocarpus elatus R. Br. ex Endl.
Podocarpus elongatus (Ait.) L'Hér. ex Pers. Most plants so called are **P. gracilior**.
Podocarpus falcatus (Thunb.) R. Br. ex Mirb.
Podocarpus gracilior Pilg. AFRICAN FERN PINE. Most plants called **P. elongatus** belong here.
Podocarpus gracilior 'Glauca'?
Podocarpus henkelii Stapf ex Dallim. & A. B. Jackson.
Podocarpus macrophyllus (Thunb.) D. Don. YEW PINE. JAPANESE YEW.
Podocarpus macrophyllus var. *maki* (Siebold & Zucc.) Endl. Syn.: *P. maki.* SHRUBBY YEW PINE. ORIENTAL YEW PINE.
Podocarpus maki Siebold & Zucc. See **P. macrophyllus** var. *maki*.
Podocarpus nagi (Thunb.) Zollinger & Moritzi ex Makino. BROAD-LEAF PODOCARPUS.
Podocarpus neriifolius D. Don.
Podocarpus nivalis Hook.
Podocarpus salignus D. Don.
Podocarpus totara D. Don. TOTARA. TOTARA PINE.
Podranea brycei (N. E. Br.) Sprague. Syn.: *Pandorea brycei. Tecoma reginae-sabae.* QUEEN-OF-SHEBA VINE.
Podranea ricasoliana (Tanfani) Sprague. Syn.: *Pandorea ricasoliana. Tecoma mackenii.* PINK TRUMPET VINE.
POINCIANA, DWARF: **Caesalpinia pulcherrima.**
  PARADISE: **Caesalpinia gilliesii.**
  ROYAL: **Delonix regia.**
Poinciana gilliesii Wallich ex Hook. See **Caesalpinia gilliesii**.
Poinciana pulcherrima L. See **Caesalpinia pulcherrima**.
Poinciana regia Bojer. See **Delonix regia**.
POINSETTIA: **Euphorbia pulcherrima.**
Poinsettia pulcherrima (Willd. ex Klotzsch) R. Graham. See **Euphorbia pulcherrima**.
POINT REYES CREEPER: **Ceanothus gloriosus.**
POKER PLANT: **Kniphofia uvaria.**
POLKA-DOT PLANT, PINK: **Hypoestes phyllostachys.**
Polyandrococos caudescens (Mart.) Barb.-Rodr. Syn.: *Diplothemium caudescens.* CAUDATE FEATHER PALM.
Polygala apopetala Brandegee.
Polygala chamaebuxus L.
Polygala chamaebuxus 'Atropurpurea'. See **P. chamaebuxus** var. *grandiflora*.
Polygala chamaebuxus var. *grandiflora* Gaudin. Syn.: *P. chamaebuxus* var. *purpurea*.
Polygala chamaebuxus var. *purpurea* Neilreich. See **P. chamaebuxus** var. *grandiflora*.
Polygala × dalmaisiana L. H. Bailey. [*P. oppositifolia* L. var. *cordata* Harv. × *P. myrtifolia* var. *grandiflora*]. SWEET PEA SHRUB.
Polygala myrtifolia L. var. *grandiflora* Hook.
Polygala virgata Thunb.
Polygonum aubertii L. Henry. CHINESE FLEECE VINE. SILVER LACE VINE.
Polygonum baldschuanicum Regel. BUKHARA FLEECE FLOWER.
Polygonum capitatum Buch.-Ham. ex D. Don.
Polygonum cuspidatum Siebold & Zucc. JAPANESE KNOTWEED. MEXICAN BAMBOO.
Polygonum cuspidatum var. *compactum* (Hook. f.) L. H. Bailey. Syn.: *P. reynoutria*.
Polygonum reynoutria Hort. not Makino. See **P. cuspidatum** var. *compactum*.
Polygonum vacciniifolium Wallich ex Meissn. ROSE-CARPET KNOTWEED.
Polyscias filicifolia (C. Moore ex Fournier) L. H. Bailey. FERN-LEAF ARALIA.
Polyscias guilfoylei (Bull) L. H. Bailey. Syn.: *Aralia guilfoylei. Nothopanax guilfoylei.* COFFEE TREE. WILD COFFEE. GERANIUM-LEAF ARALIA.
POMEGRANATE: **Punica granatum.**
  DOUBLE DWARF: **Punica granatum 'Nana Plena'.**
  DWARF: **Punica granatum 'Nana'.**
  DWARF CARNATION-FLOWERED: **Punica granatum 'Chico'.**
  FLOWERING: **Punica granatum.**
POMELO: **Citrus maxima.**
POMMELO: **Citrus maxima.**
POMPON TREE: **Dais cotinifolia.**

Poncirus trifoliata (L.) Raf. TRIFOLIATE ORANGE.
Poncirus trifoliata 'Monstrosa'. FLYING DRAGON TRIFOLIATE
ORANGE. HIRYU.
PONYTAIL: *Beaucarnea recurvata.*
POPLAR, BALSAM: *Populus balsamifera.*
   BLACK: *Populus nigra.*
   BOLLE'S: *Populus alba* 'Pyramidalis'.
   CANADIAN: *Populus* × *canadensis.*
   CAROLINA: *Populus* × *canadensis.*
   CHINESE WHITE: *Populus tomentosa.*
   EASTERN: *Populus deltoides.*
   EUGENE: *Populus* × *canadensis* 'Eugenei'.
   GRAY: *Populus canescens.*
   HEARTLEAF BALSAM: *Populus balsamifera* var. *subcordata.*
   LOMBARDY: *Populus nigra* 'Italica'.
   NECKLACE: *Populus deltoides.*
   QUEENSLAND: *Homalanthus populifolius.*
   SILVER: *Populus alba* 'Nivea'.
   SILVER-LEAVED: *Populus alba.*
   THEVES: *Populus nigra* 'Afghanica'.
   WESTERN BALSAM: *Populus trichocarpa.*
   WHITE: *Populus alba.*
   YELLOW: *Liriodendron tulipifera.*
POPPY, BUSH: *Dendromecon rigida.*
   ISLAND TREE: *Dendromecon rigida* subsp. *harfordii.*
   MATILIJA: *Romneya coulteri.*
   TREE: *Dendromecon rigida.*
Populus × *acuminata* Rydb. [*P. angustifolia* James ×
P. sargentii Dode]. MOUNTAIN COTTONWOOD.
Populus alba L. ABELE. SILVER-LEAVED POPLAR. WHITE POPLAR.
Populus alba var. *argentea* Hort. See *P. alba* 'Nivea'.
Populus alba var. *bolleana* (Lauche) Otto. See *P. alba*
'Pyramidalis'.
Populus alba 'Nivea'. Syn.: *P. alba* var. *argentea.*
SILVER POPLAR.
Populus alba 'Pyramidalis'. Syn.: *P. alba* var. *bolleana.*
P. bolleana. BOLLE'S POPLAR.
Populus balsamifera L. Syn.: *P. candicans.* BALSAM POPLAR.
HACKMATACK. TACAMAHAC.
Populus balsamifera var. *subcordata* Hylander. HEARTLEAF
BALSAM POPLAR.
Populus bolleana Lauche. See *P. alba* 'Pyramidalis'.
Populus × *canadensis* Moench. [*P. deltoides* × *P. nigra*].
CANADIAN POPLAR. CAROLINA POPLAR. Some plants called
P. caroliniana belong here.
Populus × *canadensis* 'Eugenei'. Syn.: *P. eugenei.* EUGENE
POPLAR.
Populus candicans Ait. See *P. balsamifera.*
Populus canescens (Ait.) Sm. GRAY POPLAR.
Populus canescens 'Macrophylla'.
Populus caroliniana. Listed name. Plants so called are
P. × *canadensis* or other spp.
Populus deltoides Bartr. ex Marsh. COTTONWOOD. EASTERN
COTTONWOOD. NECKLACE POPLAR. EASTERN POPLAR.
Populus eugenei Simon-Louis ex K. Koch.
See *P.* × *canadensis* 'Eugenei'.
Populus euphratica G. A. Oliver.
Populus fremontii S. Wats. WESTERN COTTONWOOD. FREMONT
COTTONWOOD.
Populus × *gileadensis* Rouleau. [?*P. balsamifera* ×
P. deltoides]. BALM-OF-GILEAD.
Populus idahoensis?
Populus macrophylla. Listed name. May be *P. canescens*
'Macrophylla'.
Populus nigra L. BLACK POPLAR.
Populus nigra 'Afghanica'. Syn.: *P. nigra* var. *thevestina.*
THEVES POPLAR.
Populus nigra 'Italica'. LOMBARDY POPLAR.
Populus nigra var. *thevestina* (Dode) Bean. See *P. nigra*
'Afghanica'.
Populus sargentii Dode. GREAT PLAINS COTTONWOOD.
Populus simonii Carrière.
Populus tomentosa Carrière. CHINESE WHITE POPLAR.
Populus tremula L. EUROPEAN ASPEN.
Populus tremuloides Michx. QUAKING ASPEN. TREMBLING ASPEN.
QUIVER-LEAF.
Populus trichocarpa Torr. & Gray. BLACK COTTONWOOD.
WESTERN BALSAM POPLAR.
PORK-AND-BEANS: *Sedum* × *rubrotinctum.*
Portulacaria afra (L.) Jacq. ELEPHANT BUSH. ELEPHANT'S FOOD.
SPEKBOOM.
POSSUMWOOD: *Diospyros virginiana.*
POTATO BUSH, BLUE: *Lycianthes rantonnei.*
POTATO CREEPER, GIANT: *Solanum wendlandii.*
POTATO VINE: *Solanum jasminoides.*
POTATO VINE: *Solanum wendlandii.*
Potentilla 'Farreri'. See *P. parvifolia* 'Gold Drop'.
Potentilla fruticosa L. SHRUBBY CINQUEFOIL. GOLDEN HARDTACK.
Some cvs. listed under *P. fruticosa* may be *P. parvifolia* cvs.
Potentilla fruticosa 'Gold Drop'. See *P. parvifolia* 'Gold Drop'.
Potentilla fruticosa 'Katherine Dykes'. See *P. parvifolia*
'Katherine Dykes'.
Potentilla fruticosa 'Parvifolia'. See *P. parvifolia.*
Potentilla fruticosa 'Snowflake'.
Potentilla fruticosa 'Sutter's Gold'.
Potentilla 'Katherine Dykes'. See *P. parvifolia* 'Katherine
Dykes'.
Potentilla parvifolia Lehm. Some cvs. listed under *P. fruticosa*
may belong here.
Potentilla parvifolia 'Farreri'. See *P. parvifolia* 'Gold Drop'.
Potentilla parvifolia 'Gold Drop'.
Potentilla parvifolia 'Katherine Dykes'.
POTHOS, GOLDEN: *Epipremnum aureum.*
Pothos aureus Linden & André. See *Epipremnum aureum.*
POWDER PUFF, PINK: *Calliandra haematocephala.*
PRIDE-OF-CHINA: *Melia azedarach.*
PRIDE-OF-INDIA: *Lagerstroemia speciosa.*
PRIDE-OF-INDIA: *Melia azedarach.*
PRIDE-OF-MADEIRA: *Echium fastuosum.*
PRIDE-OF-TENERIFFE: *Echium pininana.*
PRIMROSE TREE: *Lagunaria patersonii.*
PRINCE'S PLUME: *Stanleya pinnata.*
PRINCESS FLOWER: *Tibouchina urvilleana.*
PRINCESS TREE: *Paulownia tomentosa.*
Prinsepia sinensis (D. Oliver) D. Oliver ex Bean.
Prinsepia uniflora Batal.
Pritchardia beccariana Rock.
Pritchardia hillebrandii Becc.
PRIVET: *Ligustrum vulgare.*
   CALIFORNIA: *Ligustrum ovalifolium.*
   CHINESE: *Ligustrum sinense.*
   ENGLISH: *Ligustrum vulgare.*
   GLOSSY: *Ligustrum lucidum.*
   GOLDEN CALIFORNIA: *Ligustrum ovalifolium* 'Aureum'.
   HENRY'S: *Ligustrum henryi.*
   IBOLIUM: *Ligustrum* × *ibolium.*
   JAPANESE: *Ligustrum japonicum.*
   LUSTER-LEAF: *Ligustrum japonicum.*
   SUWANNEE RIVER: *Ligustrum* 'Suwannee River'.
   TEXAS: *Ligustrum japonicum.*
   VICARY: *Ligustrum* × *vicaryi.*
   VICARY GOLDEN: *Ligustrum* × *vicaryi.*

PRIVET, continued
   WAX-LEAF: **Ligustrum japonicum.**
Prosopis chilensis (Molina) Stuntz. Probably not cult. Plants so called may be **P. glandulosa.**
Prosopis glandulosa Torr. MESQUITE.
Prosopis glandulosa var. *torreyana* (L. Benson) M. C. Johnst. WESTERN HONEY MESQUITE.
Prosopis pubescens Benth. SCREW BEAN MESQUITE. TORNILLO. SCREW BEAN.
Prostanthera nivea A. Cunn. ex Benth. SNOWY MINTBUSH.
Prostanthera rotundifolia R. Br. ROUND-LEAVED MINTBUSH.
PROTEA, GIANT: **Protea cynaroides.**
   KING: **Protea cynaroides.**
   OLEANDER LEAVED: **Protea neriifolia.**
   PEACH: **Protea grandiceps.**
   QUEEN: **Protea barbigera.**
Protea barbigera Meissn. QUEEN PROTEA.
Protea compacta R. Br.
Protea cynaroides L. KING PROTEA. GIANT PROTEA.
Protea eximia (J. Knight) Fourcade. Syn.: *P. latifolia.*
Protea grandiceps Tratt. PEACH PROTEA.
Protea latifolia R. Br. See **P. eximia.**
Protea lepidocarpodendron (L.) L.
Protea longiflora Lam.
Protea longifolia Andr.
Protea macrocephala Thunb.
Protea neriifolia R. Br. OLEANDER LEAVED PROTEA.
Protea obtusifolia Buek.
Protea rouppelliae Meissn.
Protea scolymocephala (L.) Reichard.
Protea speciosa L.
Protea susannae Phillips.
Prunus 'Accolade'. [?*P. sargentii* × *P. subhirtella*].
Prunus 'Akebono'. See **P. × yedoensis 'Akebono'.**
Prunus 'Alma Stultz'. See **P. persica 'Alma Stultz'.**
Prunus 'Amanogawa'. See **P. serrulata 'Amanogawa'.**
Prunus amygdalus Batsch. See **P. dulcis.**
Prunus amygdalus 'Nana'. See **P. dulcis 'Nana'.**
Prunus andersonii Gray. DESERT PEACH.
Prunus armeniaca L. APRICOT.
Prunus armeniaca 'Charles Abraham'. CHARLES ABRAHAM FLOWERING APRICOT.
Prunus autumnalis Koehne ex Sarg. See **P. subhirtella 'Autumnalis'.**
Prunus avium L. BIRD CHERRY. MAZZARD. SWEET CHERRY.
Prunus avium 'Plena'. DOUBLE-FLOWERED MAZZARD CHERRY.
Prunus azorica. Listed name. May be **P. lusitanica** subsp. **azorica.**
Prunus 'Beni Hoshi'. See **P. serrulata 'Beni Hoshi'.**
Prunus besseyi L. H. Bailey. SAND CHERRY. WESTERN SAND CHERRY.
Prunus × blireiana André. [*P. cerasifera* 'Atropurpurea' × *P. mume*]. Syn.: *P. cerasifera* var. *blireiana.* FLOWERING PLUM.
Prunus 'Burbank'. See **P. persica 'Burbank'** or **P. salicina 'Burbank'.**
Prunus 'Camelliaeflora'. See **P. persica 'Camelliaeflora'.**
Prunus campanulata Maxim. TAIWAN CHERRY.
Prunus caroliniana (Mill.) Ait. AMERICAN CHERRY LAUREL. CAROLINA CHERRY LAUREL.
Prunus caroliniana 'Bright 'n Tight'.
Prunus caroliniana 'Compacta'.
Prunus cerasifera J. F. Ehrh. CHERRY PLUM. MYROBALAN PLUM.
Prunus cerasifera 'Atropurpurea'. Syn.: *P. cerasifera* 'Pissardii'. *P. pissardii.* PURPLE-LEAF PLUM. PISSARD PLUM.
Prunus cerasifera var. *blireiana* Bean. See **P. × blireiana.**
Prunus cerasifera 'Hollywood'. HOLLYWOOD FLOWERING PLUM.
Prunus cerasifera 'Krauter Vesuvius'. May be the same as **P. cerasifera 'Vesuvius'.**
Prunus cerasifera 'Newport'.
Prunus cerasifera 'Oregon Trail'.
Prunus cerasifera 'Pendula'. WEEPING FLOWERING PLUM.
Prunus cerasifera 'Pissardii'. See **P. cerasifera 'Atropurpurea'.**
Prunus cerasifera 'Thundercloud'. THUNDERCLOUD FLOWERING PLUM.
Prunus cerasifera 'Vesuvius'. VESUVIUS FLOWERING PLUM.
Prunus cerasoides D. Don.
Prunus cerasus L. SOUR CHERRY. PIE CHERRY.
Prunus cerasus 'Akebono'. See **P. × yedoensis 'Akebono'.**
Prunus × cistena Hansen. [*P. cerasifera* 'Atropurpurea' × *P. pumila*]. DWARF RED-LEAF PLUM. PURPLE-LEAF SAND CHERRY.
Prunus 'Daybreak'. See **P. × yedoensis 'Akebono'.**
Prunus depressa Pursh. Syn.: *P. pumila* var. *depressa.* SAND CHERRY.
Prunus domestica L. EUROPEAN PLUM.
Prunus 'Double White'. See **P. persica 'Double White'.**
Prunus dulcis (Mill.) D. A. Webb. Syn.: *P. amygdalus.* SWEET ALMOND. ALMOND.
Prunus dulcis var. *amara* (DC.) H. E. Moore. BITTER ALMOND.
Prunus dulcis 'Nana'. DWARF SWEET ALMOND.
Prunus 'Early Double Pink'. See **P. persica 'Early Double Pink'.**
Prunus 'Early Double Red'. See **P. persica 'Early Double Red'.**
Prunus fasciculata (Torr.) Gray. DESERT ALMOND. WILD PEACH.
Prunus glandulosa Thunb. DWARF FLOWERING ALMOND.
Prunus glandulosa 'Alboplena'.
Prunus glandulosa 'Sinensis'.
Prunus 'Hally Jolivette'. [(*P. subhirtella* × *P. × yedoensis*) × *P. subhirtella*].
Prunus 'Helen Borchers'. See **P. persica 'Helen Borchers'.**
Prunus 'Hollywood'. See **P. cerasifera 'Hollywood'.**
Prunus 'Icicle'. See **P. persica 'Icicle'.**
Prunus ilicifolia (Nutt. ex Hook. & Arn.) Walp. ISLAY. CALIFORNIA CHERRY. HOLLY-LEAF CHERRY.
Prunus integrifolia Sarg. See **P. lyonii.**
Prunus jacquemontii Hook. f. FLOWERING ALMOND. JACQUEMONT CHERRY.
Prunus 'Kanzan'. See **P. serrulata 'Kwanzan'.**
Prunus 'Krauter Vesuvius'. May be the same as **P. cerasifera 'Vesuvius'.**
Prunus 'Kwanzan'. See **P. serrulata 'Kwanzan'.**
Prunus 'Late Double Red'. See **P. persica 'Late Double Red'.**
Prunus laurocerasus L. Syn.: *Laurocerasus officinalis.* CHERRY LAUREL. ENGLISH LAUREL.
Prunus laurocerasus 'Nana'.
Prunus laurocerasus 'Otto Luyken'.
Prunus laurocerasus 'Schipkaensis'.
Prunus laurocerasus 'Zabeliana'. ZABEL CHERRY LAUREL.
Prunus lusitanica L. Syn.: *Laurocerasus lusitanica.* PORTUGAL LAUREL.
Prunus lusitanica subsp. **azorica** (Mouillefert) Franco. AZORES CHERRY LAUREL.
Prunus lyonii (Eastw.) Sarg. Syn.: *P. integrifolia.* CATALINA CHERRY.
Prunus 'Mandarin'. See **P. persica 'Dwarf Mandarin'.**
Prunus 'Mount Fuji'. See **P. serrulata 'Shirotae'.**
Prunus mume Siebold & Zucc. JAPANESE FLOWERING APRICOT. JAPANESE APRICOT.
Prunus mume 'Bonita'.
Prunus mume 'Dawn'.
Prunus mume 'Peggy Clarke'.

Prunus mume 'Rosemary Clarke'. ROSEMARY CLARKE FLOWERING APRICOT.
Prunus 'Naden'. See **P. serrulata** 'Takasago'.
Prunus 'Newport'. See **P. cerasifera** 'Newport'.
Prunus pendula. Listed name for **P. serrulata** 'Kiku-shidare-zakura' or **P. subhirtella** 'Pendula'.
Prunus pendula plena rosea. Listed name for **P. subhirtella** 'Pendula Plena Rosea'.
Prunus pendula rosea. Listed name for **P. subhirtella** 'Pendula'.
Prunus persica (L.) Batsch. Syn.: *Amygdalus persica*. PEACH.
Prunus persica 'Alma Stultz'. AZALEA-FLOWERED NECTARINE.
Prunus persica 'Burbank'. BURBANK FLOWERING PEACH.
Prunus persica 'Camelliaeflora'.
Prunus persica 'Daily News One Star'.
Prunus persica 'Double White'.
Prunus persica 'Dwarf Mandarin'. DWARF FLOWERING PEACH.
Prunus persica 'Early Double Pink'.
Prunus persica 'Early Double Red'.
Prunus persica 'Golden Treasure'?
Prunus persica 'Helen Borchers'.
Prunus persica 'Icicle'.
Prunus persica 'Late Double Red'.
Prunus persica var. *nectarina* (Ait. f.) Maxim. See **P. persica** var. *nucipersica*.
Prunus persica var. *nucipersica* (Borkh.) Schneid. Syn.: *P. persica* var. *nectarina*. NECTARINE.
Prunus persica 'Peppermint Stick'.
Prunus persica 'Royal Red Leaf'.
Prunus persica 'San Jose Pink'.
Prunus persica 'Weeping Double Pink'.
Prunus persica 'Weeping Double Red'.
Prunus pissardii Carrière. See **P. cerasifera** 'Atropurpurea'.
Prunus pumila L. SAND CHERRY. DWARF CHERRY.
Prunus pumila var. *depressa* (Pursh) Bean. See **P. depressa**.
Prunus rosea. Listed name for **P. subhirtella** 'Rosea'.
Prunus 'Royal Red Leaf'. See **P. persica** 'Royal Red Leaf'.
Prunus salasii Standl.
Prunus salicina Lindl. JAPANESE PLUM.
Prunus salicina 'Burbank'. BURBANK PLUM.
Prunus 'San Jose Pink'. See **P. persica** 'San Jose Pink'.
Prunus sargentii Rehd. NORTH JAPANESE HILL CHERRY. SARGENT CHERRY.
Prunus serrula Franch. BIRCH BARK CHERRY.
Prunus serrulata Lindl. JAPANESE FLOWERING CHERRY.
Prunus serrulata 'Amanogawa'. AMANOGAWA CHERRY.
Prunus serrulata 'Amayadori'.
Prunus serrulata 'Beni Higan'. See **P. subhirtella** 'Rosea'.
Prunus serrulata 'Beni Hoshi'.
Prunus serrulata 'Botan-zakura'.
Prunus serrulata 'Cheal's Weeping'. Not cult. Plants so called are **P. serrulata** 'Kiku-shidare-zakura'.
Prunus serrulata 'Fugenzo'.
Prunus serrulata 'Hi-zakura'.
Prunus serrulata 'Kiku-shidare'. See **P. serrulata** 'Kiku-shidare-zakura'.
Prunus serrulata 'Kiku-shidare-zakura'. DOUBLE WEEPING FLOWERING CHERRY. ORIENTAL WEEPING CHERRY. Plants called **P. serrulata** 'Cheal's Weeping' belong here.
Prunus serrulata 'Kirin'.
Prunus serrulata 'Kwanzan'.
Prunus serrulata 'Mikuruma-gaeshi'.
Prunus serrulata 'Mount Fuji'. See **P. serrulata** 'Shirotae'.
Prunus serrulata 'Ojochin'.
Prunus serrulata 'Pendula'. See **P. serrulata** 'Kiku-shidare-zakura'.
Prunus serrulata 'Sekiyama'. See **P. serrulata** 'Kwanzan'.
Prunus serrulata 'Shirofugen'.
Prunus serrulata 'Shirotae'. MOUNT FUJI CHERRY.
Prunus serrulata 'Shogetsu'.
Prunus serrulata 'Sieboldii'. See **P. serrulata** 'Takasago'.
Prunus serrulata 'Tai Haku'. GREAT WHITE CHERRY.
Prunus serrulata 'Takasago'. Syn.: *P. sieboldii*. TAKASAGO CHERRY. NADEN CHERRY.
Prunus serrulata 'Tanko Shinju'. TANKO SHINJU CHERRY.
Prunus serrulata 'Ukon'. UKON CHERRY.
Prunus serrulata 'Washino-o'. WASHINO-O CHERRY.
Prunus 'Shirofugen'. See **P. serrulata** 'Shirofugen'.
Prunus 'Shogetsu'. See **P. serrulata** 'Shogetsu'.
Prunus sieboldii (Carrière) Wittm. See **P. serrulata** 'Takasago'.
Prunus subcordata Benth. SIERRA PLUM.
Prunus subhirtella Miq. ROSEBUD CHERRY. SPRING CHERRY. HIGAN CHERRY.
Prunus subhirtella 'Autumnalis'. Syn.: *P. autumnalis*. AUTUMN FLOWERING HIGAN CHERRY.
Prunus subhirtella 'Beni-higan'. Plants so called are **P. subhirtella** 'Rosea'.
Prunus subhirtella 'Hally Jolivette'. See **P.** 'Hally Jolivette'.
Prunus subhirtella 'Pendula'. SINGLE WEEPING CHERRY. SINGLE WEEPING ROSEBUD CHERRY.
Prunus subhirtella 'Pendula Plena Rosea'. DOUBLE WEEPING ROSEBUD CHERRY. YAE-SHIDARE-HIGAN.
Prunus subhirtella 'Pendula Rosea'. See **P. subhirtella** 'Pendula'.
Prunus subhirtella 'Rosea'. PINK SPRING CHERRY. PINK HIGAN CHERRY.
Prunus subhirtella 'Shiro-higan'. Probably **P. subhirtella** or a selected cv., such as **P. subhirtella** 'Rosea'.
Prunus subhirtella 'Whitcombii'.
Prunus subhirtella 'Yae-shidare-higan'. See **P. subhirtella** 'Pendula Plena Rosea'.
Prunus tenella Batsch. DWARF RUSSIAN ALMOND.
Prunus tenella 'Alba'.
Prunus 'Thundercloud'. See **P. cerasifera** 'Thundercloud'.
Prunus tomentosa Thunb. DOWNY CHERRY. NANKING CHERRY. MANCHU CHERRY.
Prunus 'Trailblazer'. TRAILBLAZER PLUM.
Prunus triloba Lindl. FLOWERING ALMOND.
Prunus 'Ukon'. See **P. serrulata** 'Ukon'.
Prunus 'Vesuvius'. See **P. cerasifera** 'Vesuvius'.
Prunus 'Whitcombii'. See **P. subhirtella** 'Whitcombii'.
Prunus × yedoensis Matsumura. [?**P. serrulata** × **P. subhirtella**]. JAPANESE CHERRY. YOSHINO CHERRY. POTOMAC CHERRY.
Prunus × yedoensis 'Akebono'. AKEBONO CHERRY. Plants called *P. cerasus* 'Akebono' belong here.
Prunus 'Zabeliana'. See **P. laurocerasus** 'Zabeliana'.
Pseudobombax ellipticum (HBK) Dug. Syn.: *Bombax ellipticum*. SHAVING-BRUSH TREE.
Pseudolarix amabilis (J. Nels.) Rehd. See **P. kaempferi**.
Pseudolarix kaempferi Gord. Syn.: *P. amabilis*. *Chrysolarix amabilis*. GOLDEN LARCH.
Pseudopanax crassifolium (A. Cunn.) K. Koch. LANCEWOOD.
Pseudopanax discolor (T. Kirk) Harms.
Pseudopanax ferox (T. Kirk) T. Kirk.
Pseudopanax lessonii (DC.) K. Koch.
Pseudosasa japonica (Siebold & Zucc. ex Steud.) Makino. See **Sasa japonica**.
Pseudosasa japonica 'Tsutsumiana'. See **Sasa japonica** var. *tsutsumiana*.
Pseudotsuga macrocarpa (Vasey) Mayr. BIGCONE SPRUCE.
Pseudotsuga menziesii (Mirb.) Franco. Syn.: *P. taxifolia*. DOUGLAS FIR. OREGON PINE.
Pseudotsuga menziesii var. *glauca* (Beissner) Franco.

Pseudotsuga taxifolia (Lamb.) Britt. See **P. menziesii**.
Psidium cattleianum Sabine. Syn.: *P. littorale*. STRAWBERRY GUAVA.
Psidium cattleianum 'Lucidum'. YELLOW STRAWBERRY GUAVA.
Psidium guajava L. GUAVA. LEMON GUAVA.
Psidium guajava 'Peruvian Pink'.
Psidium littorale Raddi. See **P. cattleianum**.
Psidium 'Lucidum'. See **P. cattleianum 'Lucidum'**.
Psoralea pinnata L.
Ptelea trifoliata L. STINKING ASH. WATER ASH.
Pterocarya stenoptera C. DC. CHINESE WINGNUT.
Pterostyrax hispida Siebold & Zucc. EPAULETTE TREE.
Ptychosperma alexandrae F. J. Muell. See **Archontophoenix alexandrae**.
Ptychosperma cunninghamianum H. Wendl. See **Archontophoenix cunninghamiana**.
Ptychosperma elegans (R. Br.) Blume. Syn.: *Seaforthia elegans*. ALEXANDER PALM. SOLITAIRE PALM.
Ptychosperma macarthurii (H. Wendl.) Nichols. Syn.: *Actinophloeus macarthurii*. MACARTHUR PALM. MACARTHUR FEATHER PALM. HURRICANE PALM.
PUKA: **Meryta sinclairii**.
PUMELO: **Citrus maxima**.
PUMMELO: **Citrus maxima**.
   CHANDLER'S: **Citrus maxima 'Chandler'**.
Punica 'Chico'. See **P. granatum 'Chico'**.
Punica granatum L. POMEGRANATE. FLOWERING POMEGRANATE.
Punica granatum 'Chico'. DWARF CARNATION-FLOWERED POMEGRANATE.
Punica granatum 'Legrelle'.
Punica granatum 'Nana'. DWARF POMEGRANATE.
Punica granatum 'Nana Plena'. DOUBLE DWARF POMEGRANATE.
Punica granatum 'Variegatum'.
Punica granatum 'Wonderful'.
Punica 'Legrelle'. See **P. granatum 'Legrelle'**.
PUNK TREE: **Melaleuca quinquenervia**.
PURPLE GLORY PLANT: **Sutera grandiflora**.
Purshia glandulosa Curran.
Purshia tridentata (Pursh) DC. ANTELOPE BRUSH.
PYRACANTHA, LALANDE: **Pyracantha coccinea 'Lalandei'**.
   ROSEDALE: **Pyracantha koidzumii 'Rosedale'**.
   SAN JOSE: **Pyracantha koidzumii 'San Jose'**.
   SANTA CRUZ: **Pyracantha koidzumii 'Santa Cruz'**.
   STRIBLING'S: **Pyracantha koidzumii 'Striblingii'**.
   VICTORY: **Pyracantha koidzumii 'Victory'**.
   WALDER: **Pyracantha koidzumii 'Walderi Prostrata'**.
Pyracantha angustifolia (Franch.) Schneid.
Pyracantha angustifolia 'Gnome'.
Pyracantha atalantioides (Hance) Stapf. Syn.: *P. gibbsii*.
Pyracantha atalantioides 'Aurea'.
Pyracantha 'Aurea'. See **P. atalantioides 'Aurea'** or **P. coccinea 'Fructo Luteo'**.
Pyracantha 'Boxwood'.
Pyracantha coccinea M. J. Roemer.
Pyracantha coccinea 'Aurea'. See **P. coccinea 'Fructo Luteo'**.
Pyracantha coccinea 'Fructo Luteo'.
Pyracantha coccinea 'Government Red'.
Pyracantha coccinea 'Kasan'.
Pyracantha coccinea 'Lalandei'. LALANDE PYRACANTHA.
Pyracantha coccinea 'Lalandei Monrovia'.
Pyracantha coccinea 'Lalandei Thornless'.
Pyracantha coccinea 'Pauciflora'.
Pyracantha coccinea 'Pyra-Box'.
Pyracantha coccinea 'Wyattii'.
Pyracantha crenato-serrata (Hance) Rehd. Syn.: *P. fortuneana. P. yunnanensis*.
Pyracantha crenato-serrata 'Graberi'.
Pyracantha crenato-serrata 'Lewisii'. See **P. koidzumii 'Lewisii'**.
Pyracantha crenato-serrata 'Lynnii'.
Pyracantha crenato-serrata 'Rosedale'. See **P. koidzumii 'Rosedale'**.
Pyracantha crenato-serrata 'Santa Ana'.
Pyracantha crenulata (D. Don) M. J. Roemer.
Pyracantha crenulata var. *rogersiana* A. B. Jackson. See **P. rogersiana**.
Pyracantha crenulata 'Rubra'.
Pyracantha 'Duvalii'. [*P. crenato-serrata* × *P. koidzumii*].
Pyracantha formosana. Listed name for **P. koidzumii**.
Pyracantha fortuneana (Maxim.) Li. See **P. crenato-serrata**.
Pyracantha fortuneana 'Graberi'. See **P. crenato-serrata 'Graberi'**.
Pyracantha fortuneana 'Lynii'. See **P. crenato-serrata 'Lynnii'**.
Pyracantha fortuneana 'Orange Glow'. See **P. 'Orange Glow'**.
Pyracantha fortuneana 'Rosedale'. See **P. koidzumii 'Rosedale'**.
Pyracantha fortuneana 'Santa Ana'. See **P. crenato-serrata 'Santa Ana'**.
Pyracantha 'Fructo Luteo'. See **P. coccinea 'Fructo Luteo'**.
Pyracantha gibbsii A. B. Jackson. See **P. atalantioides**.
Pyracantha 'Gnome'. See **P. angustifolia 'Gnome'**.
Pyracantha 'Golden Delicious'.
Pyracantha 'Government Orange'. See **P. koidzumii 'Government Orange'**.
Pyracantha 'Government Red'. See **P. coccinea 'Government Red'**.
Pyracantha 'Graberi'. See **P. crenato-serrata 'Graberi'**.
Pyracantha 'Kasan'. See **P. coccinea 'Kasan'**.
Pyracantha koidzumii (Hayata) Rehd.
Pyracantha koidzumii 'Government Orange'.
Pyracantha koidzumii 'Government Red'. See **P. coccinea 'Government Red'**.
Pyracantha koidzumii 'Lewisii'.
Pyracantha koidzumii 'Low-Dense'.
Pyracantha koidzumii 'Rosedale'. ROSEDALE PYRACANTHA.
Pyracantha koidzumii 'San Jose'. SAN JOSE PYRACANTHA.
Pyracantha koidzumii 'Santa Cruz'. SANTA CRUZ PYRACANTHA.
Pyracantha koidzumii 'Striblingii'. STRIBLING'S PYRACANTHA.
Pyracantha koidzumii 'Victory'. VICTORY PYRACANTHA.
Pyracantha koidzumii 'Walderi Prostrata'. WALDER PYRACANTHA.
Pyracantha 'Lalandei'. See **P. coccinea 'Lalandei'**.
Pyracantha 'Lalandei Monrovia'. See **P. coccinea 'Lalandei Monrovia'**.
Pyracantha 'Lalandei Thornless'. See **P. coccinea 'Lalandei Thornless'**.
Pyracantha 'Lewisii'. See **P. koidzumii 'Lewisii'**.
Pyracantha 'Low-Dense'. See **P. koidzumii 'Low-Dense'**.
Pyracantha 'Lynnii'. See **P. crenato-serrata 'Lynnii'**.
Pyracantha 'Mohave'. [*P. koidzumii* × *P. coccinea*].
Pyracantha 'Orange Glow'. [?*P. crenato-serrata* × *P. coccinea*].
Pyracantha 'Pauciflora'. See **P. coccinea 'Pauciflora'**.
Pyracantha 'Pyra-Box'. See **P. coccinea 'Pyra-Box'**.
Pyracantha 'Radiance'. RADIANT PYRACANTHA.
Pyracantha 'Red Elf'.
Pyracantha rogersiana (A. B. Jackson) Bean. Syn.: *P. crenulata* var. *rogersiana*.
Pyracantha 'Rosedale'. See **P. koidzumii 'Rosedale'**.
Pyracantha 'Rubra'. See **P. crenulata 'Rubra'**.
Pyracantha 'San Jose'. See **P. koidzumii 'San Jose'**.
Pyracantha 'Santa Ana'. See **P. crenato-serrata 'Santa Ana'**.
Pyracantha 'Santa Cruz'. See **P. koidzumii 'Santa Cruz'**.
Pyracantha 'Shawnee'. [*P. crenato-serrata* × *P. koidzumii*].
Pyracantha 'Striblingii'. See **P. koidzumii 'Striblingii'**.
Pyracantha 'Tiny Tim'.

Pyracantha 'Victory'. See **P. koidzumii 'Victory'**.
Pyracantha 'Walderi'. See **P. koidzumii 'Walderi Prostrata'**.
Pyracantha 'Walderi Prostrata'. See **P. koidzumii 'Walderi Prostrata'**.
Pyracantha 'Watereri'. [?**P. coccinea** 'Lalandei' × **P. rogersiana**].
Pyracantha 'Wyattii'. See **P. coccinea 'Wyattii'**.
Pyracantha yunnanensis (M. L. Vilm. ex Mottet) Chitt. See **P. crenato-serrata**.
× *Pyracomeles vilmorinii* Rehd. ex Guillaum. [**Pyracantha atalantioides** × **Osteomeles subrotunda** K. Koch].
Pyrostegia ignea (Vell.) K. Presl. See **P. venusta**.
Pyrostegia venusta (Ker-Gawl.) Miers. Syn.: *Bignonia venusta. Pyrostegia ignea*. FLAME VINE. FLAMING TRUMPET. FLAME FLOWER.
Pyrus baccata L. See **Malus baccata**.
Pyrus betulifolia Bunge.
Pyrus calleryana Decne. CALLERY PEAR.
Pyrus calleryana 'Aristocrat'.
Pyrus calleryana 'Bradford'. BRADFORD PEAR.
Pyrus communis L. PEAR.
Pyrus coronaria L. See **Malus coronaria**.
Pyrus faurei Schneid.
Pyrus ioensis (Wood) L. H. Bailey. See **Malus ioensis**.
Pyrus japonica Thunb. See **Chaenomeles japonica**.
Pyrus kawakamii Hayata. EVERGREEN PEAR.
Pyrus malus L. See **Malus domestica**.
Pyrus 'Okusankichi'. LATE JAPANESE PEAR. OCTOBER JAPANESE PEAR. NIHON NASHI. Plants called *P. pyrifolia* 'Nihon Nashi' belong here.
Pyrus pyrifolia (Burm. f.) Nakai. Syn.: *P. serotina*. SAND PEAR. JAPANESE SAND PEAR.
Pyrus pyrifolia var. *culta* (Makino) Nakai. NASHI.
Pyrus pyrifolia 'Nihon Nashi'. See **P. 'Okusankichi'**.
Pyrus serotina Rehd. See **P. pyrifolia**.

Q

QUAIL-BUSH: **Atriplex lentiformis**.
QUARTER VINE: **Bignonia capreolata**.
QUEEN-OF-SHEBA VINE: **Podranea brycei**.
QUEENSLAND NUT: **Macadamia integrifolia**.
  SMALL-FRUITED: **Macadamia ternifolia**.
QUEEN'S TEARS: **Bilbergia nutans**.
QUEEN'S WREATH: **Antigonon leptopus**.
Quercus acutissima Carruthers. CHESTNUT-LEAVED OAK. SAW-TOOTH OAK.
Quercus agrifolia Née. COAST LIVE OAK. CALIFORNIA LIVE OAK.
Quercus alba L. AMERICAN WHITE OAK.
Quercus bicolor Willd. SWAMP WHITE OAK.
Quercus borealis Michx. f. See **Q. rubra**.
Quercus borealis var. *maxima* Marsh. See **Q. rubra**.
Quercus cerris L. TURKEY OAK.
Quercus chrysolepis Liebm. CANYON OAK. MAUL OAK. GOLDEN CUP OAK.
Quercus coccinea Muenchh. SCARLET OAK.
Quercus douglasii Hook. & Arn. BLUE OAK.
Quercus dumosa Nutt. CALIFORNIA SCRUB OAK.
Quercus dumosa var. *turbinella* (Greene) Jeps. See **Q. turbinella**.
Quercus durata Jeps. LEATHER OAK.
Quercus engelmannii Greene. MESA OAK. ENGELMANN OAK.

Quercus garryana Dougl. ex Hook. OREGON WHITE OAK. GARRY OAK.
Quercus ilex L. HOLLY OAK. HOLM OAK. TRUFFLE OAK.
Quercus kelloggii Newb. KELLOGG OAK. CALIFORNIA BLACK OAK.
Quercus lobata Née. VALLEY OAK. CALIFORNIA WHITE OAK.
Quercus lyrata Walter. OVERCUP OAK. SWAMP POST OAK.
Quercus macrocarpa Michx. BURR OAK. MOSSY-CUP OAK.
Quercus macrolepis Kotschy.
Quercus macrophylla Née. LARGE-LEAVED OAK.
Quercus myrsinifolia Blume.
Quercus nigra L. WATER OAK. POSSUM OAK.
Quercus palustris Muenchh. PIN OAK.
Quercus phellos L. WILLOW OAK.
Quercus robur L. ENGLISH OAK. TRUFFLE OAK.
Quercus robur 'Fastigiata'. PYRAMIDAL ENGLISH OAK.
Quercus rubra L. Syn.: *Q. borealis. Q. borealis* var. *maxima*. RED OAK. EASTERN RED OAK.
Quercus suber L. CORK OAK.
Quercus turbinella Greene. Syn.: *Q. dumosa* var. *turbinella*. SHRUB LIVE OAK.
Quercus turbinella subsp. *californica* Tucker. CALIFORNIA SCRUB OAK.
Quercus velutina Lam. EASTERN BLACK OAK. YELLOW-BARKED OAK.
Quercus virginiana Mill. SOUTHERN LIVE OAK.
Quercus virginiana var. *eximia* Sarg. See **Q. virginiana** var. *maritima*.
Quercus virginiana var. *maritima* (Michx.) Sarg. Syn.: *Q. virginiana* var. *eximia*. SAND LIVE OAK.
Quercus wislizeni A. DC. INTERIOR LIVE OAK.
Quercus wislizeni var. *frutescens* Engelm. DWARF INTERIOR LIVE OAK.
Quillaja saponaria Molina. SOAP-BARK TREE.
QUINCE: **Cydonia oblonga**.
  CHINESE: **Cydonia sinensis**.
  FLOWERING: **Chaenomeles speciosa**.
  JAPANESE: **Chaenomeles speciosa**.
  JAPANESE FLOWERING: **Chaenomeles japonica**.
QUIVER-LEAF: **Populus tremuloides**.

R

RABBITBUSH: **Chrysothamnus nauseosus**.
RAIN TREE, BRAZIL: **Brunfelsia pauciflora** var. **calycina**.
  EVERGREEN GOLDEN: **Koelreuteria bipinnata**.
  GOLDEN: **Koelreuteria paniculata**.
RAISIN TREE, JAPANESE: **Hovenia dulcis**.
RAMBUTAN, SMOOTH: **Alectryon subcinereus**.
RANGPUR: **Citrus × limonia**.
Raoulia australis Hook. f.
Raphidophora aurea (Linden & André) Birdsey. See **Epipremnum aureum**.
Raphidophora decursiva (Roxb.) Schott.
Raphiolepis. See **Rhaphiolepis**.
Raphionacme velutina Schlechter.
RASPBERRY, EUROPEAN: **Rubus idaeus**.
  RED: **Rubus idaeus**.
RATA, NORTH ISLAND: **Metrosideros robustus**.
  SOUTHERN: **Metrosideros umbellatus**.
RATAMA: **Parkinsonia aculeata**.
RAT-STRIPPER: **Paxistima canbyi**.
Rauvolfia samarensis Merrill. Some plants called **Alstonia scholaris** belong here.

Rauwolfia. See **Rauvolfia.**
Ravenala madagascariensis Sonn. TRAVELER'S TREE. TRAVELER'S PALM.
REDBERRY: **Rhamnus crocea.**
   HOLLYLEAF: **Rhamnus crocea** subsp. *ilicifolia.*
   ISLAND: **Rhamnus crocea** subsp. *pyrifolia.*
REDBUD, CALIFORNIA: **Cercis occidentalis.**
   CHINESE: **Cercis chinensis.**
   EASTERN: **Cercis canadensis.**
   OKLAHOMA: **Cercis reniformis 'Oklahoma'.**
   WESTERN: **Cercis occidentalis.**
RED-HEART: **Ceanothus spinosus.**
RED-HOT-POKER: **Kniphofia uvaria.**
REDWOOD, COAST: **Sequoia sempervirens.**
   DAWN: **Metasequoia glyptostroboides.**
   JAPANESE: **Cryptomeria japonica.**
   SIERRA: **Sequoiadendron giganteum.**
REED, GIANT: **Arundo donax.**
Reinwardtia indica Dumort. Syn.: *R. trigyna.* YELLOW FLAX BUSH. Some plants called **Linum flavum** belong here.
Reinwardtia trigyna (Roxb.) Planch. See **R. indica.**
Relhania quinquenervis Thunb.
Retinospora. See **Chamaecyparis** and **Thuja.**
RETINOSPORA, DWARF GOLDEN THREAD-BRANCH: **Chamaecyparis pisifera 'Filifera Aurea Nana'.**
   GOLDEN THREAD-BRANCH: **Chamaecyparis pisifera 'Filifera Aurea'.**
   SAWARA: **Chamaecyparis pisifera.**
   THREAD-BRANCH: **Chamaecyparis pisifera 'Filifera'.**
Rhamnus alaternus L. ITALIAN BUCKTHORN.
Rhamnus alaternus 'Argenteo-variegata'. VARIEGATED ITALIAN BUCKTHORN.
Rhamnus alaternus 'John Edwards'.
Rhamnus alaternus 'Variegata'. See **R. alaternus 'Argenteo-variegata'.**
Rhamnus californica Eschsch. COFFEEBERRY.
Rhamnus californica 'Compacta'.
Rhamnus californica subsp. *crassifolia* (Jeps.) C. B. Wolf.
Rhamnus californica 'Eve Case'.
Rhamnus californica 'Sea View'.
Rhamnus californica subsp. *tomentella* (Benth.) C. B. Wolf. CHAPARRAL COFFEEBERRY.
Rhamnus cathartica L. EUROPEAN BUCKTHORN.
Rhamnus crocea Nutt. BUCKTHORN. REDBERRY.
Rhamnus crocea subsp. *ilicifolia* (Kell.) C. B. Wolf. HOLLYLEAF REDBERRY.
Rhamnus crocea subsp. *pyrifolia* (Greene) C. B. Wolf. ISLAND REDBERRY.
Rhamnus frangula L. ALDER BUCKTHORN.
Rhamnus frangula 'Columnaris'.
Rhamnus purshiana DC. CASCARA SAGRADA. CASCARA.
Rhamnus rubra Greene.
Rhaphidophora. See **Raphidophora.**
Rhaphiolepis 'Clara'. See **R. indica 'Clara'.**
Rhaphiolepis 'Coates Crimson'. See **R. indica 'Coates Crimson'.**
Rhaphiolepis × *delacourii* André. [**R. indica** × **R. umbellata**].
Rhaphiolepis indica Lindl. Syn.: *R. rubra.* INDIAN HAWTHORN.
Rhaphiolepis indica 'Apple Blossom'.
Rhaphiolepis indica 'Ballerina'.
Rhaphiolepis indica 'Bill Evans'.
Rhaphiolepis indica 'Clara'.
Rhaphiolepis indica 'Coates Crimson'.
Rhaphiolepis indica 'Compacta'.
Rhaphiolepis indica 'Coolidge Primrose'.
Rhaphiolepis indica 'Enchantress'.
Rhaphiolepis indica 'Fascination'.
Rhaphiolepis indica 'Flamingo'.
Rhaphiolepis indica 'Jack Evans'.
Rhaphiolepis indica 'Pink Cloud'.
Rhaphiolepis indica 'Pinkie'.
Rhaphiolepis indica 'Pink Lady'.
Rhaphiolepis indica 'Rosea'.
Rhaphiolepis indica 'Santa Barbara'.
Rhaphiolepis indica 'Snow White'.
Rhaphiolepis indica 'Springtime'.
Rhaphiolepis indica 'W. B. Clarke'.
Rhaphiolepis 'Majestic Beauty'.
Rhaphiolepis ovata Briot. See **R. umbellata.**
Rhaphiolepis 'Pinkie'. See **R. indica 'Pinkie'.**
Rhaphiolepis 'Pink Lady'. See **R. indica 'Pink Lady'.**
Rhaphiolepis 'Rosea Compacta'. See **R. indica 'Compacta'.**
Rhaphiolepis rubra Lindl. See **R. indica.**
Rhaphiolepis umbellata (Thunb.) Makino. Syn.: *R. ovata. R. umbellata* forma *ovata.* YEDDO HAWTHORN.
Rhaphiolepis umbellata forma *ovata* (Briot) Schneid. See **R. umbellata.**
Rhapidophyllum hystrix (Pursh) H. Wendl. & Drude. NEEDLE PALM. DWARF SAW PALMETTO.
Rhapis excelsa (Thunb.) Henry. LADY PALM.
Rhapis humilis Blume. RATTAN PALM. SLENDER LADY PALM.
RHODODENDRON, CAROLINA: **Rhododendron carolinianum.**
   CATAWBA: **Rhododendron catawbiense.**
   FRINGED: **Rhododendron ciliatum.**
   POOR MAN'S: **Impatiens sodenii.**
   TREE: **Rhododendron arboreum.**
   WEST COAST: **Rhododendron macrophyllum.**
   WESTERN: **Rhododendron macrophyllum.**
   YUNNAN: **Rhododendron yunnanense.**
Rhododendron 'A. Bedford'. See **R. 'Arthur Bedford'.**
Rhododendron 'Alaska'.
Rhododendron 'Albert and Elizabeth'.
Rhododendron 'Alice'.
Rhododendron 'Anah Kruschke'.
Rhododendron 'Anchorite'.
Rhododendron 'Anna Rose Whitney'.
Rhododendron 'Antoon van Welie'.
Rhododendron anwheiense Wils.
Rhododendron 'Aphrodite'.
Rhododendron arboreum Sm. TREE RHODODENDRON.
Rhododendron 'Arthur Bedford'.
Rhododendron augustinii Hemsl.
Rhododendron augustinii 'Towercourt'.
Rhododendron 'Avenir'.
Rhododendron barbatum Wallich.
Rhododendron 'Belle Heller'.
Rhododendron 'Beni-Kirishima'.
Rhododendron 'Betty Wormald'.
Rhododendron 'Blaney's Blue'. [R. 'Blue Diamond' × R. augustinii 'Towercourt'].
Rhododendron 'Blue Diamond'.
Rhododendron 'Blue Peter'.
Rhododendron 'Blue Tit'.
Rhododendron 'Bo-Peep'.
Rhododendron 'Bow Bells'.
Rhododendron brachyanthum Hutchinson.
Rhododendron 'Brazil'.
Rhododendron 'Bric-a-brac'.
Rhododendron 'Bridesmaid'.
Rhododendron 'Brilliant'.
Rhododendron 'Britannia'.
Rhododendron 'Broughtonii Aureum'.
Rhododendron 'Buccaneer'.

Rhododendron 'Bunkwa'.
Rhododendron burmanicum Hutchinson.
Rhododendron caeruleum Lév.
Rhododendron calendulaceum (Michx.) Torr. FLAME AZALEA.
Rhododendron 'California Sunset'.
Rhododendron calophytum Franch.
Rhododendron campylocarpum Hook. f.
Rhododendron campylogynum Franch.
Rhododendron 'Carita'.
Rhododendron 'Caroline Gable'.
Rhododendron carolinianum Rehd. CAROLINA RHODODENDRON.
Rhododendron catawbiense Michx. CATAWBA RHODODENDRON. MOUNTAIN ROSEBAY.
Rhododendron 'Cecile'.
Rhododendron 'Charles Encke'.
Rhododendron chartophyllum Franch.
Rhododendron 'Chikor'.
Rhododendron 'Chimes'.
Rhododendron 'Christmas Cheer'.
Rhododendron chryseum Balf. f. & F. K. Ward.
Rhododendron ciliatum Hook. f. FRINGED RHODODENDRON.
Rhododendron 'Cilpinense'.
Rhododendron cinnabarinum Hook. f.
Rhododendron 'C. I. S.'.
Rhododendron 'Coccinea'. See R. 'Formosa'.
Rhododendron 'Coccinea Speciosa'.
Rhododendron 'Conemaugh'.
Rhododendron 'Coral Bells'.
Rhododendron 'Cornubia'.
Rhododendron 'Cotton Candy'.
Rhododendron 'Countess of Derby'.
Rhododendron 'Countess of Haddington'.
Rhododendron 'Countess of Sefton'.
Rhododendron 'Crest'.
Rhododendron 'Cunningham's White'.
Rhododendron 'Cynthia'.
Rhododendron dauricum L.
Rhododendron 'David'.
Rhododendron davidsonianum Rehd. & Wils.
Rhododendron degronianum Carrière.
Rhododendron 'Doctor Bergmann'.
Rhododendron 'Dorothy Gish'.
Rhododendron 'Duc de Rohan'.
Rhododendron 'Earl of Athlone'.
Rhododendron 'Easter Bonnet'.
Rhododendron 'Easter Parade'.
Rhododendron edgeworthii Hook. f.
Rhododendron 'Elegans Superba'.
Rhododendron 'Elizabeth'.
Rhododendron 'Else Frye'.
Rhododendron 'Everest'.
Rhododendron 'Exbury'.
Rhododendron 'Fabia'.
Rhododendron 'Fabia Tangerine'.
Rhododendron falconeri Hook. f.
Rhododendron fastigiatum Franch.
Rhododendron 'Fastuosum Flore Pleno'.
Rhododendron 'Fedora'.
Rhododendron ferrugineum L. ALPINE ROSE.
Rhododendron 'Fielder's White'.
Rhododendron 'Flame Creeper'.
Rhododendron 'Flamingo'.
Rhododendron flavum (Hoffmannsegg) G. Don. See R. luteum.
Rhododendron 'Formosa'.
Rhododendron forrestii Balf. f. ex Diels.

Rhododendron forrestii var. *repens* (Balf. f. & Forrest) Cowan & Davidian.
Rhododendron 'Forsteranum'.
Rhododendron 'Fragrantissimum'.
Rhododendron 'Fred Sanders'. See R. 'Mrs. Frederick Sanders'.
Rhododendron fulgens Hook. f.
Rhododendron 'Geisha'.
Rhododendron 'George Lindley Taber'.
Rhododendron 'Gibraltar'.
Rhododendron 'Glacier'.
Rhododendron 'Glamour'.
Rhododendron glaucophyllum Rehd.
Rhododendron 'Glory of Sunninghill'.
Rhododendron 'Gomer Waterer'.
Rhododendron 'Gumpo'.
Rhododendron hanceanum Hemsl. 'Nanum'.
Rhododendron 'Harvest Moon'.
Rhododendron 'Helene Schiffner'.
Rhododendron 'Herbert'.
Rhododendron 'Hexe'.
Rhododendron 'Hino-crimson'.
Rhododendron 'Hinodegiri'.
Rhododendron 'Hinomayo'.
Rhododendron hippophaeoides Balf. f. & W. W. Sm.
Rhododendron 'Homebush'.
Rhododendron hyperythrum Hayata.
Rhododendron impeditum Balf. f. & W. W. Sm.
Rhododendron 'Indica Alba'. See R. mucronatum.
Rhododendron 'Indica Rosea'. See R. mucronatum 'Sekidera'.
Rhododendron indicum (L.) Sweet.
Rhododendron indicum 'Balsaminiflorum'.
Rhododendron intricatum Franch.
Rhododendron 'Iveryana'.
Rhododendron 'Jan Dekens'.
Rhododendron japonicum (Gray) Suringar.
Rhododendron 'John Cairns'.
Rhododendron 'Judge Solomon'. See R. 'Southern Charm'.
Rhododendron kaempferi Planch. TORCH AZALEA. KAEMPFER AZALEA.
Rhododendron keiskei Miq.
Rhododendron keleticum Balf. f. & Forrest.
Rhododendron kiusianum Makino. KYUSHU AZALEA.
Rhododendron kotschyi Simonkai. Syn.: R. *myrtifolium*.
Rhododendron 'Ledifolia Alba'. See R. mucronatum.
Rhododendron 'Ledifolia Rosea'. See R. mucronatum 'Sekidera'.
Rhododendron 'Leo'.
Rhododendron lepidostylum Balf. f. & Forrest.
Rhododendron 'Letty Edwards'.
Rhododendron leucaspis Tagg.
Rhododendron 'L. J. Bobbink'.
Rhododendron lochae F. J. Muell.
Rhododendron 'Loderi'.
Rhododendron 'Loderi King George'.
Rhododendron 'Loderi Venus'.
Rhododendron 'Loder's White'.
Rhododendron 'Louise Gable'.
Rhododendron lutescens Franch.
Rhododendron luteum Sweet. Syn.: R. *flavum*. PONTIC AZALEA.
Rhododendron macrophyllum D. Don ex G. Don. WESTERN RHODODENDRON. WEST COAST RHODODENDRON. CALIFORNIA ROSEBAY.
Rhododendron 'Madame August Mestdagh'.
Rhododendron 'Madame Mason'.
Rhododendron 'Madame Pericat'.
Rhododendron 'Madame Petrick'.
Rhododendron 'Madame Petrick Alba'.

Rhododendron 'Madonna'.
Rhododendron 'Marinus Koster'.
Rhododendron 'Mars'.
Rhododendron 'Mme. Mestdagh'. See **R. 'Madame August Mestdagh'**.
Rhododendron molle (Blume) G. Don. CHINESE AZALEA.
Rhododendron 'Moonstone'.
Rhododendron 'Mother of Pearl'.
Rhododendron moupinense Franch.
Rhododendron 'Mrs. A. T. de la Mare'.
Rhododendron 'Mrs. Betty Robertson'.
Rhododendron 'Mrs. Charles E. Pearson'.
Rhododendron 'Mrs. E. C. Stirling'.
Rhododendron 'Mrs. Frederick Sanders'.
Rhododendron 'Mrs. Furnival'.
Rhododendron 'Mrs. G. W. Leak'.
Rhododendron mucronatum (Blume) G. Don.
Rhododendron mucronatum 'Indica Rosea'. See **R. mucronatum 'Sekidera'**.
Rhododendron mucronatum 'Ledifolia Rosea'. See **R. mucronatum 'Sekidera'**.
Rhododendron mucronatum 'Sekidera'.
Rhododendron mucronulatum Turcz.
Rhododendron myrtifolium Schott & Kotschy. See **R. kotschyi**.
Rhododendron 'Naomi Stella Maris'.
Rhododendron 'Noyo Chief'.
Rhododendron obtusum (Lindl.) Planch.
Rhododendron obtusum 'Amoenum'.
Rhododendron occidentale (Torr. & Gray) Gray. WESTERN AZALEA.
Rhododendron 'Orchidiflora'.
Rhododendron 'Palestrina'.
Rhododendron 'Paul Schame'.
Rhododendron pemakoense F. K. Ward.
Rhododendron 'Phoenicia'. See **R. 'Formosa'**.
Rhododendron 'Pink Pearl'.
Rhododendron ponticum L.
Rhododendron poukhanense Lév. See **R. yedoense** var. **poukhanense**.
Rhododendron 'Praecox'.
Rhododendron 'Pride of Dorking'.
Rhododendron 'Pride of Mobile'. See **R. 'Elegans Superba'**.
Rhododendron 'Prince of Wales'.
Rhododendron 'Professeur Wolters'.
Rhododendron 'Ptarmigan'.
Rhododendron 'Purple Splendor'. [**R.** 'Hexe' × **R. poukhanense**].
Rhododendron 'Purple Splendour'. [**R. ponticum** × **R.** sp.].
Rhododendron quinquefolium Bisset & S. Moore.
Rhododendron racemosum Franch.
Rhododendron 'Racil'.
Rhododendron 'Radium'.
Rhododendron 'Rainbow'.
Rhododendron 'Red Poppy'.
Rhododendron 'Redwing'.
Rhododendron 'Roman Pottery'.
Rhododendron 'Rosaeflora'. See **R. indicum 'Balsaminiflorum'**.
Rhododendron 'Rosebud'.
Rhododendron 'Rose Elf'.
Rhododendron 'Rose Greeley'.
Rhododendron 'Rose Parade'.
Rhododendron roseum Rehd.
Rhododendron rupicola W. W. Sm.
Rhododendron 'Saffron Queen'.
Rhododendron sanctum Nakai.
Rhododendron 'Sapphire'.
Rhododendron 'Sappho'.
Rhododendron 'Scarlet King'.
Rhododendron schlippenbachii Maxim. ROYAL AZALEA.
Rhododendron 'Scintillation'.
Rhododendron 'Shamrock'. [**R. keiskei** × **R. hanceanum** 'Nanum'].
Rhododendron 'Sherwood Orchid'.
Rhododendron 'Sherwood Red'.
Rhododendron 'Shinnyo-No-Tsuki'.
Rhododendron 'Snow'.
Rhododendron 'Snowbank'. See **R. 'Alaska'**.
Rhododendron 'Snow Lady'.
Rhododendron 'Southern Charm'.
Rhododendron strigillosum Franch.
Rhododendron 'Sun Valley'.
Rhododendron sutchuenense Franch.
Rhododendron 'Sweetheart Supreme'.
Rhododendron temenium Balf. f. & Forrest.
Rhododendron thayeranum Rehd. & Wils.
Rhododendron 'The Hon. Jean Marie de Montague'.
Rhododendron 'Trilby'.
Rhododendron 'Twenty Grand'.
Rhododendron 'Unique'. [Ghent hybrid].
Rhododendron 'Unique'. [**R. campylocarpum** hybrid].
Rhododendron 'Unknown Warrior'.
Rhododendron 'Van Nes Sensation'.
Rhododendron vaseyi Gray. PINK-SHELL AZALEA.
Rhododendron viscosum (L.) Torr.
Rhododendron 'Vivid'.
Rhododendron 'Vulcan'.
Rhododendron wardii W. W. Sm.
Rhododendron 'Ward's Ruby'.
Rhododendron 'White April'.
Rhododendron 'White Orchid'.
Rhododendron 'William van Orange'.
Rhododendron williamsii Rehd. & Wils.
Rhododendron wilsonae Hemsl. & Wils.
Rhododendron yakusimanum Nakai.
Rhododendron yedoense Maxim. YODOGAWA AZALEA.
Rhododendron yedoense var. *poukhanense* (Lév.) Nakai. Syn.: R. poukhanense. KOREAN AZALEA.
Rhododendron yunnanense Franch. YUNNAN RHODODENDRON.
Rhododendron zoelleri Warb.
Rhodosphaera rhodanthema (F. J. Muell.) Engler. YELLOWWOOD.
Rhodotypos kerrioides Siebold & Zucc. See **R. scandens**.
Rhodotypos scandens (Thunb.) Makino. Syn.: R. kerrioides. JETBEAD. WHITE KERRIA.
Rhoicissus capensis (Burm.) Planch. Syn.: *Cissus capensis. Vitis capensis.* CAPE GRAPE. EVERGREEN GRAPE.
Rhopalostylis baueri H. Wendl. & Drude. NIKAU PALM.
Rhopalostylis cheesemanii Becc. NIKAU PALM.
Rhopalostylis sapida H. Wendl. & Drude. FEATHER-DUSTER PALM. SHINING BRUSH PALM. NIKAU PALM.
Rhus aromatica Ait. FRAGRANT SUMAC. LEMON SUMAC.
Rhus chinensis Mill. NUTGALL TREE.
Rhus copallina L. SHINING SUMAC.
Rhus cotinus L. See **Cotinus coggygria**.
Rhus glabra L. SMOOTH SUMAC. SCARLET SUMAC.
Rhus integrifolia (Nutt.) Benth & Hook. LEMONADE BERRY. LEMONADE SUMAC.
Rhus lancea L. f. AFRICAN SUMAC.
Rhus lanceolata (Gray) Britt. PRAIRIE SUMAC.
Rhus laurina Nutt. LAUREL SUMAC.
Rhus ovata S. Wats. SUGARBUSH.
Rhus succedanea L. WAX TREE.

Rhus trilobata Nutt. SQUAWBUSH. SKUNKBUSH.
Rhus typhina L. STAGHORN SUMAC. VELVET SUMAC.
Rhynchospermum asiaticum. Listed name for **Trachelospermum asiaticum.**
Rhynchospermum jasminoides Lindl. See **Trachelospermum jasminoides.**
Rhyticocos amara (Jacq.) Becc. Syn.: *Cocos amara.*
OVERTOP PALM.
RIBBONBUSH: **Homalocladium platycladum.**
RIBBONWOOD: **Adenostoma sparsifolium.**
RIBBONWOOD: **Hoheria sexstylosa.**
MOUNTAIN: **Hoheria glabrata.**
Ribes aureum Pursh. GOLDEN CURRANT. BUFFALO CURRANT.
Ribes aureum var. *gracillimum* (Cov. & Britt.) Jeps.
GOLDEN CURRANT.
Ribes indecorum Eastw. WHITE-FLOWERED CURRANT.
Ribes malvaceum Sm. CHAPARRAL CURRANT.
Ribes menziesii Pursh. CANYON GOOSEBERRY.
Ribes roezlii Regel. SIERRA GOOSEBERRY.
Ribes sanguineum Pursh. RED-FLOWERING CURRANT. PINK WINTER CURRANT.
Ribes sanguineum var. *glutinosum* (Benth.) Loud.
FLOWERING CURRANT.
Ribes speciosum Pursh. FUCHSIA-FLOWERED CURRANT. FUCHSIA-FLOWERED GOOSEBERRY.
Ribes viburnifolium Gray. CATALINA CURRANT. CATALINA PERFUME. EVERGREEN CURRANT.
RICEFLOWER, ROSY: **Pimelea ferruginea.**
RICE-PAPER PLANT: **Tetrapanax papyriferus.**
Ricinus communis L. CASTOR BEAN. CASTOR-OIL PLANT. PALMA CHRISTI.
Ricinus communis 'San Diego Red'.
Robinia × *ambigua* Poir. [*R. pseudoacacia* × *R. viscosa*]. Syn.: *R.* × *hybrida.* HYBRID LOCUST.
Robinia × *ambigua* 'Decaisneana'. Syn.: *R. decaisneana.*
PINK LOCUST. PINK-FLOWERING LOCUST.
Robinia × *ambigua* 'Idahoensis'. IDAHO LOCUST. PINK IDAHO LOCUST.
Robinia 'Contorta'. See **R. pseudoacacia 'Tortuosa'.**
Robinia decaisneana Hort. See **R.** × **ambigua 'Decaisneana'.**
Robinia 'Decaisneana Idaho'. See **R.** × **ambigua 'Idahoensis'.**
Robinia 'Fastigiata'. See **R. pseudoacacia 'Pyramidalis'.**
Robinia fertilis Ashe.
Robinia fertilis 'Monument'. MONUMENT LOCUST.
Robinia globosa. Listed name for **R. pseudoacacia 'Rehderi'.**
Robinia hispida L. ROSE ACACIA. MOSSY LOCUST. BRISTLY LOCUST.
Robinia hispida var. *inermis* Kirchn. See **R. hispida** var. **macrophylla.** Some plants called *R. hispida* var. *inermis* may be **R. pseudoacacia 'Inermis'.**
Robinia hispida var. *macrophylla* DC. Syn.: *R. hispida* var. *inermis.* SMOOTH ROSE ACACIA.
Robinia hispida 'Monument'. See **R. fertilis 'Monument'.**
Robinia × *hybrida* Audibert ex DC. See **R.** × **ambigua.**
Robinia × *hybrida* 'Monument'. See **R. fertilis 'Monument'.**
Robinia 'Idaho'. See **R.** × **ambigua 'Idahoensis'.**
Robinia 'Idahoensis'. See **R.** × **ambigua 'Idahoensis'.**
Robinia 'Idaho Pink'. See **R.** × **ambigua 'Idahoensis'.**
Robinia pseudoacacia L. BLACK LOCUST. FALSE ACACIA.
Robinia pseudoacacia 'Decaisneana'. See **R.** × **ambigua 'Decaisneana'.**
Robinia pseudoacacia 'Fastigiata'. See **R. pseudoacacia 'Pyramidalis'.**
Robinia pseudoacacia 'Frisia'.
Robinia pseudoacacia 'Idahoensis'. See **R.** × **ambigua 'Idahoensis'.**
Robinia pseudoacacia 'Inermis'. Some plants called *R. hispida* var. *inermis* may belong here.
Robinia pseudoacacia 'Purple Robe'.
Robinia pseudoacacia 'Pyramidalis'.
Robinia pseudoacacia 'Rehderi'.
Robinia pseudoacacia 'Tortuosa'.
Robinia pseudoacacia 'Umbraculifera'.
Robinia 'Purple Robe'. See **R. pseudoacacia 'Purple Robe'.**
Robinia viscosa Vent. CLAMMY LOCUST.
ROBIN-REDBREAST BUSH: **Melaleuca lateritia.**
Rochea coccinea (L.) DC.
ROCKROSE or ROCK-ROSE: See ROSE, ROCK.
Romneya coulteri Harv. MATILIJA POPPY.
Romneya coulteri var. *trichocalyx* (Eastw.) Jeps. Syn.: *R. trichocalyx.*
Romneya coulteri 'White Cloud'.
Romneya trichocalyx Eastw. See **R. coulteri** var. **trichocalyx.**
Rondeletia amoena (Planch.) Hemsl.
Rondeletia cordata Benth.
Rondeletia odorata Jacq.
Rosa banksiae Ait. f. LADY BANKS' ROSE. BANKSIA ROSE.
Rosa banksiae 'Alba Plena'.
Rosa banksiae 'Lutea'. BANKS' YELLOW ROSE.
Rosa bengalensis Pers. See **R. chinensis.**
Rosa bracteata J. C. Wendl. MACARTNEY ROSE.
Rosa bracteata 'Mermaid'.
Rosa brunonii Lindl. HIMALAYAN MUSK ROSE.
Rosa californica Cham. & Schlechtend. CALIFORNIA WILD ROSE.
Rosa canina L. DOG ROSE. BRIER ROSE. DOG BRIER.
Rosa centifolia L. CABBAGE ROSE.
Rosa chinensis Jacq. Syn.: *R. bengalensis. R. mutabilis.*
CHINA ROSE.
Rosa chinensis 'Gloire des Rosomanes'. RAGGED ROBIN ROSE.
Rosa chinensis 'Minima'. Syn.: *R. lawranceana. R. roulettii.*
FAIRY ROSE. PYGMY ROSE. ROULETT ROSE.
Rosa damascena Mill. DAMASK ROSE.
Rosa 'Doctor Huey'. DOCTOR HUEY ROSE.
Rosa eglanteria L. See **R. rubiginosa.**
Rosa foetida Herrmann. Syn.: *R. lutea.* AUSTRIAN BRIER. AUSTRIAN BRIER ROSE.
Rosa foetida 'Bicolor'. AUSTRIAN COPPER BRIER. AUSTRIAN COPPER ROSE.
Rosa foetida 'Persiana'.
Rosa gallica L. FRENCH ROSE.
Rosa gallica 'Rosa Mundi'. See **R. gallica 'Versicolor'.**
Rosa gallica 'Versicolor'.
Rosa gymnocarpa Nutt. WOOD ROSE.
Rosa × *harisonii* Rivers. [*R. foetida* × *R. pimpinellifolia*]. HARISON'S YELLOW ROSE.
Rosa hugonis Hemsl. FATHER HUGO'S ROSE. GOLDEN ROSE-OF-CHINA.
Rosa laevigata Michx. CHEROKEE ROSE.
Rosa lawranceana Sweet. See **R. chinensis 'Minima'.**
Rosa lutea Mill. See **R. foetida.**
Rosa moschata Herrmann. MUSK ROSE.
Rosa moschata 'Plena'. DOUBLE MUSK ROSE.
Rosa multiflora Thunb. ex J. Murr. Syn.: *R. polyantha* Siebold & Zucc. BABY ROSE. MULTIFLORA ROSE. JAPANESE ROSE.
Rosa mutabilis Correvon. See **R. chinensis.**
Rosa × *noisettiana* Thory. [*R. chinensis* × *R. moschata*]. NOISETTE ROSE. CHAMPNEY ROSE.
Rosa × *noisettiana* 'Manettii'. MANETTI ROSE.
Rosa nutkana K. Presl. NOOTKA ROSE.
Rosa odorata (Andr.) Sweet. TEA ROSE.
Rosa odorata 'Pseudindica'. GOLD ROSE. OPHIR ROSE.
Rosa omeiensis Rolfe.

Rosa omeiensis forma *chrysocarpa* Rehd.
Rosa omeiensis forma *pterocantha* (Franch.) Rehd. & Wils.
Rosa persica Michx. ex Juss.
Rosa pimpinellifolia L. Syn.: *R. spinosissima*.
    SCOTCH ROSE. BURNET ROSE.
Rosa polyantha Siebold & Zucc., not Carrière. See *R. multiflora*.
Rosa × *polyantha* Carrière, not Siebold & Zucc.
    See *R.* × *rehderana*.
Rosa × *rehderana* Blackburn. [*R. chinensis* × *R. multiflora*].
    Syn.: *R.* × *polyantha* Carrière. POLYANTHA ROSE.
Rosa roulettii Correvon. See *R. chinensis* 'Minima'.
Rosa roxburghii Tratt. CHESTNUT ROSE. CHINQUAPIN ROSE.
Rosa rubiginosa L. Syn.: *R. eglanteria*. EGLANTINE. EGLANTINE
    ROSE. SWEET BRIER.
Rosa rubrifolia Vill.
Rosa rugosa Thunb. RUGOSA ROSE. TURKESTAN ROSE. JAPANESE
    ROSE.
Rosa soulieana Crépin.
Rosa spinosissima L. See *R. pimpinellifolia*.
Rosa spithamea S. Wats. GROUND ROSE.
Rosa wichuraiana Crépin. MEMORIAL ROSE.
Rosa woodsii Lindl.
Rosa woodsii var. *ultramontana* (S. Wats.) Jeps.
ROSA DE MONTAÑA: **Antigonon leptopus.**
ROSARY VINE: **Ceropegia woodii.**
ROSE, ALPINE: **Rhododendron ferrugineum.**
    AUSTRIAN BRIER: **Rosa foetida.**
    AUSTRIAN COPPER: **Rosa foetida 'Bicolor'.**
    BABY: **Rosa multiflora.**
    BANKSIA: **Rosa banksiae.**
    BANKS' YELLOW: **Rosa banksiae 'Lutea'.**
    BRIER: **Rosa canina.**
    BROWN-EYED ROCK: **Cistus ladanifer.**
    BURNET: **Rosa pimpinellifolia.**
    CABBAGE: **Rosa centifolia.**
    CALIFORNIA WILD: **Rosa californica.**
    CHAMPNEY: **Rosa** × **noisettiana.**
    CHEROKEE: **Rosa laevigata.**
    CHESTNUT: **Rosa roxburghii.**
    CHINA: **Rosa chinensis.**
    CHINQUAPIN: **Rosa roxburghii.**
    CHRISTMAS: **Helleborus niger.**
    CLIFF: **Cowania mexicana** var. **stansburiana.**
    CONFEDERATE: **Hibiscus mutabilis.**
    DAMASK: **Rosa damascena.**
    DOCTOR HUEY: **Rosa 'Doctor Huey'.**
    DOG: **Rosa canina.**
    DOUBLE MUSK: **Rosa moschata 'Plena'.**
    EGLANTINE: **Rosa rubiginosa.**
    FAIRY: **Rosa chinensis 'Minima'.**
    FATHER HUGO'S: **Rosa hugonis.**
    FRENCH: **Rosa gallica.**
    GOLD: **Rosa odorata 'Pseudindica'.**
    GROUND: **Rosa spithamea.**
    GUELDER: **Viburnum opulus.**
    HARISON'S YELLOW: **Rosa** × **harisonii.**
    HIMALAYAN MUSK: **Rosa brunonii.**
    JAPANESE: **Kerria japonica.**
    JAPANESE: **Rosa multiflora.**
    JAPANESE: **Rosa rugosa.**
    LADY BANKS': **Rosa banksiae.**
    LAUREL ROCK: **Cistus laurifolius.**
    LENTEN: **Helleborus orientalis.**
    MACARTNEY: **Rosa bracteata.**
    MANETTI: **Rosa** × **noisettiana 'Manettii'.**
    MEMORIAL: **Rosa wichuraiana.**
    MULTIFLORA: **Rosa multiflora.**
    MUSK: **Rosa moschata.**
    NOISETTE: **Rosa** × **noisettiana.**
    NOOTKA: **Rosa nutkana.**
    OPHIR: **Rosa odorata 'Pseudindica'.**
    ORCHID-SPOT ROCK: **Cistus** × **purpureus.**
    POLYANTHA: **Rosa** × **rehderana.**
    PYGMY: **Rosa chinensis 'Minima'.**
    RAGGED ROBIN: **Rosa chinensis 'Gloire des Rosomanes'.**
    ROULETT: **Rosa chinensis 'Minima'.**
    RUGOSA: **Rosa rugosa.**
    RUSH: **Helianthemum scoparium.**
    SAGE-LEAF ROCK: **Cistus salvifolius.**
    SCOTCH: **Rosa pimpinellifolia.**
    SPOTTED ROCK: **Cistus ladanifer.**
    STURT'S DESERT: **Gossypium sturtianum.**
    SUN: **Helianthemum nummularium.**
    TEA: **Rosa odorata.**
    TURKESTAN: **Rosa rugosa.**
    WHITE ROCK: **Cistus** × **hybridus.**
    WOOD: **Rosa gymnocarpa.**
ROSEBAY, CALIFORNIA: **Rhododendron macrophyllum.**
    MOUNTAIN: **Rhododendron catawbiense.**
ROSEMARY: **Rosmarinus officinalis.**
    BOG: **Andromeda polifolia.**
    CREEPING: **Rosmarinus officinalis 'Lockwood de Forest'.**
    SANTA BARBARA: **Rosmarinus officinalis 'Lockwood de Forest'.**
ROSE-OF-CHINA: **Hibiscus rosa-sinensis.**
    GOLDEN: **Rosa hugonis.**
ROSE-OF-SHARON: **Hibiscus syriacus.**
ROSE-OF-SHARON: **Hypericum calycinum.**
    PINK: **Hibiscus syriacus 'Pulcherrimus'.**
    PURPLE: **Hibiscus syriacus 'Ardens'.**
    WHITE: **Hibiscus syriacus 'Albus'.**
Rosmarinus 'Forestii'. See *R. officinalis* **'Lockwood de Forest'.**
Rosmarinus 'Ingrami'. See *R. officinalis* **'Collingwood Ingram'.**
Rosmarinus 'Lockwoodii'. See *R. officinalis* **'Lockwood de Forest'.**
Rosmarinus officinalis L. ROSEMARY.
Rosmarinus officinalis 'Collingwood Ingram'.
Rosmarinus officinalis 'Lockwood de Forest'.
    CREEPING ROSEMARY. SANTA BARBARA ROSEMARY.
Rosmarinus officinalis 'Prostratus'.
Rosmarinus officinalis 'Tuscan Blue'.
Rosmarinus 'Tuscan Blue'. See *R. officinalis* **'Tuscan Blue'.**
ROWAN: **Sorbus aucuparia.**
ROYAL CLIMBER: **Oxera pulchella.**
Roystonea elata (Bartr.) F. Harper. FLORIDA ROYAL PALM.
Roystonea regia (HBK) O. F. Cook. Syn.: *Oreodoxa regia*.
    CUBAN ROYAL PALM.
RUBBER PLANT: **Ficus elastica.**
    BROAD-LEAVED INDIAN: **Ficus elastica 'Decora'.**
RUBBER TREE, INDIA: **Ficus elastica.**
    LITTLE-LEAF: **Ficus rubiginosa.**
    MYSORE: **Ficus drupacea.**
Rubus calycinoides Hayata.
Rubus fockeanus Kurz.
Rubus idaeus L. RED RASPBERRY. EUROPEAN RASPBERRY.
    FRAMBOISE.
Rubus parviflorus Nutt. THIMBLEBERRY.
Rubus procerus P. J. Muell. HIMALAYA BERRY.
Rubus rosifolius Sm.
Rubus rosifolius 'Coronarius'.
Rubus spectabilis Pursh. SALMONBERRY.
Rubus 'Tridel'. [*R. deliciosus* Torr. × *R. trilobus*].
Rubus trilobus Ser.

RUE: **Ruta graveolens.**
Ruellia amoena Nees. See **R. graecizans.**
Ruellia graecizans Backer. Syn.: *R. amoena.*
Ruellia macrantha Mart. ex Nees.
Ruellia tuberosa L. Some plants so called are **Mackaya bella.**
Ruscus aculeatus L. BUTCHER'S BROOM.
Ruscus hypoglossum L.
Ruscus racemosus L. See **Danae racemosa.**
Russelia equisetiformis Schlechtend. & Cham. Syn.: *R. juncea.*
    CORAL PLANT. FOUNTAIN PLANT. FOUNTAIN BUSH.
Russelia juncea Zucc. See **R. equisetiformis.**
Ruta bracteosa DC. See **R. chapalensis.**
Ruta chapalensis L. Syn.: *R. bracteosa.* RUE.
Ruta graveolens L. RUE. HERB-OF-GRACE.

S

Sabal blackburniana Glazebr. ex Schult. & Schult. f. Syn.: *S. umbraculifera.* HISPANIOLAN PALMETTO. BLACKBURN PALMETTO.
Sabal deeringiana Small. See **S. minor.**
Sabal exul (O. F. Cook) L. H. Bailey. See **S. mexicana.**
Sabal glaucescens Lodd. ex H. E. Moore. See **S. mauritiiformis.**
Sabal louisiana Bomhard. See **S. minor.**
Sabal mauritiiformis (Karst.) Griseb. & H. Wendl. Syn.: *S. glaucescens.*
Sabal mexicana Mart. Syn.: *S. exul. S. texana.* OAXACANA PALMETTO. TEXAS PALMETTO. RIO GRANDE PALMETTO. MEXICAN PALMETTO. VICTORIA PALMETTO.
Sabal minor (Jacq.) Pers. Syn.: *S. deeringiana. S. louisiana.* BUSH PALMETTO. SAVANNAH PALMETTO.
Sabal palmetto (Walter) Lodd. ex Schult. & Schult. f. CABBAGE PALMETTO. PALMETTO. BLUE PALMETTO.
Sabal rosei (O. F. Cook) Becc.
Sabal texana (O. F. Cook) Becc. See **S. mexicana.**
Sabal umbraculifera (Jacq.) Mart. See **S. blackburniana.**
Sabal uresana Trel. SONORAN PALM.
SACRED-FLOWER-OF-THE-INCAS: **Cantua buxifolia.**
Sadleria cyatheoides Kaulfuss. HAWAIIAN TREE FERN. DWARF HAWAIIAN TREE FERN.
SAGE, AUTUMN: **Salvia greggii.**
    BLACK: **Salvia mellifera.**
    BLUE: **Eranthemum pulchellum.**
    BLUE: **Salvia azurea.**
    BLUE: **Salvia dorrii.**
    CLEVELAND: **Salvia clevelandii.**
    CREEPING: **Salvia sonomensis.**
    FRAGRANT: **Salvia clevelandii.**
    GARDEN: **Salvia officinalis.**
    GENTIAN: **Salvia patens.**
    GRAY: **Salvia leucophylla.**
    HUMMINGBIRD: **Salvia spathacea.**
    JERUSALEM: **Phlomis fruticosa.**
    MEALY-CUP: **Salvia farinacea.**
    MEXICAN BUSH: **Salvia leucantha.**
    PINEAPPLE: **Salvia elegans.**
    PINEAPPLE-SCENTED: **Salvia elegans.**
    PITCHER: **Lepechinia calycina.**
    PITCHER: **Salvia spathacea.**
    PURPLE: **Salvia dorrii.**
    PURPLE: **Salvia leucophylla.**
    ROSY-LEAF: **Salvia involucrata.**
    SAN MIGUEL MOUNTAIN: **Salvia munzii.**
    SANDHILL: **Artemisia pycnocephala.**
    SCARLET: **Salvia coccinea.**
    SCARLET: **Salvia splendens.**
    TEXAS: **Leucophyllum frutescens.**
    TEXAS: **Salvia coccinea.**
    WHITE: **Salvia apiana.**
    YELLOW: **Lantana camara.**
SAGEBRUSH: **Artemisia tridentata.**
    BEACH: **Artemisia pycnocephala.**
    CALIFORNIA: **Artemisia californica.**
    GREAT BASIN: **Artemisia tridentata.**
    SANDHILL: **Artemisia pycnocephala.**
SAGO, QUEEN: **Cycas circinalis.**
SAGUARO: **Carnegiea gigantea.**
SAHUARA: **Carnegiea gigantea.**
SAINT CATHERINE'S LACE: **Eriogonum giganteum.**
SALAL: **Gaultheria shallon.**
Salix alba L. WHITE WILLOW.
Salix alba var. *tristis* (Ser.) Gaudin. GOLDEN WEEPING WILLOW.
Salix alba 'Vitellina'. See **S. alba** var. **vitellina.**
Salix alba var. *vitellina* (L.) Stokes. YELLOW WILLOW. GOLDEN WILLOW.
Salix alba var. *vitellina* 'Pendula'. See **S. alba** var. **tristis.**
Salix arctica Pall. ARCTIC WILLOW.
Salix babylonica L. WEEPING WILLOW. Some plants so called are **S.** × **blanda** or **S.** × **elegantissima.**
Salix babylonica forma *annularis* Aschers. See **S. babylonica 'Crispa'.**
Salix babylonica 'Aurea'. See **S. alba** var. **tristis.**
Salix babylonica 'Crispa'. Syn.: *S. babylonica* forma *annularis.* RINGLEAF WILLOW.
Salix × *blanda* Anderss. [?*S. babylonica* × *S. fragilis*]. WISCONSIN WEEPING WILLOW. NIOBE WILLOW. WEEPING WILLOW. Some plants called **S. babylonica** belong here.
Salix caprea L. GOAT WILLOW. FRENCH PUSSY WILLOW. PINK PUSSY WILLOW. FLORIST'S WILLOW.
Salix chilensis Molina. Syn.: *S. humboldtiana.*
Salix 'Contorta'. See **S. matsudana 'Tortuosa'.**
Salix discolor Muhlenb. PUSSY WILLOW.
Salix × *elegantissima* K. Koch. [?*S. babylonica* × *S.* × *blanda*]. THURLOW WEEPING WILLOW. WEEPING WILLOW. Some plants called **S. babylonica** belong here.
Salix fragilis L. CRACK WILLOW. BRITTLE WILLOW.
Salix gooddingii Ball.
Salix gracilistyla Miq. ROSE-GOLD PUSSY WILLOW.
Salix humboldtiana Willd. See **S. chilensis.**
Salix lasiandra Benth.
Salix lasiolepis Benth. ARROYO WILLOW.
Salix magnifica Hemsl.
Salix matsudana G. Koidz. PEKIN WILLOW. HANKOW WILLOW.
Salix matsudana 'Navajo'. GLOBE NAVAJO WILLOW.
Salix matsudana 'Tortuosa'. DRAGON-CLAW WILLOW. CORKSCREW WILLOW. TWISTED HANKOW WILLOW.
Salix matsudana 'Umbraculifera'. GLOBE WILLOW.
Salix × *multinervis* Döll. [*S. aurita* L. × *S. cinerea* L.].
Salix 'Navajo'. See **S. matsudana 'Navajo'.**
Salix pentandra L. BAY WILLOW. BAY-LEAF WILLOW. LAUREL WILLOW. LAUREL-LEAF WILLOW.
Salix purpurea L. PURPLE WILLOW. PURPLE OSIER. ALASKA BLUE WILLOW.
Salix purpurea var. *lambertiana* (Sm.) W. D. J. Koch.
Salix sachalinensis Fr. Schmidt.
Salix sachalinensis 'Sekka'.
Salix sachalinensis 'Setsuka'. See **S. sachalinensis 'Sekka'.**
Salix scouleriana Barratt ex Hook. SCOULER WILLOW.

Salix × *sepulcralis* Simonkai. [*S. alba* × *S. babylonica*].
Salix 'Tortuosa'. See *S. matsudana* 'Tortuosa'.
Salix 'Umbraculifera'. See *S. matsudana* 'Umbraculifera'.
SALLY, BLACK: **Eucalyptus stellulata**.
   NARROW-LEAVED: **Eucalyptus moorei**.
SALMONBERRY: **Rubus spectabilis**.
SALTBUSH: **Atriplex lentiformis**.
   BREWER: **Atriplex lentiformis** subsp. **breweri**.
   CALIFORNIA: **Atriplex californica**.
   COAST: **Atriplex californica**.
   FOUR-WING: **Atriplex canescens**.
   GARDNER VALLEY: **Atriplex nuttallii** var. **gardneri**.
   MEDITERRANEAN: **Atriplex halimus**.
Salvia apiana Jeps. WHITE SAGE.
Salvia azurea Lam. BLUE SAGE.
Salvia azurea var. *grandiflora* Benth. Syn.: *S. azurea* subsp. *pitcheri*. *S. pitcheri*.
Salvia azurea subsp. *pitcheri* (Torr. ex Benth.) Epling. See **S. azurea** var. **grandiflora**.
Salvia clevelandii (Gray) Greene. FRAGRANT SAGE. CLEVELAND SAGE.
Salvia coccinea Juss. ex J. Murr. TEXAS SAGE. SCARLET SAGE. Some plants called *S. gracilistyla* and *S. rutilans* belong here.
Salvia dorrii (Kell.) Abrams. BLUE SAGE. PURPLE SAGE.
Salvia elegans Vahl. PINEAPPLE SAGE. PINEAPPLE-SCENTED SAGE. Some plants called *S. gracilistyla* and *S. rutilans* belong here.
Salvia farinacea Benth. MEALY-CUP SAGE.
Salvia gesneraeflora Lindl. & Paxt.
Salvia gracilistyla. Listed name. Plants so called are **S. coccinea** or **S. elegans**.
Salvia grahamii Benth. See **S. microphylla**.
Salvia greggii Gray. AUTUMN SAGE.
Salvia involucrata Cav. ROSY-LEAF SAGE.
Salvia leucantha Cav. MEXICAN BUSH SAGE.
Salvia leucophylla Greene. PURPLE SAGE. GRAY SAGE.
Salvia mellifera Greene. BLACK SAGE.
Salvia microphylla HBK. Syn.: *S. grahamii*.
Salvia munzii Epling. SAN MIGUEL MOUNTAIN SAGE.
Salvia officinalis L. GARDEN SAGE.
Salvia patens Cav. GENTIAN SAGE.
Salvia pitcheri Torr. ex Benth. See **S. azurea** var. **grandiflora**.
Salvia pratensis L. MEADOW CLARY.
Salvia purpurea Cav.
Salvia rutilans Carrière. Plants so called are **S. coccinea** or **S. elegans**.
Salvia sonomensis Greene. CREEPING SAGE.
Salvia spathacea Greene. PITCHER SAGE. HUMMINGBIRD SAGE.
Salvia splendens F. Sellow ex Roemer & Schult. SCARLET SAGE.
Salvia wagnerana Polak.
Sambucus caerulea Raf. Syn.: *S. glauca*. BLUE ELDERBERRY.
Sambucus caerulea var. *mexicana* (K. Presl) L. Benson. See **S. mexicana**.
Sambucus callicarpa Greene. Syn.: *S. racemosa* var. *callicarpa*. COAST RED ELDERBERRY.
Sambucus glauca Nutt. See **S. caerulea**.
Sambucus melanocarpa Gray. BLACK ELDERBERRY.
Sambucus mexicana K. Presl. Syn.: *S. caerulea* var. *mexicana*. MEXICAN ELDERBERRY.
Sambucus microbotrys Rydb. MOUNTAIN RED ELDERBERRY.
Sambucus racemosa L. EUROPEAN RED ELDERBERRY. RED ELDERBERRY.
Sambucus racemosa var. *callicarpa* (Greene) Jeps. See **S. callicarpa**.
SANDALWOOD, BASTARD: **Myoporum sandwicense**.
SANSEVIERIA, BIRD'S-NEST: **Sansevieria trifasciata** 'Hahnii'.
   HAHN'S: **Sansevieria trifasciata** 'Hahnii'.
Sansevieria trifasciata Prain. SNAKE PLANT. MOTHER-IN-LAW'S TONGUE. Some plants called **S. zeylanica** may belong here.
Sansevieria trifasciata 'Hahnii'. BIRD'S-NEST SANSEVIERIA. HAHN'S SANSEVIERIA.
Sansevieria zeylanica (L.) Willd. CEYLON BOWSTRING HEMP. Some plants so called are **S. trifasciata**.
Santolina chamaecyparissus L. LAVENDER COTTON. CYPRESS LAVENDER COTTON.
Santolina virens Mill. GREEN LAVENDER COTTON.
Sapindus drummondii Hook. & Arn. DRUMMOND SOAPBERRY.
Sapium indicum Willd.
Sapium japonicum (Siebold & Zucc.) Pax & K. Hoffm. JAPANESE TALLOW TREE.
Sapium sebiferum (L.) Roxb. CHINESE TALLOW TREE.
SAPOTE, WHITE: **Casimiroa edulis**.
SAPPHIRE BERRY: **Symplocos paniculata**.
SARCOCOCCA, FRAGRANT: **Sarcococca ruscifolia**.
   WILLOWLEAF: **Sarcococca saligna**.
Sarcococca confusa Sealy. Some plants called *S. ruscifolia* belong here.
Sarcococca hookerana Baill.
Sarcococca hookerana var. *digyna* Franch.
Sarcococca hookerana var. *humilis* Rehd. & Wils. Syn.: *S. humilis*.
Sarcococca humilis (Rehd. & Wils.) Stapf. See **S. hookerana** var. **humilis**.
Sarcococca ruscifolia Stapf. FRAGRANT SARCOCOCCA. Some plants so called are **S. confusa**.
Sarcococca saligna (D. Don) Muell.-Arg. WILLOWLEAF SARCOCOCCA.
SARVICEBERRY: See SERVICEBERRY.
Sasa disticha (Mitf.) E. G. Camus. See **Arundinaria disticha**.
Sasa fortunei. Listed name, probably for **Arundinaria variegata**.
Sasa humilis (Mitf.) E. G. Camus. See **Arundinaria humilis**.
Sasa japonica (Siebold & Zucc. ex Steud.) Makino. Syn.: *Arundinaria japonica*. *Pseudosasa japonica*. ARROW BAMBOO. JAPANESE ARROW BAMBOO. METAKE. YADAKE.
Sasa japonica 'Tsutsumiana'.
Sasa kumasasa. Listed name for **Shibataea kumasasa**.
Sasa megalophylla Makino & Uchida.
Sasa megalophylla 'Nobilis'?
Sasa palmata (Burbidge) E. G. Camus. Syn.: *S. paniculata*. PALMATE BAMBOO. Plants called *Sasa senanensis* belong here.
Sasa palmata 'Nebulosa'.
Sasa paniculata (Fr. Schmidt) Makino. See **S. palmata**.
Sasa pygmaea (Miq.) E. G. Camus. See **Arundinaria pygmaea**.
Sasa senanensis (Franch. & Sav.) Rehd. Not cult. Plants so called are **S. palmata**.
Sasa tessellata (Munro) Makino & Shibata.
Sasa variegata (Miq.) E. G. Camus. See **Arundinaria variegata**.
Sasa veitchii (Carrière) Rehd.
Sassafras albidum (Nutt.) Nees.
SATSUMA: **Citrus reticulata**.
   OWARI: **Citrus reticulata** 'Owari'.
Satureja douglasii (Benth.) Briq. Syn.: *Micromeria chamissonis*. YERBA BUENA.
Satureja montana L. WINTER SAVORY.
SAUSAGE TREE: **Kigelia africana**.
SAVORY, WINTER: **Satureja montana**.
SCARLET-BUGLER: **Penstemon centranthifolius**.
SCARLET BUSH: **Hamelia patens**.
SCHEFFLERA, NEW ZEALAND: **Schefflera digitata**.

Schefflera actinophylla (Endl.) Harms. See **Brassaia actinophylla.**
Schefflera arboricola Hayata. Plants called *Brassaia arboricola* belong here.
Schefflera delavayi (Franch.) Harms.
Schefflera digitata J. R. Forst. & G. Forst. NEW ZEALAND SCHEFFLERA.
Schefflera venulosa (Wight & Arn.) Harms.
Schinus dependens Ort. See **S. polygamus.**
Schinus molle L. CALIFORNIA PEPPER TREE. PERUVIAN MASTIC TREE.
Schinus polygamus (Cav.) Cabrera. Syn.: *S. dependens.*
Schinus terebinthifolius Raddi. BRAZILIAN PEPPER TREE. CHRISTMAS-BERRY TREE.
Schizocentron elegans (Schlechtend.) Rose. See **Heterocentron elegans.**
Schizophragma hydrangeoides Siebold & Zucc. JAPANESE HYDRANGEA VINE.
SCHOLAR TREE, CHINESE: **Sophora japonica.**
Schotia afra (L.) Bodin. Syn.: *S. speciosa.* KAFFIR BEAN.
Schotia brachypetala Sond.
Schotia speciosa Jacq. See **S. afra.**
Schrebera alata (Hochst.) Welw. Syn.: *Schrebera saundersiae.*
Schrebera saundersiae Harv. See **S. alata.**
Sciadopitys verticillata (Thunb.) Siebold & Zucc. UMBRELLA PINE.
SCIMITAR SHRUB: **Brachysema lanceolatum.**
Scindapsus aureus (Linden & André) Engler. See **Epipremnum aureum.**
Scolopia ecklonii (Arn.) Harv.
SCOTCH BROOM HYBRIDS: **Cytisus (Dallimore Hybrids).**
Seaforthia elegans R. Br. See **Ptychosperma elegans.** Some plants called *Seaforthia elegans* are **Archontophoenix cunninghamiana.**
SEA-URCHIN TREE: **Hakea laurina.**
Securidaca diversifolia (L.) S. F. Blake. Syn.: *S. volubilis.*
Securidaca volubilis L. See **S. diversifolia.**
Sedum adolphii Hamet.
Sedum confusum Hemsl.
Sedum decumbens R. T. Clausen.
Sedum dendroideum Moç. & Sessé ex DC.
Sedum dendroideum subsp. **praealtum** (DC.) R. T. Clausen. Syn.: *S. praealtum.*
Sedum guatemalense Hemsl. Probably not cult. Some plants so called may be **S. × rubrotinctum.**
Sedum morganianum Walther. DONKEY-TAIL. BURRO-TAIL.
Sedum nussbaumeranum Bitter.
Sedum oxypetalum HBK.
Sedum pachyphyllum Rose. JELLY-BEAN PLANT. JELLY-BEANS.
Sedum praealtum DC. See **S. dendroideum** subsp. *praealtum.*
Sedum × rubrotinctum R. T. Clausen. [Parentage unknown]. PORK-AND-BEANS. CHRISTMAS-CHEER. Some plants called **S. guatemalense** belong here.
Sedum stahlii Solms-Laub. CORAL-BEADS.
Semele androgyna (L.) Kunth. CLIMBING BUTCHER'S BROOM.
Semiarundinaria fastuosa (Mitf.) Makino. NARIHIRA BAMBOO.
SEMINOLE-BREAD: **Zamia pumila.**
Senecio angulatus L. f.
Senecio cineraria DC. Syn.: *Cineraria maritima.* DUSTY-MILLER.
Senecio compactus T. Kirk. Plants called **S. greyi** probably belong here.
Senecio confusus (DC.) Britten. MEXICAN FLAME VINE.
Senecio glastifolius L. f.
Senecio greyi Hook. f. May not be cult. Plants so called are probably **S. compactus.**
Senecio haworthii Schultz-Bip. Syn.: *Kleinia tomentosa.*
Senecio leucostachys Baker. See **S. vira-vira.**
Senecio mandraliscae (Tineo) Jacobsen. Syn.: *Kleinia mandraliscae.*
Senecio mikanioides Otto. Syn.: *S. scandens.* GERMAN IVY.
Senecio petasitis (Sims) DC. VELVET GROUNDSEL. CALIFORNIA GERANIUM.
Senecio praecox (Cav.) DC.
Senecio scandens DC. See **S. mikanioides.**
Senecio vira-vira Hieron. Syn.: *Cineraria candidissima. C. maritima* var. *candidissima. Senecio leucostachys.* DUSTY-MILLER.
SENNA, FLOWERY: **Cassia corymbosa.**
  GOLDEN-WONDER: **Cassia splendida.**
  SCORPION: **Coronilla emerus.**
  SHRUBBY BLADDER: **Colutea arborescens.**
  WOOLLY: **Cassia tomentosa.**
SENSITIVE PLANT: **Mimosa pudica.**
SEQUOIA, GIANT: **Sequoiadendron giganteum.**
Sequoia gigantea (Lindl.) Decne. See **Sequoiadendron giganteum.**
Sequoia gigantea 'Pendula'. See **Sequoiadendron giganteum 'Pendulum'.**
Sequoia sempervirens (D. Don) Endl. COAST REDWOOD.
Sequoia sempervirens 'Aptos Blue'.
Sequoia sempervirens 'Santa Cruz'.
Sequoia sempervirens 'Soquel'.
Sequoiadendron giganteum (Lindl.) Buchh. Syn.: *Sequoia gigantea.* GIANT SEQUOIA. BIG TREE. SIERRA REDWOOD.
Sequoiadendron giganteum 'Pendulum'.
Serenoa repens (Bartr.) Small. SAW PALMETTO. SCRUB PALMETTO.
Serissa foetida (L. f.) Lam. Syn.: *S. japonica.*
Serissa foetida 'Variegata'.
Serissa japonica Thunb. See **S. foetida.**
SERRISA: **Leucophyllum frutescens.**
Serruria florida J. Knight. BLUSHING-BRIDE.
SERVICEBERRY, SASKATOON: **Amelanchier alnifolia.**
  SHADBLOW: **Amelanchier canadensis.**
  WESTERN: **Amelanchier alnifolia.**
SERVICE TREE: **Sorbus domestica.**
Sesbania grandiflora (L.) Poir. Syn.: *Agati grandiflora.* Probably not cult. in Calif. Some plants so called are **S. punicea.**
Sesbania punicea (Cav.) Benth. Syn.: *Daubentonia punicea. D. tripetiana. D. tripetii. Sesbania tripetii.* SCARLET WISTERIA TREE. GLORY PEA. Plants called **Sesbania grandiflora** probably belong here.
Sesbania tripetii (Poit.) Hort. ex F. T. Hubbard. See **S. punicea.**
SHADBUSH: **Amelanchier laevis.**
SHADDOCK: **Citrus maxima.**
SHAVING-BRUSH TREE: **Pseudobombax ellipticum.**
SHELLFLOWER: **Alpinia zerumbet.**
Shepherdia argentea (Pursh) Nutt. BUFFALO BERRY.
Shibataea kumasasa (Zollinger ex Steud.) Nakai. RUSCUS-LEAVED BAMBOO. Plants called *Sasa kumasasa* belong here.
SHRIMP BUSH: **Justicia brandegeana.**
SHRIMP PLANT: **Justicia brandegeana.**
Sibiraea laevigata (L.) Maxim.
Sidalcea neomexicana Gray. var. **thurberi** (B. L. Robinson ex Gray) C. L. Hitchc.
SILK TREE: **Albizia julibrissin.**
SILK-COTTON TREE, RED: **Bombax ceiba.**
SILK-TASSEL, COAST: **Garrya elliptica.**
  FREMONT: **Garrya fremontii.**
  PALE: **Garrya flavescens** var. **pallida.**
SILVER-BELL: **Halesia carolina.**

SILVER-BELL, continued
    MOUNTAIN: **Halesia monticola.**
SILVERBERRY: **Elaeagnus angustifolia.**
SILVERBERRY: **Elaeagnus pungens.**
    EBBINGE'S: **Elaeagnus × ebbingei.**
    FRUITLAND: **Elaeagnus pungens 'Fruitlandii'.**
SILVER BUSH: **Convolvulus cneorum.**
SILVER-DOLLAR TREE: **Eucalyptus cinerea.**
SILVER-DOLLAR TREE: **Eucalyptus polyanthemos.**
SILVER-DOLLAR TREE: **Eucalyptus pulverulenta.**
SILVER-LEAF, TEXAS: **Leucophyllum frutescens.**
SILVER TREE: **Leucodendron argenteum.**
Simmondsia californica (Link) Nutt. See **S. chinensis.**
Simmondsia chinensis (Link) Schneid. Syn.: *S. californica.*
    GOAT NUT. JOJOBA.
Sinarundinaria murielae (Gamble) Nakai. Syn.: *Arundinaria murielae.*
Sinarundinaria nitida (Mitf.) Nakai. Syn.: *Arundinaria nitida.*
    BLUE BAMBOO.
Sinocalamus beecheyanus (Munro) McClure. See **Bambusa beecheyana.**
Sinocalamus oldhamii (Munro) McClure. See **Bambusa oldhamii.**
Siphonosmanthus delavayi (Franch.) Stapf. See **Osmanthus delavayi.**
SISKIYOU-MAT: **Ceanothus pumilus.**
SKIMMIA, FOREMAN: **Skimmia × foremanii.**
    JAPANESE: **Skimmia japonica.**
Skimmia × foremanii H. Knight. [*S. japonica × S. reevesiana*]. FOREMAN SKIMMIA.
Skimmia fortunei M. T. Mast. See **S. reevesiana.**
Skimmia fragrans Carrière. See **S. japonica.**
Skimmia japonica Thunb. Syn.: *S. fragrans.* JAPANESE SKIMMIA.
Skimmia japonica 'Macrophylla'.
Skimmia japonica 'Nana'?
Skimmia reevesiana Fortune. Syn.: *S. fortunei.*
Skimmia repens Nakai.
SKUNKBUSH: **Rhus trilobata.**
SKY FLOWER: **Duranta repens.**
    BLUE: **Thunbergia grandiflora.**
    BRAZILIAN: **Duranta stenostachya.**
SKY VINE, BLUE: **Thunbergia grandiflora.**
SMOKE TREE: **Cotinus coggygria.**
SMOKE TREE: **Dalea spinosa.**
SNAIL FLOWER: **Vigna caracalla.**
SNAILSEED, LAUREL-LEAF: **Cocculus laurifolius.**
SNAIL VINE: **Vigna caracalla.**
SNAKE PLANT: **Sansevieria trifasciata.**
SNAKE VINE: **Hibbertia scandens.**
SNAPDRAGON, BUSH: **Galvezia speciosa.**
SNAPWEED, OLIVER'S: **Impatiens sodenii.**
SNOWBALL: **Viburnum opulus 'Roseum'.**
    CHINESE: **Viburnum macrocephalum.**
    EUROPEAN: **Viburnum opulus 'Roseum'.**
    FRAGRANT: **Viburnum × carlcephalum.**
    JAPANESE: **Viburnum plicatum.**
SNOWBELL: **Styrax officinalis.**
    FRAGRANT: **Styrax obassia.**
    JAPANESE: **Styrax japonicus.**
SNOWBERRY: **Symphoricarpos albus.**
    CREEPING: **Symphoricarpos mollis.**
    SPREADING: **Symphoricarpos mollis.**
SNOWBUSH: **Ceanothus cordulatus.**
SNOWDROP BUSH: **Styrax officinalis** var. **californicus.**
    HAIRY: **Styrax officinalis** var. **fulvescens.**
SNOWDROP TREE: **Halesia carolina.**

SNOWFLAKE, TROPICAL: **Trevesia palmata 'Micholitzii'.**
SNOWFLAKE TREE: **Trevesia palmata 'Micholitzii'.**
SOAP-BARK TREE: **Quillaja saponaria.**
SOAPBERRY, DRUMMOND: **Sapindus drummondii.**
SOAP WEED: **Yucca glauca.**
SOAP WEED: **Yucca elata.**
Solandra guttata D. Don ex Lindl. Some plants so called are **S. maxima.**
Solandra hartwegii N. E. Br. See **S. maxima.**
Solandra longiflora Tussac. COPA-DE-LECHE.
Solandra maxima (Sessé & Moç.) P. S. Green. Syn.: *S. hartwegii.* COPA-DE-ORO. CHALICE VINE. CUP-OF-GOLD. Some plants called **S. guttata** belong here.
SOLANUM, BLUE: **Lycianthes rantonnei.**
Solanum aculeatissimum Jacq. Syn.: *S. ciliatum.* LOVE APPLE.
Solanum aviculare G. Forst. Some plants so called are **S. laciniatum.**
Solanum capsicastrum Link ex Schauer. See **S. diflorum.**
Solanum ciliatum Lam. See **S. aculeatissimum.**
Solanum diflorum Vell. Syn.: *S. capsicastrum.* FALSE JERUSALEM CHERRY.
Solanum dulcamaroides Poir. Syn.: *S. macrantherum.*
Solanum hispidum Pers. Syn.: *S. warszewiczii.* DEVIL'S FIG.
Solanum jasminoides Paxt. POTATO VINE. JASMINE NIGHTSHADE.
Solanum laciniatum Ait. KANGAROO APPLE. Some plants called **S. aviculare** belong here.
Solanum lanceolatum Cav.
Solanum laurifolium Mill. See **S. subinerme.**
Solanum macrantherum Dunal. See **S. dulcamaroides.**
Solanum marginatum L. f.
Solanum pseudocapsicum L. CHRISTMAS CHERRY. JERUSALEM CHERRY.
Solanum quitoense Lam. NARANJILLA. LULO.
Solanum rantonnetii Carrière ex Lescuyer. See **Lycianthes rantonnei.**
Solanum robustum Wendl.
Solanum seaforthianum Andr.
Solanum subinerme Jacq. Syn.: *S. laurifolium.*
Solanum wallacei (Gray) Parish. CATALINA NIGHTSHADE.
Solanum warszewiczii Weick ex Lambertye. See **S. hispidum.**
Solanum wendlandii Hook. f. POTATO VINE. GIANT POTATO CREEPER. PARADISE FLOWER. COSTA RICAN NIGHTSHADE.
Sollya fusiformis (Labill.) Briq. See **S. heterophylla.**
Sollya heterophylla Lindl. Syn.: *S. fusiformis.* AUSTRALIAN BLUEBELL.
Sophora japonica L. JAPANESE PAGODA TREE. CHINESE SCHOLAR TREE.
Sophora japonica 'Pendula'. WEEPING JAPANESE PAGODA TREE.
Sophora japonica 'Regent'.
Sophora microphylla Ait.
Sophora secundiflora (Ort.) Lag. ex DC. MESCAL BEAN. TEXAS MOUNTAIN LAUREL.
Sophora tetraptera J. Mill.
Sorbaria aitchisonii (Hemsl.) Hemsl. ex Rehd. FALSE SPIRAEA.
Sorbaria arborea Schneid. FALSE SPIRAEA.
Sorbaria sorbifolia (L.) A. Braun. FALSE SPIRAEA.
Sorbus aria (L.) Crantz. WHITE BEAM.
Sorbus aucuparia L. ROWAN. EUROPEAN MOUNTAIN ASH.
Sorbus aucuparia 'Columbia Queen'.
Sorbus aucuparia 'Leonard Springer'.
Sorbus aucuparia 'Red Top'.
Sorbus aucuparia 'Wilsonii'.
Sorbus californica Greene.
Sorbus cascadensis G. N. Jones.
Sorbus domestica L. SERVICE TREE.
Sorbus hybrida L.

Sorbus intermedia (J. F. Ehrh.) Pers.
Sorbus scopulina Greene.
SORREL TREE: **Oxydendrum arboreum.**
SOURWOOD: **Oxydendrum arboreum.**
SOUTHERNWOOD: **Artemisia abrotanum.**
SPANISH-BAYONET: **Yucca aloifolia.**
    VARIEGATED: **Yucca aloifolia 'Variegata'.**
SPANISH-DAGGER: **Yucca gloriosa.**
SPANISH-SHAWL: **Heterocentron elegans.**
SPANISH-STOPPER: **Eugenia foetida.**
Sparmannia africana L. f. GERMAN LINDEN. AFRICAN LINDEN. AFRICAN HEMP.
Spartium junceum L. WEAVER'S BROOM. SPANISH BROOM.
Spathodea campanulata Beauvois. AFRICAN TULIP-TREE. FLAME-OF-THE-FOREST.
SPEAR FLOWER: **Doryanthes palmeri.**
SPEKBOOM: **Portulacaria afra.**
Sphaeralcea ambigua Gray. APRICOT MALLOW.
Sphaeralcea grossulariifolia (Hook. & Arn.) Rydb.
Sphaeralcea umbellata (Cav.) G. Don. See **Phymosia umbellata.**
Sphaeropteris cooperi (F. J. Muell.) R. Tryon. Syn.: *Alsophila cooperi.* AUSTRALIAN TREE FERN. Some plants called **Alsophila australis** belong here.
Sphaeropteris medullaris (G. Forst.) Bernh. Syn.: *Cyathea medullaris.* BLACK-STEM TREE FERN. SAGO TREE FERN.
SPICEBUSH: **Lindera benzoin.**
    WESTERN: **Calycanthus occidentalis.**
SPIDER NET: **Grevillea thelemanniana.**
SPINACH, CATTLE: **Atriplex polycarpa.**
SPINDLE TREE, EUROPEAN: **Euonymus europaea.**
    WINGED: **Euonymus alata.**
SPIRAEA, ANTHONY WATERER: **Spiraea bumalda 'Anthony Waterer'.**
    BLUE: **Caryopteris incana.**
    BRIDAL-WREATH: **Spiraea prunifolia.**
    BRIDAL-WREATH: **Spiraea × vanhouttei.**
    DOUBLE BRIDAL-WREATH: **Spiraea cantoniensis 'Lanceata'.**
    DOUGLAS: **Spiraea douglasii.**
    DWARF PINK: **Spiraea × bumalda 'Anthony Waterer'.**
    DWARF PINK BRIDAL-WREATH: **Spiraea × bumalda 'Anthony Waterer'.**
    FALSE: **Sorbaria aitchisonii.**
    FALSE: **Sorbaria arborea.**
    FALSE: **Sorbaria sorbifolia.**
    JAPANESE: **Spiraea japonica.**
    MOUNTAIN: **Spiraea densiflora.**
    REEVES: **Spiraea cantoniensis.**
    SHOE-BUTTON: **Spiraea prunifolia.**
    THUNBERG: **Spiraea thunbergii.**
    TOSA: **Spiraea nipponica.**
    TOSA: **Spiraea nipponica** var. **tosaensis.**
    VAN HOUTTE: **Spiraea × vanhouttei.**
    WESTERN: **Spiraea douglasii.**
Spiraea albiflora (Miq.) Zabel.
Spiraea 'Anthony Waterer'. See **S. × bumalda 'Anthony Waterer'.**
Spirea × arguta Zabel. [**S. × multiflora** Zabel × **S. thunbergii**]. BABY'S BREATH.
Spiraea betulifolia Pall.
Spiraea × billardii Herincq. [**S. douglasii × S. salicifolia** L.].
Spiraea bullata Maxim.
Spiraea × bumalda Burv. [**S. japonica × S. albiflora**].
Spiraea × bumalda 'Alpina'.
Spiraea × bumalda 'Anthony Waterer'. DWARF PINK BRIDAL-WREATH. DWARF PINK BRIDAL-WREATH SPIRAEA. DWARF PINK SPIRAEA. ANTHONY WATERER SPIRAEA.
Spiraea × bumalda 'Froebelii'.
Spiraea callosa Thunb. See **S. japonica.**
Spiraea cantoniensis Lour. Syn.: **S. reevesiana.** REEVES SPIRAEA.
Spiraea cantoniensis 'Lanceata'. DOUBLE BRIDAL-WREATH. DOUBLE BRIDAL-WREATH SPIRAEA.
Spiraea densiflora Nutt. ex Rydb. MOUNTAIN SPIRAEA.
Spiraea douglasii Hook. DOUGLAS SPIRAEA. WESTERN SPIRAEA.
Spiraea japonica L. f. Syn.: **S. callosa.** JAPANESE SPIRAEA.
Spiraea japonica 'Alpina'.
Spiraea lucida Dougl. ex Greene.
Spiraea menziesii Hook.
Spiraea nipponica Maxim. TOSA SPIRAEA.
Spiraea nipponica var. *tosaensis* (Yatabe) Makino. TOSA SPIRAEA.
Spiraea opulifolia L. See **Physocarpus opulifolius.**
Spiraea opulifolia 'Nana'. See **Physocarpus opulifolius 'Nanus'.**
Spiraea prunifolia Siebold & Zucc. BRIDAL-WREATH. SHOE-BUTTON SPIRAEA. BRIDAL-WREATH SPIRAEA.
Spiraea prunifolia 'Plena'. See **S. prunifolia.**
Spiraea reevesiana Lindl. See **S. cantoniensis.**
Spiraea reevesii. Listed name for **S. cantoniensis.**
Spiraea thunbergii Siebold ex Blume. THUNBERG SPIRAEA.
Spiraea trilobata L.
Spiraea × vanhouttei (Briot) Zabel. [**S. cantoniensis × S. trilobata**]. BRIDAL-WREATH. BRIDAL-WREATH SPIRAEA. VAN HOUTTE SPIRAEA.
SPRUCE, ALBERTA: **Picea glauca** var. **albertiana.**
    BIGGONE: **Pseudotsuga macrocarpa.**
    BLACK HILLS: **Picea glauca 'Densata'.**
    BREWER'S WEEPING: **Picea brewerana.**
    COLORADO: **Picea pungens.**
    COLORADO BLUE: **Picea pungens.**
    COLORADO BLUE: **Picea pungens 'Glauca'.**
    DWARF: **Picea abies 'Pumila'.**
    DWARF ALBERTA: **Picea glauca 'Conica'.**
    DWARF WHITE: **Picea glauca 'Conica'.**
    EASTERN: **Picea orientalis.**
    ENGELMANN: **Picea engelmannii.**
    HEMLOCK: **Tsuga canadensis.**
    HIMALAYAN: **Picea smithiana.**
    HONDO: **Picea jezoensis** var. **hondoensis.**
    KOSTER BLUE: **Picea pungens 'Koster'.**
    KOSTER'S WEEPING BLUE: **Picea pungens 'Glauca Pendula'.**
    MONTGOMERY: **Picea pungens 'Montgomery'.**
    MOERHEIM BLUE: **Picea pungens 'Moerheimii'.**
    NEST: **Picea abies 'Nidiformis'.**
    NORWAY: **Picea abies.**
    ORIENTAL: **Picea orientalis.**
    RED: **Picea rubens.**
    SERBIAN: **Picea omorika.**
    SHERWOOD GEM: **Picea abies 'Sherwood Gem'.**
    SHERWOOD MULTNOMAH: **Picea abies 'Sherwoodii'.**
    SITKA: **Picea sitchensis.**
    TIDELAND: **Picea sitchensis.**
    TIGER-TAIL: **Picea torano.**
    WHITE: **Picea glauca.**
    WILSON: **Picea wilsonii.**
    YEDDO: **Picea jezoensis.**
SPURGE, JAPANESE: **Pachysandra terminalis.**
    VARIEGATED JAPANESE: **Pachysandra terminalis 'Variegata'.**
SQUAWBUSH: **Rhus trilobata.**
SQUAW CARPET: **Ceanothus prostratus.**
SQUAW GRASS: **Xerophyllum tenax.**
ST. JOHN'S-BREAD: **Ceratonia siliqua.**

ST. JOHN'S-WORT, CREEPING: **Hypericum calycinum.**
   HENRY: **Hypericum beanii.**
Stachys byzantina K. Koch. Syn.: *S. lanata. S. olympica.*
   LAMB'S EARS.
Stachys lanata Jacq. See **S. byzantina.**
Stachys olympica of authors, not Poir. See **S. byzantina.**
Stachytarpheta frantzii Polakowsky.
Stachytarpheta speciosa Pohl.
Stachytarpheta urticaefolia (Salisb.) Sims.
Stachyurus chinensis Franch.
Stachyurus praecox Siebold & Zucc.
STAGBUSH: **Viburnum prunifolium.**
Stanleya pinnata (Pursh) Britt. PRINCE'S PLUME.
Stapelia gigantea N. E. Br. STARFISH FLOWER. CARRION FLOWER. GIANT TOAD PLANT.
Stapelia variegata L. TOAD PLANT. TOAD CACTUS. STARFISH PLANT.
Staphylea pinnata L. EUROPEAN BLADDERNUT.
Staphylea trifolia L. AMERICAN BLADDERNUT.
STAR BUSH: **Turraea obtusifolia.**
STAR-CLUSTER: **Pentas lanceolata.**
   EGYPTIAN: **Pentas lanceolata.**
STARFISH FLOWER: **Stapelia gigantea.**
STARFISH PLANT: **Stapelia variegata.**
STAR FLOWER, LAVENDER: **Grewia caffra.**
STAR PLANT, LAVENDER: **Grewia caffra.**
Statice pectinata Ait. See **Limonium pectinatum.**
Statice perezii Stapf. See **Limonium perezii.**
Stauntonia hexaphylla (Thunb.) Decne.
Stenocarpus salignus R. Br. BEEFWOOD.
Stenocarpus sinuatus (A. Cunn.) Endl. FIREWHEEL TREE.
Stenolobium alatum (DC.) Sprague. See **Tecoma guarume.**
Stenolobium stans (L.) Seem. See **Tecoma stans.**
Stephanotis floribunda Brongn. MADAGASCAR JASMINE. CORSAGE VINE.
Sterculia acerifolia A. Cunn. See **Brachychiton acerifolius.**
Sterculia bidwillii (Hook.) Hook. ex Benth. See **Brachychiton bidwillii.**
Sterculia discolor (F. J. Muell.) F. J. Muell. ex Benth. See **Brachychiton discolor.**
Sterculia diversifolia G. Don. See **Brachychiton populneus.**
Sterculia foetida L. INDIAN ALMOND.
Sterculia lurida (C. Moore ex F. J. Muell.) F. J. Muell. ex Benth. See **Brachychiton discolor.**
Sterculia platanifolia L. f. See **Firmiana simplex.**
Sterculia rupestris (Lindl.) Benth. See **Brachychiton rupestris.**
Sterculia sextonii. Listed name for **Brachychiton × hybridus.**
Sterculia trichosiphon Benth. See **Brachychiton australis.**
STEWARTIA, JAPANESE: **Stewartia pseudo-camellia.**
   KOREAN: **Stewartia koreana.**
   MOUNTAIN: **Stewartia ovata.**
   TALL: **Stewartia monadelpha.**
Stewartia koreana Nakai ex Rehd. KOREAN STEWARTIA.
Stewartia monadelpha Siebold & Zucc. TALL STEWARTIA.
Stewartia ovata (Cav.) Weatherby. MOUNTAIN STEWARTIA.
Stewartia pseudo-camellia Maxim. JAPANESE STEWARTIA.
Stigmaphyllon ciliatum (Lam.) A. Juss. AMAZON VINE. BRAZILIAN GOLDEN VINE. ORCHID VINE.
Stigmaphyllon littorale A. Juss.
STORAX: **Styrax officinalis.**
Stranvaesia davidiana Decne.
Stranvaesia davidiana var. *undulata* (Decne.) Rehd. & Wils. Syn.: *S. undulata.*
Stranvaesia glaucescens Lindl. See **S. nussia.**
Stranvaesia nussia Decne. Syn.: *S. glaucescens.*
Stranvaesia undulata Decne. See **S. davidiana** var. **undulata.**
STRAWBERRY TREE: **Arbutus unedo.**
   COMPACT: **Arbutus unedo 'Compacta'.**
Strelitzia alba (L. f.) Skeels. Syn.: *S. augusta.*
Strelitzia augusta Thunb. See **S. alba.**
Strelitzia nicolai Regel & Koern. BIRD-OF-PARADISE TREE. GIANT BIRD-OF-PARADISE.
Strelitzia parvifolia Ait. f. See **S. reginae.**
Strelitzia reginae Ait. Syn.: *S. parvifolia.* BIRD-OF-PARADISE. BIRD-OF-PARADISE FLOWER. QUEEN'S BIRD-OF-PARADISE. CRANE FLOWER.
Streptosolen jamesonii (Benth.) Miers. FIREBUSH. MARMALADE BUSH.
STRINGYBARK, MEATY: **Eucalyptus cinerea.**
   RED: **Eucalyptus macrorhyncha.**
   WHITE: **Eucalyptus globoidea.**
Styrax californicus Torr. See **S. officinalis** var. **californicus.**
Styrax japonicus Siebold & Zucc. JAPANESE SNOWBELL.
Styrax obassia Siebold & Zucc. FRAGRANT SNOWBELL.
Styrax officinalis L. STORAX. SNOWBELL.
Styrax officinalis var. *californicus* (Torr.) Rehd. Syn.: *S. californicus.* SNOWDROP BUSH.
Styrax officinalis var. *fulvescens* (Eastw.) Munz & I. M. Johnst. HAIRY SNOWDROP BUSH.
SUGARBERRY: **Celtis laevigata.**
SUGARBERRY: **Ehretia anacua.**
SUGARBUSH: **Rhus ovata.**
SULPHUR FLOWER: **Eriogonum umbellatum.**
SUMAC, AFRICAN: **Rhus lancea.**
   FRAGRANT: **Rhus aromatica.**
   LAUREL: **Rhus laurina.**
   LEMON: **Rhus aromatica.**
   LEMONADE: **Rhus integrifolia.**
   PRAIRIE: **Rhus lanceolata.**
   SCARLET: **Rhus glabra.**
   SHINING: **Rhus copallina.**
   SMOOTH: **Rhus glabra.**
   STAGHORN: **Rhus typhina.**
   VELVET: **Rhus typhina.**
SUMMERSWEET: **Clethra alnifolia.**
SUNFLOWER, ACTON: **Encelia virginensis** subsp. **actonii.**
   BUSH: **Encelia californica.**
   MEXICAN: **Tithonia rotundifolia.**
SUPORA SHRUB: **Hauya elegans.**
Sutera grandiflora (Galpin) Hiern. PURPLE GLORY PLANT.
Sutherlandia frutescens (L.) R. Br.
Swainsonia galegifolia (Andr.) R. Br. WINTER SWEET PEA. DARLING PEA. SWAN FLOWER.
SWAN FLOWER: **Swainsonia galegifolia.**
SWEET BAY: **Magnolia virginiana.**
SWEET GUM: See GUM, SWEET.
SWEETLEAF, ASIATIC: **Symplocos paniculata.**
SWEETSHADE: **Hymenosporum flavum.**
SWEET SHRUB: **Calycanthus floridus.**
SWEETSPIRE: **Itea virginiana.**
   HOLLY-LEAF: **Itea ilicifolia.**
Syagrus coronata (Mart.) Becc. LICURY PALM. LICURI PALM.
Syagrus weddelliana (H. Wendl.) Becc. See **Microcoelum weddellianum.**
SYCAMORE, AMERICAN: **Platanus occidentalis.**
   ARIZONA: **Platanus wrightii.**
   CALIFORNIA: **Platanus racemosa.**
   EUROPEAN: **Platanus orientalis.**
   WESTERN: **Platanus racemosa.**
Symingtonia populnea (R. Br.) Steenis. See **Exbucklandia populnea.**
Symphoricarpos albus (L.) S. F. Blake. Syn.: *S. racemosus.* SNOWBERRY. WAXBERRY.

Symphoricarpos albus, of authors, not (L.) S. F. Blake.
See **S. rivularis**.
Symphoricarpos × chenaultii Rehd. [?**S. microphyllus** HBK × **S. orbiculatus**]. CHENAULT CORALBERRY.
Symphoricarpos mollis Nutt. CREEPING SNOWBERRY. SPREADING SNOWBERRY.
Symphoricarpos orbiculatus Moench. Syn.: *S. vulgaris*. CORALBERRY. INDIAN CURRANT.
Symphoricarpos racemosus Michx. See **S. albus** (L.) S. F. Blake.
Symphoricarpos rivularis Suksd. Syn.: *S. albus* of authors, not (L.) S. F. Blake.
Symphoricarpos vulgaris Michx. See **S. orbiculatus**.
Symplocos paniculata (Thunb.) Miq. SAPPHIRE BERRY. ASIATIC SWEETLEAF.
Synadenium grantii Hook. f. AFRICAN MILK BUSH.
Syncarpia glomulifera (Sm.) Niedenzu. Syn.: *S. laurifolia*. TURPENTINE TREE.
Syncarpia laurifolia Ten. See **S. glomulifera**.
SYRINGA, EVERGREEN: **Philadelphus mexicanus**.
Syringa amurensis Rupr. See **S. reticulata** var. **mandschurica**.
Syringa amurensis var. *japonica* (Maxim.) Franch. & Sav. See **S. reticulata**.
Syringa 'Angel White'. See **S. vulgaris 'Angel White'**.
Syringa 'Belle de Nancy'. See **S. vulgaris 'Belle de Nancy'**.
Syringa 'Charles Joly'. See **S. vulgaris 'Charles Joly'**.
Syringa 'Charles X'. See **S. vulgaris 'Charles X'**.
Syringa × chinensis Willd. [*S. persica* × *S. vulgaris*]. Syn.: *S. rothomagensis*. CHINESE LILAC. ROUEN LILAC.
Syringa × chinensis 'Hathaway'.
Syringa 'Congo'. See **S. vulgaris 'Congo'**.
Syringa 'Eximia'. See **S. josikaea 'Eximia'**.
Syringa × hyacinthiflora (Lemoine) Rehd. [**S. oblata** Lindl. × **S. vulgaris**].
Syringa × hyacinthiflora 'Blue Hyacinth'.
Syringa × hyacinthiflora 'Clarke's Giant'.
Syringa × hyacinthiflora 'Esther Staley'.
Syringa × hyacinthiflora 'Kate Sessions'.
Syringa × hyacinthiflora 'Purple Glory'.
Syringa × hyacinthiflora 'Purple Heart'.
Syringa × hyacinthiflora 'Splendor'.
Syringa × hyacinthiflora 'Spring Glory'.
Syringa × hyacinthiflora 'Sunset'.
Syringa 'James Macfarlane'. See **S. × prestoniae 'James Macfarlane'**.
Syringa × josiflexa Preston. [*S. josikaea* × *S. reflexa*].
Syringa josikaea Jacq. f. ex Rchb. HUNGARIAN LILAC.
Syringa josikaea 'Eximia'.
Syringa josikaea 'Sylvia'. See **S. vulgaris 'Sylvia'**.
Syringa laciniata Mill. Syn.: *S. × persica* var. *laciniata*. CUT-LEAF PERSIAN LILAC. CUT-LEAF LILAC.
Syringa 'Lavender Lady'.
Syringa 'Ludwig Spaeth'. See **S. vulgaris 'Ludwig Spaeth'**.
Syringa 'Mme. Casimir Périer'. See **S. vulgaris 'Madame Casimer Périer'**.
Syringa 'Mme. F. Morel'. See **S. vulgaris 'Madame F. Morel'**.
Syringa 'Mme. Lemoine'. See **S. vulgaris 'Madame Lemoine'**.
Syringa 'Negro'. See **S. vulgaris 'Negro'**.
Syringa palibiniana Nakai. See **S. patula**.
Syringa patula (Palibin) Nakai. Syn.: *S. palibiniana*. *S. velutina*.
Syringa × persica L. [**S. afghanica** Schneid. × **S. laciniata**]. PERSIAN LILAC.
Syringa × persica 'Alba'.
Syringa × persica 'Hathaway'. See **S. × chinensis 'Hathaway'**.
Syringa × persica var. *laciniata* (Mill.) Weston. See **S. laciniata**.
Syringa 'President Grevy'. See **S. vulgaris 'President Grevy'**.
Syringa × prestoniae McKelvey. [*S. reflexa* × *S. villosa*].
Syringa × prestoniae 'James Macfarlane'.
Syringa reflexa Schneid.
Syringa reticulata (Blume) Hara. Syn.: *S. amurensis* var. *japonica*. JAPANESE TREE LILAC.
Syringa reticulata var. **mandschurica** (Maxim.) Hara. Syn.: *S. amurensis*.
Syringa rothomagensis Mordant de Launay. See **S. × chinensis**.
Syringa × swegiflexa Hesse. [*S. reflexa* × *S. sweginzowii* Koehne & Lingelsh.].
Syringa 'Sylvia'. See **S. vulgaris 'Sylvia'**.
Syringa velutina Kom. See **S. patula**.
Syringa villosa Vahl. LATE LILAC.
Syringa vulgaris L. COMMON LILAC.
Syringa vulgaris 'Adelaide Dunbar'.
Syringa vulgaris 'Ami Schott'.
Syringa vulgaris 'Angel White'.
Syringa vulgaris 'Belle de Nancy'.
Syringa vulgaris 'Blue Hyacinth'. See **S. × hyacinthiflora 'Blue Hyacinth'**.
Syringa vulgaris 'Capitaine Perrault'.
Syringa vulgaris 'Charles Joly'.
Syringa vulgaris 'Charles X'.
Syringa vulgaris 'Clarke's Giant'. See **S. × hyacinthiflora 'Clarke's Giant'**.
Syringa vulgaris 'Congo'.
Syringa vulgaris 'Edith Cavell'.
Syringa vulgaris 'Ellen Willmott'. See **S. vulgaris 'Miss Ellen Willmott'**.
Syringa vulgaris 'Esther Staley'. See **S. × hyacinthiflora 'Esther Staley'**.
Syringa vulgaris 'Firmament'.
Syringa vulgaris 'Gismonda'.
Syringa vulgaris 'Henri Martin'.
Syringa vulgaris 'Jan van Tol'.
Syringa vulgaris 'Jeanne d'Arc'.
Syringa vulgaris 'Kate Sessions'. See **S. × hyacinthiflora 'Kate Sessions'**.
Syringa vulgaris 'Katherine Havemeyer'.
Syringa vulgaris 'Lucie Baltet'.
Syringa vulgaris 'Ludwig Spaeth'.
Syringa vulgaris 'Madame Casimer Périer'.
Syringa vulgaris 'Madame Charles Souchet'.
Syringa vulgaris 'Madame F. Morel'.
Syringa vulgaris 'Madame Lemoine'.
Syringa vulgaris 'Marceau'.
Syringa vulgaris 'Maréchal Foch'.
Syringa vulgaris 'Marie Legraye'.
Syringa vulgaris 'Maximowicz'.
Syringa vulgaris 'Michel Buchner'.
Syringa vulgaris 'Miss Ellen Willmott'.
Syringa vulgaris 'Monument'.
Syringa vulgaris 'Mrs. Edward Harding'.
Syringa vulgaris 'Mrs. W. E. Marshall'.
Syringa vulgaris 'Negro'.
Syringa vulgaris 'Paul Thirion'.
Syringa vulgaris 'President Grevy'.
Syringa vulgaris 'President Lincoln'.
Syringa vulgaris 'Primrose'.
Syringa vulgaris 'Prodige'.
Syringa vulgaris 'Professor E. H. Wilson'.
Syringa vulgaris 'Purple Glory'. See **S. × hyacinthiflora 'Purple Glory'**.

Syringa vulgaris 'Purple Heart'. See **S. × hyacinthiflora 'Purple Heart'**.
Syringa vulgaris 'Souvenir de Georges Truffaut'.
Syringa vulgaris 'Souvenir de Louis Chasset'.
Syringa vulgaris 'Splendor'. See **S. × hyacinthiflora 'Splendor'**.
Syringa vulgaris 'Spring Glory'. See **S. × hyacinthiflora 'Spring Glory'**.
Syringa vulgaris 'Sunset'. See **S. × hyacinthiflora 'Sunset'**.
Syringa vulgaris 'Sweetheart'.
Syringa vulgaris 'Sylvia'.
Syringa vulgaris 'Violetta'.
Syringa vulgaris 'Volcan'.
Syzygium jambos (L.) Alston. Syn.: *Eugenia jambos*.
  ROSE APPLE.
Syzygium paniculatum Gaertner. Syn.: *Eugenia australis. E. myrtifolia. E. paniculata. E. paniculata* var. *australis*.
  AUSTRALIAN BUSH CHERRY.
Syzygium paniculatum 'Antone Dwarf'.
Syzygium paniculatum 'Compactum'. DWARF BUSH CHERRY.
Syzygium paniculatum 'Flaming Sphere'.
Syzygium paniculatum 'Globulus'.
Syzygium paniculatum 'Nanum'.
Syzygium paniculatum 'Newport'.
Syzygium paniculatum 'Red Flame'. BRIGHT RED BUSH CHERRY.

T

Tabebuia argentea (Bur. & K. Schum.) Britt. See **Tabebuia caraiba**.
Tabebuia avellanedae Lorentz ex Griseb. See **Tabebuia impetiginosa**.
Tabebuia caraiba (Mart.) Bur. Syn.: *T. argentea. Tecoma argentea*. SILVER TRUMPET TREE.
Tabebuia chrysotricha (Mart. ex DC.) Standl. Syn.: *Tecoma chrysotricha*. GOLDEN TRUMPET TREE.
Tabebuia impetiginosa (Mart. ex DC.) Standl. Syn.: *T. avellanedae. T. ipe. Tecoma impetiginosa*.
Tabebuia ipe Mart. See **T. impetiginosa**.
Tabebuia palmeri Rose.
Tabebuia pulcherrima Sandwith.
Tabebuia umbellata (Sond.) Sandwith.
TACAMAHAC: **Populus balsamifera**.
TALI: **Gigantochloa apus**.
TALLOW TREE, CHINESE: **Sapium sebiferum**.
  JAPANESE: **Sapium japonicum**.
TALLOW WOOD, BASTARD: **Eucalyptus planchoniana**.
TAMARACK: **Larix laricina**.
  WESTERN: **Larix occidentalis**.
TAMARISK: See **Tamarix** spp.
TAMARIX, FRENCH: **Tamarix gallica**.
Tamarix africana Poir.
Tamarix aphylla (L.) Karst. Syn.: *T. articulata*. ATHEL.
Tamarix articulata Vahl. See **T. aphylla**.
Tamarix 'Cheyanne'?
Tamarix chinensis Lour. Syn.: *T. japonica. T. juniperina. T. plumosa*.
Tamarix gallica L. MANNA PLANT. FRENCH TAMARIX.
Tamarix hispida Willd.
Tamarix hispida 'Coolidgei'.
Tamarix japonica Dipp. See **T. chinensis**.
Tamarix juniperina Bunge. See **T. chinensis**.
Tamarix parviflora DC. Plants cult. as **T. tetrandra** belong here.
Tamarix pentandra Pall. See **T. ramosissima**.
Tamarix plumosa Carrière. See **T. chinensis**.
Tamarix ramosissima Ledeb. Syn.: *T. pentandra*.
Tamarix tetrandra Pall. Not cult. Plants so called are **T. parviflora**.
TANGELO: **Citrus × tangelo**.
  MINNEOLA: **Citrus × tangelo 'Minneola'**.
  SAMPSON: **Citrus × tangelo 'Sampson'**.
TANGERINE: **Citrus reticulata**.
  DANCY: **Citrus reticulata 'Dancy'**.
TANGOR: **Citrus × nobilis**.
  KING: **Citrus × nobilis 'King'**.
  TEMPLE: **Citrus × nobilis 'Temple'**.
TANYOSHO: **Pinus densiflora 'Umbraculifera'**.
TAPEWORM PLANT: **Homalocladium platycladum**.
TARA: **Caesalpinia spinosa**.
TARAJO: **Ilex latifolia**.
TARATA: **Pittosporum eugenioides**.
TARA VINE: **Actinidia arguta**.
TARO: **Alocasia macrorrhiza**.
TARO: **Colocasia esculenta**.
  BLUE: **Xanthosma violaceum**.
TARRAGON, FRENCH: **Artemisia dracunculus**.
  TRUE: **Artemisia dracunculus**.
TAWHIWHI: **Pittosporum tenuifolium**.
Taxodium distichum (L.) L. Rich. BALD CYPRESS. SWAMP CYPRESS.
Taxodium mucronatum Ten. MONTEZUMA CYPRESS.
Taxus baccata L. ENGLISH YEW.
Taxus baccata 'Adpressa'.
Taxus baccata 'Adpressa Aurea'.
Taxus baccata 'Aurea'. GOLDEN YEW.
Taxus baccata 'Aureovariegata'.
Taxus baccata 'Dovastoniana'. DOVASTON YEW.
Taxus baccata 'Dovastonii Aurea'.
Taxus baccata 'Erecta'. UPRIGHT ENGLISH YEW. BROOM YEW.
Taxus baccata 'Fastigiata'. See **T. baccata 'Stricta'**.
Taxus baccata 'Fastigiata Aurea'. GOLDEN IRISH YEW.
Taxus baccata 'Hicksii'. See **T. × media 'Hicksii'**.
Taxus baccata 'Pendula'.
Taxus baccata 'Repandens'. SPREADING ENGLISH YEW.
Taxus baccata 'Repandens Aurea'?
Taxus baccata 'Semperaurea'.
Taxus baccata 'Silver Green'. SILVER GREEN YEW.
Taxus baccata 'Stricta'. IRISH YEW.
Taxus baccata 'Washingtonii'.
Taxus baccata 'Washingtonii Aurea'. See **T. baccata 'Washingtonii'**.
Taxus baccata 'Washingtonia Nana'. See **T. baccata 'Washingtonii'**.
Taxus brevifolia Nutt. PACIFIC YEW. WESTERN YEW.
Taxus canadensis Marsh. CANADA YEW.
Taxus canadensis 'Pyramidalis'. See **T. canadensis 'Stricta'**.
Taxus canadensis 'Stricta'.
Taxus 'Capitata'. See **T. cuspidata**.
Taxus cuspidata Siebold & Zucc. JAPANESE YEW. UPRIGHT JAPANESE YEW.
Taxus cuspidata 'Capitata'. See **T. cuspidata**.
Taxus cuspidata 'Columnaris'. COLUMNAR JAPANESE YEW.
Taxus cuspidata 'Densa'. CUSHION JAPANESE YEW.
Taxus cuspidata 'Expansa'. SPREADING JAPANESE YEW.
Taxus cuspidata 'Fastigiata'. Listed name. May be **T. cuspidata 'Columnaris'** or **T. × media 'Andorra'** or **T. × media 'Hicksii'**.

Taxus cuspidata 'Fastigiata Aurea'. Listed name. May be
  T. baccata 'Fastigiata Aurea'.
Taxus cuspidata 'Hillii'. See **T. × media 'Hillii'**.
Taxus cuspidata var. **nana** Rehd. DWARF JAPANESE YEW.
  COMPACT JAPANESE YEW.
Taxus cuspidata 'Nana Pyramidalis Hillii'. See **T. × media
  'Hillii'**.
Taxus cuspidata 'Prostrata'.
Taxus cuspidata 'Pyramidalis'. See **T. × media 'Hillii'**.
Taxus erecta nana?
Taxus 'Fastigiata'. Listed name for **T. baccata 'Stricta'**.
Taxus hibernica. Listed name for **T. baccata 'Stricta'**.
Taxus × media Rehd. [*T. baccata* × *T. cuspidata*].
Taxus × media 'Andersonii'.
Taxus × media 'Andorra'.
Taxus × media 'Brownii'. BROWN'S YEW.
Taxus × media 'Halloriana'. HALLORAN YEW.
Taxus × media 'Hatfieldii'. HATFIELD YEW.
Taxus × media 'Hicksii'. HICKS' YEW.
Taxus × media 'Hillii'. HILL'S PYRAMIDAL YEW.
Taxus × media 'Pyramidalis'. See **T. × media 'Hillii'**.
Taxus × media 'Robusta'.
Taxus 'Repandens'. See **T. baccata 'Repandens'**.
TEA, ARABIAN: **Catha edulis**.
  LABRADOR: **Ledum groenlandicum**.
  MOUNTAIN: **Gaultheria procumbens**.
  NEW JERSEY: **Ceanothus americanus**.
TEA PLANT: **Camellia sinensis**.
TEA TREE, AUSTRALIAN: **Leptospermum laevigatum**.
  COMPACT AUSTRALIAN: **Leptospermum laevigatum 'Reevesii'**.
  DWARF: **Leptospermum laevigatum 'Reevesii'**.
  MANUKA: **Leptospermum scoparium**.
  NEW ZEALAND: **Leptospermum scoparium**.
  REEVES' AUSTRALIAN: **Leptospermum laevigatum 'Reevesii'**.
  ROUND-LEAVED: **Leptospermum rotundifolium**.
  SANDERS': **Leptospermum scoparium 'Sandersii'**.
  SWAMP: **Melaleuca quinquenervia**.
  WOOLLY: **Leptospermum lanigerum**.
Tecoma alata. Listed name for **Tecoma guarume**.
Tecoma argentea Bur. & K. Schum. See **Tabebuia caraiba**.
Tecoma australis R. Br. See **Pandorea pandorana**.
Tecoma capensis Lindl. See **Tecomaria capensis**.
Tecoma chrysotricha Mart. ex DC. See **Tabebuia chrysotricha**.
Tecoma garrocha Hieron.
Tecoma guarume DC. Syn.: *Tecoma smithii*. *Stenolobium
  alatum*. Plants called *Tecoma alata* and *Tecomaria smithii*
  belong here.
Tecoma impetiginosa Mart. See **Tabebuia impetiginosa**.
Tecoma jasminoides Lindl. See **Pandorea jasminoides**.
Tecoma jasminoides 'Alba'. See **Pandorea jasminoides 'Alba'**.
Tecoma jasminoides 'Rosea'. See **Pandorea jasminoides 'Rosea'**.
Tecoma mackenii W. Wats. See **Podranea ricasoliana**.
Tecoma reginae-sabei Franceschi. See **Podranea brycei**.
Tecoma smithii W. Wats. See **Tecoma guarume**.
Tecoma stans (L.) HBK. Syn.: *Bignonia stans*. *Stenolobium
  stans*. YELLOWBELLS. YELLOW ELDER.
Tecomaria capensis (Thunb.) Spach. Syn.: *Bignonia capensis*.
  Tecoma capensis. CAPE HONEYSUCKLE.
Tecomaria capensis 'Aurea'. YELLOW CAPE HONEYSUCKLE.
Tecomaria smithii. Listed name for **Tecoma guarume**.
Templetonia retusa (Vent.) R. Br.
TEMU: **Luma apiculata**.
Ternstroemia gymnanthera (Wight & Arn.) Sprague.
  Syn.: *T. japonica* of authors, not Thunb.
Ternstroemia japonica of authors, not Thunb.
  See **T. gymnanthera**.

Testudinaria elephantipes (L'Hér.) Burchell. See **Dioscorea
  elephantipes**.
TETRACOCCUS, PARRY'S: **Tetracoccus dioicus**.
Tetracoccus dioicus Parry. PARRY'S TETRACOCCUS.
Tetrapanax papyriferus (Hook.) K. Koch. Syn.: *Aralia
  papyrifera*. RICE-PAPER PLANT.
Tetrastigma voinieranum (Baltet) Pierre ex Gagnep. Syn.:
  Cissus voinierana. *Vitis voinierana*. CHESTNUT VINE. LIZARD
  PLANT. Plants called *Cissus vomerensis* and *Tetrastigma
  vomerensis* belong here.
Tetrastigma vomerensis. Listed name for **T. voinieranum**.
Teucrium chamaedrys L. GERMANDER.
Teucrium chamaedrys 'Prostratum'.
Teucrium fruticans L. BUSH GERMANDER.
TEXAS RANGER: **Leucophyllum frutescens**.
Thamnocalamus tesselatus. Listed name for **Arundinaria
  tesselata**.
THATCH, SILVER: **Coccothrinax argentea**.
Thea sinensis L. See **Camellia sinensis**.
Thevetia neriifolia A. Juss. ex Steud. See **T. peruviana**.
Thevetia peruviana (Pers.) K. Schum. Syn.: *T. neriifolia*.
  YELLOW OLEANDER. LUCKY NUT.
Thevetia thevetioides (HBK) K. Schum.
THIMBLEBERRY: **Rubus parviflorus**.
THORN, CAMEL: **Acacia giraffae**.
  COCKSPUR: **Crataegus crus-galli**.
  GLASTONBURY: **Crataegus monogyna 'Biflora'**.
  JERUSALEM: **Parkinsonia aculeata**.
  KANGAROO: **Acacia armata**.
  LAVALLE: **Crataegus × lavallei**.
  WASHINGTON: **Crataegus phaenopyrum**.
Thrinax argentea Lodd. ex Schult. & Schult. f.
  See **Coccothrinax argentea**.
Thryallis glauca (Cav.) Kuntze. See **Galphimia glauca**.
Thuja 'Aurea Nana'. See **Platycladus orientalis 'Aureus Nanus'**.
Thuja 'Bakeri'. See **Platycladus orientalis 'Bakeri'**.
Thuja 'Berkmannii'. See **Platycladus orientalis 'Aureus Nanus'**.
Thuja 'Beverleyensis'. See **Platycladus orientalis 'Beverleyensis'**.
Thuja 'Beverleyensis Elegans'. See **Platycladus orientalis
  'Beverleyensis'**.
Thuja 'Blue Cone'. See **Platycladus orientalis 'Blue Cone'**.
Thuja 'Bonita'. See **Platycladus orientalis 'Bonita'**.
Thuja 'Chase's Golden'. See **Platycladus orientalis 'Chase's
  Golden'**.
Thuja 'Compacta'. See **T. occidentalis 'Compacta'** or
  Platycladus orientalis 'Sieboldii'.
Thuja 'Elegantissima'. See **Platycladus orientalis
  'Elegantissimus'**.
Thuja 'Fruitlandii'. See **Platycladus orientalis 'Fruitlandii'**.
Thuja 'Little Gem'. See **T. occidentalis 'Little Gem'**.
Thuja lobbii Gord. See **T. plicata**.
Thuja 'Nestoides'. Listed name. May be **Chamaecyparis
  lawsoniana 'Nestoides'**.
Thuja occidentalis L. NORTHERN WHITE CEDAR. EASTERN
  ARBORVITAE.
Thuja occidentalis 'Aureovariegata'.
Thuja occidentalis 'Brewers'?
Thuja occidentalis 'Columna'. Some plants so called may be
  T. occidentalis 'Pyramidalis Compacta'.
Thuja occidentalis 'Columnaris'. See **T. occidentalis 'Fastigiata'**.
Thuja occidentalis 'Compacta'. LITTLE GLOBE ARBORVITAE.
Thuja occidentalis 'Compacta Erecta'.
Thuja occidentalis 'Douglasii'. See **T. occidentalis 'Filiformis'**.
Thuja occidentalis 'Elegantissima'.
Thuja occidentalis 'Ellwangerana'. ELLWANGER ARBORVITAE.

Thuja occidentalis 'Ellwangerana Aurea'. ELLWANGER GOLDEN ARBORVITAE.
Thuja occidentalis 'Fastigiata'. PYRAMIDAL ARBORVITAE. Some plants so called may be **T. occidentalis 'Pyramidalis'**.
Thuja occidentalis 'Filiformis'.
Thuja occidentalis 'George Washington'?
Thuja occidentalis 'Globosa'. GLOBE ARBORVITAE.
Thuja occidentalis 'Hetz's Midget'. HETZ'S MIDGET ARBORVITAE.
Thuja occidentalis 'Little Gem'. LITTLE GEM ARBORVITAE.
Thuja occidentalis 'Lutea'.
Thuja occidentalis 'Nana'. See **T. occidentalis 'Compacta'**.
Thuja occidentalis 'Pumila'.
Thuja occidentalis 'Pumila Sudworthii'?
Thuja occidentalis 'Pyramidalis'. PYRAMIDAL ARBORVITAE. Some plants so called may be **T. occidentalis 'Columna'**, **T. occidentalis 'Fastigiata'**, or **T. occidentalis 'Pyramidalis Compacta'**.
Thuja occidentalis 'Pyramidalis Compacta'. Some plants so called may be **T. occidentalis 'Columna'**.
Thuja occidentalis 'Pyramidalis Hillii'?
Thuja occidentalis 'Pyramidalis Nana'?
Thuja occidentalis 'Recurva Nana'.
Thuja occidentalis 'Rosenthalii'.
Thuja occidentalis 'Sherwood Column'?
Thuja occidentalis 'Sherwood Frost'?
Thuja occidentalis 'Sherwood Moss'?
Thuja occidentalis 'Sherwood Plumespire'?
Thuja occidentalis 'Techny'.
Thuja occidentalis 'Umbraculifera'. UMBRELLA ARBORVITAE.
Thuja occidentalis 'Woodward Globe'. See **T. occidentalis 'Woodwardii'**.
Thuja occidentalis 'Woodwardii'. WOODWARD ARBORVITAE.
Thuja orientalis L. and all listed cvs. See **Platycladus orientalis** and cvs.
Thuja plicata J. Donn ex D. Don. Syn.: *T. lobbii*. WESTERN RED CEDAR. GIANT ARBORVITAE. CANOE CEDAR. GIANT CEDAR. RED CEDAR.
Thuja plicata 'Aurea'. GOLDSPOT ARBORVITAE.
Thuja plicata 'Aureovariegata'. See **T. plicata 'Aurea'**.
Thuja plicata 'Cuprea'.
Thuja plicata 'Fastigiata'.
Thuja plicata 'Rogersii'.
Thuja plicata 'Zebrina'. See **T. plicata 'Aurea'**.
Thuja 'Pyramidalis'. Listed name. May be **T. occidentalis 'Pyramidalis'**, **T. occidentalis 'Fastigiata'**, or **T. plicata 'Fastigiata'**.
Thuja 'Rosedale'. See **Platycladus orientalis 'Rosedale'**.
Thuja 'Westmont'. See **Platycladus orientalis 'Westmont'**.
Thuja 'Woodwardii'. See **T. occidentalis 'Woodwardii'**.
Thuja 'Zebrina'. See **T. plicata 'Aurea'**.
Thujopsis dolabrata (L. f.) Siebold & Zucc. FALSE ARBORVITAE. DEERHORN CEDAR. HIBA CEDAR. HIBA ARBORVITAE.
Thujopsis dolabrata 'Variegata'.
Thunbergia battiscombei Turrill.
Thunbergia erecta (Benth.) T. Anderson. KING'S-MANTLE. BUSH CLOCK VINE.
Thunbergia grandiflora (Roxb. ex Rottl.) Roxb. BLUE SKY VINE. BLUE SKY FLOWER. BLUE TRUMPET VINE.
Thunbergia mysorensis (Wight) T. Anderson ex Beddome.
THYME, CARAWAY-SCENTED: **Thymus herba-barona**.
  COMMON: **Thymus vulgaris**.
  LEMON: **Thymus serpyllum**.
  SILVER: **Thymus vulgaris 'Argenteus'**.
Thymus × citriodorus (Pers.) Schreber ex Schweigg. & Koerte. [*T. vulgaris* × *T. pulegioides*]. Some plants called *T. serpyllum* var. *vulgaris* belong here.

Thymus × citriodorus 'Argenteus'.
Thymus herba-barona Loisel. CARAWAY-SCENTED THYME.
Thymus lanuginosus Mill. May not be cult. Some plants so called are **T. pseudolanuginosus** or other hairy thymes.
Thymus praecox Opiz.
Thymus praecox subsp. **arcticus** (E. Durand) Jalas. MOTHER-OF-THYME. Some plants called **T. serpyllum** belong here.
Thymus pseudolanuginosus Ronn. Some plants called **T. lanuginosus** and **T. serpyllum** probably belong here.
Thymus pulegioides L. Some plants called **T. serpyllum** belong here.
Thymus serpyllum L. LEMON THYME. May not be cult. Plants so called may be **T. praecox** subsp. **arcticus**, **T. pseudolanuginosus**, or **T. pulegioides**.
Thymus serpyllum 'Argenteus'. Listed name; may refer to **T. × citriodorus 'Argenteus'**.
Thymus serpyllum var. *vulgaris*. Listed name for **T. × citriodorus** or **T. vulgaris**.
Thymus vulgaris L. COMMON THYME. Some plants called *T. serpyllum* var. *vulgaris* belong here.
Thymus vulgaris 'Argenteus'. SILVER THYME.
TI: **Cordyline terminalis**.
Tibouchina clavata (Pers.) Wurdack.
Tibouchina semidecandra (Schrank & Mart. ex DC.) Cogn. Not cult. Plants so called are **T. urvilleana**.
Tibouchina urvilleana (DC.) Cogn. PRINCESS FLOWER. PLEROMA. Plants called **T. semidecandra** belong here.
Tilia americana L. AMERICAN LINDEN. AMERICAN BASSWOOD.
Tilia cordata Mill. LITTLE-LEAF LINDEN.
Tilia cordata 'Greenspire'.
Tilia dasystyla Steven.
Tilia × euchlora K. Koch. [*T. cordata* × *T. dasystyla*]. CRIMEAN LINDEN.
Tilia × euchlora 'Redmont'.
Tilia × europaea L. See **T. × vulgaris**.
Tilia 'Greenspire'. See **T. cordata 'Greenspire'**.
Tilia platyphyllos Scop. LARGE-LEAVED LINDEN.
Tilia 'Redmont'. See **T. × euchlora 'Redmont'**.
Tilia tomentosa Moench. SILVER LINDEN.
Tilia × vulgaris Hayne. [*T. cordata* × *T. platyphyllos*]. Syn.: *T. × europaea*. LINDEN. EUROPEAN LINDEN.
Tipuana tipu (Benth.) Kuntze. TIPU TREE.
TIPU TREE: **Tipuana tipu**.
TOAD PLANT: **Stapelia variegata**.
  GIANT: **Stapelia gigantea**.
TOBACCO, PURPLE: **Iochroma cyaneum**.
  TREE: **Nicotiana glauca**.
TOBACCO BRUSH: **Ceanothus velutinus**.
TOBIRA: **Pittosporum tobira**.
  VARIEGATED: **Pittosporum tobira 'Variegated'**.
TOMATO TREE: **Cyphomandra crassifolia**.
TOOTHACHE TREE: **Zanthoxylum americanum**.
TORCH FLOWER: **Kniphofia uvaria**.
TORNILLO: **Prosopis pubescens**.
Torreya californica Torr. CALIFORNIA NUTMEG.
TOTARA: **Podocarpus totara**.
TOUCH-ME-NOT: **Impatiens sodenii**.
TOWER-OF-JEWELS: **Echium wildpretii**.
TOYON: **Heteromeles arbutifolia**.
  RED-TWIG CHINESE: **Photinia serrulata 'Aculeata'**.
Trachelospermum asiaticum (Siebold & Zucc.) Nakai. Syn.: *Rhynchospermum asiaticum*.
Trachelospermum fragrans Wallich ex Hook. f. Probably not cult. Some plants so called are **Chonemorpha fragrans**.

Trachelospermum jasminoides (Lindl.) Lem. Syn.: *Rhynchospermum jasminoides*. STAR JASMINE. CONFEDERATE VINE.
Trachelospermum jasminoides 'Bronze Star'.
Trachelospermum jasminoides 'North Star'.
Trachelospermum jasminoides 'Ryan's Twist Leaf'?
Trachelospermum jasminoides 'Variegatum'. VARIEGATED CONFEDERATE VINE.
Trachycarpus fortunei (Hook.) H. Wendl. Syn.: *Chamaerops fortunei*. WINDMILL PALM. CHINESE WINDMILL PALM. HEMP PALM. Plants called *Chamaerops excelsa* belong here.
Trachycarpus martianus (Wallich) H. Wendl. Syn.: *Chamaerops martiana*.
Trachycarpus nanus Becc. May not be cult. Some plants so called are *T. wagneranus*.
Trachycarpus takil Becc. Perhaps not cult. Some plants so called are *T. wagneranus*.
Trachycarpus wagneranus Hort. ex Roster. Some plants called *T. takil* and *T. nanus* belong here.
TRAVELER'S JOY: **Clematis vitalba**.
TRAVELER'S PALM: **Ravenala madagscariensis**.
TRAVELER'S TREE: **Ravenala madagascariensis**.
TREEBINE, KANGAROO: **Cissus antarctica**.
    VENEZUELA: **Cissus rhombifolia**.
TREE-OF-HEAVEN: **Ailanthus altissima**.
TREE TOMATO: **Cyphomandra crassifolia**.
TREFOIL, BIRD'S FOOT: **Lotus corniculatus**.
Trevesia burckii Boerlage. Plants called *T. sanderi* belong here.
Trevesia micholitzii. Listed name for *T. palmata* 'Micholitzii'.
Trevesia palmata (Roxb.) Vis.
Trevesia palmata 'Micholitzii'. SNOWFLAKE TREE. TROPICAL SNOWFLAKE. SNOWFLAKE ARALIA.
Trevesia sanderi. Listed name for *T. burckii*.
Trichosporum lobbianum Kuntze. See **Aeschynanthus radicans**.
Trichostema lanatum Benth. WOOLLY BLUE-CURLS.
Trichostema parishii Vasey. MOUNTAIN BLUE-CURLS.
Tricuspidaria dependens Ruiz & Pav. See **Crinodendron patagua**.
Triplochlamys multiflora (Juss.) Ulbrich. Syn.: *Pavonia multiflora*.
Tristania conferta R. Br. BRISBANE BOX.
Tristania laurina (Sm.) R. Br.
Trithrinax acanthocoma Drude.
Trithrinax campestris (H. Burmeister) Drude & Griseb.
Tritoma uvaria (L.) Ker-Gawl. See **Kniphofia uvaria**.
Trochodendron aralioides Siebold & Zucc.
TRUMPET CREEPER: **Campsis radicans**.
    CHINESE: **Campsis grandiflora**.
TRUMPET FLOWER: **Bignonia capreolata**.
TRUMPET TREE, GOLDEN: **Tabebuia chrysotricha**.
    SILVER: **Tabebuia caraiba**.
TRUMPET VINE, BLOOD-RED: **Distictis buccinatoria**.
    BLUE: **Thunbergia grandiflora**.
    LAVENDER: **Clytostoma callistegioides**.
    PINK: **Pandorea jasminoides** 'Rosea'.
    PINK: **Podranea ricasoliana**.
    ROYAL: **Distictis** 'Rivers'.
    VANILLA: **Distictis laxiflora**.
    VIOLET: **Clytostoma callistegioides**.
    YELLOW: **Anemopaegma chamberlaynii**.
    YELLOW: **Macfadyena unguis-cati**.
Tsuga canadensis (L.) Carrière. CANADIAN HEMLOCK. EASTERN HEMLOCK. HEMLOCK SPRUCE.
Tsuga canadensis 'Lewisii'?
Tsuga canadensis 'Nana'.

Tsuga canadensis 'Pendula'. SARGENT WEEPING HEMLOCK. CANADIAN WEEPING HEMLOCK.
Tsuga canadensis 'Prostrata'.
Tsuga caroliniana Engelm. CAROLINA HEMLOCK.
Tsuga diversifolia (Maxim.) M. T. Mast. JAPANESE HEMLOCK.
Tsuga heterophylla (Raf.) Sarg. WESTERN HEMLOCK. COAST HEMLOCK.
Tsuga mertensiana (Bong.) Carrière. MOUNTAIN HEMLOCK.
Tsuga sargentii. Listed name for *T. canadensis* 'Pendula'.
TUCKEROO: **Cupaniopsis anacardioides**.
TULIP TREE: **Liriodendron tulipifera**.
TULIP TREE: **Magnolia × soulangiana**.
    AFRICAN: **Spathodea campanulata**.
    CHINESE: **Liriodendron chinense**.
TUNA: **Opuntia ficus-indica**.
TUNG-OIL TREE: **Aleurites fordii**.
TUPELO: **Nyssa sylvatica**.
Tupidanthus calyptratus Hook. f. & Thoms.
TURK'S CAP: **Malvaviscus arboreus**.
TURK'S CAP: **Malvaviscus arboreus** var. **mexicanus**.
TURPENTINE TREE: **Syncarpia glomulifera**.
Turraea obtusifolia Hochst. STAR BUSH. SOUTH AFRICAN HONEYSUCKLE.
Turricula parryi (Gray) Macb.
TUTSAN: **Hypericum androsaemum**.
TWINBERRY: **Lonicera involucrata**.
TWINBERRY: **Mitchella repens**.
TWINFLOWER: **Linnaea borealis**.

U

Ugni molinae Turcz. Syn.: *Myrtus ugni*. CHILEAN GUAVA.
Ulex europaeus L. GORSE. FURZE. WHIN.
Ulmus americana L. AMERICAN ELM.
Ulmus americana 'Cool Shade'. See *U.* 'Coolshade'.
Ulmus 'Brea'. See *U. parvifolia* 'Drake'.
Ulmus 'Camperdownii'. See *U. glabra* 'Camperdownii'.
Ulmus carpinifolia Ruppius ex Suckow. SMOOTH-LEAF ELM.
Ulmus 'Christine Buisman'. BUISMAN ELM.
Ulmus 'Coolshade'.
Ulmus crassifolia Nutt. CEDAR ELM.
Ulmus 'Drake'. See *U. parvifolia* 'Drake'.
Ulmus glabra Huds. SCOTCH ELM. WYCH ELM.
Ulmus glabra 'Camperdownii'. CAMPERDOWN ELM.
Ulmus × hollandica Mill. [*U. carpinifolia* × *U. glabra* × *U. plottii* Druce]. DUTCH ELM. HOLLAND ELM.
Ulmus × hollandica 'Belgica'. BELGIAN ELM.
Ulmus parvifolia Jacq. EVERGREEN CHINESE ELM. EVERGREEN ELM. CHINESE ELM.
Ulmus parvifolia 'Brea'. See *U. parvifolia* 'Drake'.
Ulmus parvifolia 'Drake'. DRAKE EVERGREEN ELM. BREA ELM.
Ulmus parvifolia 'Sempervirens'. See *U. parvifolia*.
Ulmus parvifolia 'Sempervirens Brea'. See *U. parvifolia* 'Drake'.
Ulmus parvifolia 'Sempervirens Drake'. See *U. parvifolia* 'Drake'.
Ulmus procera Salisb. ENGLISH ELM.
Ulmus pumila L. SIBERIAN ELM.
Ulmus 'Sempervirens'. See *U. parvifolia*.
Ulmus 'Sempervirens Drake'. See *U. parvifolia* 'Drake'.
Ulmus × vegeta (Loud.) Lindl. [*U. carpinifolia* × *U. glabra*]. HUNTINGDON ELM. CHICHESTER ELM.

Ulmus × *vegeta* 'Camperdownii'. See **U. glabra 'Camperdownii'**.
Umbellularia californica Nutt. CALIFORNIA BAY. CALIFORNIA LAUREL. OREGON MYRTLE. PEPPERWOOD.
UMBRELLA TREE: **Magnolia tripetala**.
  AUSTRALIAN: **Brassaia actinophylla**.
  QUEENSLAND: **Brassaia actinophylla**.
  TEXAS: **Melia azedarach 'Umbraculifera'**.
UMBU: **Phytolacca dioica**.
Urbinia agavoides (Lem.) Rose. See **Echeveria agavoides**.

V

Vaccinium corymbosum L. HIGHBUSH BLUEBERRY.
Vaccinium cuneata. Listed name. Plants so called may be **Gaultheria cuneata**.
Vaccinium floribundum HBK.
Vaccinium moupinense Franch.
Vaccinium ovatum Pursh. EVERGREEN HUCKLEBERRY. CALIFORNIA HUCKLEBERRY.
Vaccinium oxycoccus L. SMALL CRANBERRY. EUROPEAN CRANBERRY.
Vaccinium parvifolium Sm. RED HUCKLEBERRY.
Vaccinium vitis-idaea L. FOXBERRY. COWBERRY. CRANBERRY.
Vaccinium vitis-idaea var. **minus** Lodd. MOUNTAIN CRANBERRY. LINGBERRY. LINGONBERRY. ROCK CRANBERRY.
VARNISH TREE: **Koelreuteria paniculata**.
  JAPANESE: **Firmiana simplex**.
Veitchia merrillii (Becc.) H. E. Moore. CHRISTMAS PALM. MANILLA PALM.
VELVET LEAF: **Kalanchoe beharensis**.
VERBENA, LEMON: **Aloysia triphylla**.
Veronica × *andersonii* Lindl. & Paxt. See **Hebe** × **andersonii**.
Veronica 'Autumn Glory'. See **Hebe 'Autumn Glory'**.
Veronica × *bidwillii* Hook., not Hook. f. See **Parahebe** × **bidwillii**.
Veronica bidwillii Hook. f., not Hook. See **Parahebe decora**.
Veronica buxifolia Benth. See **Hebe buxifolia**. Plants called *Veronica buxifolia* are probably **Hebe odora**.
Veronica canescens T. Kirk. See **Parahebe canescens**.
Veronica × *carnea* J. B. Armstr. See **Hebe** × **carnea**.
Veronica 'Carnea'. See **Hebe** × **carnea 'Carnea'**.
Veronica catarractae G. Forst. See **Parahebe catarractae**.
Veronica chathamica Buchanan. See **Hebe chathamica**.
Veronica 'Co-ed'. See **Hebe 'Co-ed'**.
Veronica decussata Ait. See **Hebe elliptica**.
Veronica 'Desilor'. See **Hebe 'Desilor'**.
Veronica elliptica G. Forst. See **Hebe elliptica**.
Veronica 'Evansii'?
Veronica 'Evansii Rubra'?
Veronica glaucophylla Cockayne. See **Hebe glaucophylla**.
Veronica 'Imperialis'. See **Hebe speciosa 'Imperialis'**.
Veronica imperialis Boucharlat. See **Hebe speciosa 'Imperialis'**.
Veronica 'Lake'?
Veronica lyallii Hook. f. See **Parahebe lyallii**.
Veronica menziesii Benth. See **Hebe menziesii**. Plants called *Veronica menziesii* are **Hebe divaricata**.
Veronica 'Patty's Purple'. See **Hebe 'Patty's Purple'**.
Veronica 'Purpurea'?
Veronica salicifolia G. Forst. See **Hebe salicifolia**.
Veronica speciosa R. Cunn. ex A. Cunn. See **Hebe speciosa**.
Veronica traversii Hook. f. See **Hebe traversii**.

Vestia lycioides Willd.
VIBURNUM, BODNANT: **Viburnum** × **bodnantense**.
  BURKWOOD: **Viburnum** × **burkwoodii**.
  CHENAULT: **Viburnum** × **burkwoodii 'Chenault'**.
  CINNAMON: **Viburnum cinnamomifolium**.
  DAVID: **Viburnum davidii**.
  DOUBLE-FILE: **Viburnum plicatum** forma **tomentosum**.
  DWARF FRAGRANT: **Viburnum farreri 'Nanum'**.
  ICHANG: **Viburnum ichangense**.
  JAPANESE: **Viburnum japonicum**.
  JUDD: **Viburnum** × **juddii**.
  KOREAN SPICE: **Viburnum carlesii**.
  LEATHERLEAF: **Viburnum rhytidophyllum**.
  LINDEN: **Viburnum dilatatum**.
  SANDANKWA: **Viburnum suspensum**.
  SWEET: **Viburnum odoratissimum**.
  TEA: **Viburnum setigerum**.
  WRIGHT: **Viburnum wrightii**.
Viburnum awabuki K. Koch. See **V. odoratissimum** var. **awabuki**.
Viburnum bitchiuense Makino.
Viburnum × **bodnantense** Aberconway. [*V. farreri* × *V. grandiflorum*]. BODNANT VIBURNUM.
Viburnum buddleifolium C. H. Wright.
Viburnum × **burkwoodii** Burkw. & Skipw. [*V. carlesii* × *V. utile*]. BURKWOOD VIBURNUM.
Viburnum × **burkwoodii 'Chenault'**. CHENAULT VIBURNUM.
Viburnum × **carlcephalum** Burkw. ex Pike. [*V. carlesii* × *V. macrocephalum*]. FRAGRANT SNOWBALL.
Viburnum carlesii Hemsl. KOREAN SPICE VIBURNUM.
Viburnum 'Carlotta'.
Viburnum 'Chenault'. See **V.** × **burkwoodii 'Chenault'**.
Viburnum cinnamomifolium Rehd. CINNAMON VIBURNUM.
Viburnum cylindricum D. Don.
Viburnum davidii Franch. DAVID VIBURNUM.
Viburnum dentatum L. ARROWWOOD. SOUTHERN ARROWWOOD.
Viburnum dilatatum Thunb. LINDEN VIBURNUM.
Viburnum erubescens Wallich.
Viburnum farreri W. T. Stearn. Syn.: *V. fragrans*.
Viburnum farreri 'Nanum'. DWARF FRAGRANT VIBURNUM.
Viburnum fragrans Bunge. See **V. farreri**.
Viburnum grandiflorum Wallich.
Viburnum ichangense (Hemsl.) Rehd. ICHANG VIBURNUM.
Viburnum japonicum (Thunb.) K. Spreng. JAPANESE VIBURNUM.
Viburnum × **juddii** Rehd. [*V. bitchiuense* × *V. carlesii*]. JUDD VIBURNUM.
Viburnum lantana L. WAYFARING TREE.
Viburnum lucidum Mill. See **V. tinus 'Lucidum'**.
Viburnum macrocephalum Fortune. Syn.: *V. macrocephalum* 'Sterile'. CHINESE SNOWBALL.
Viburnum macrocephalum 'Sterile'. See **V. macrocephalum**.
Viburnum 'Mariesii'. See **V. plicatum** forma **tomentosum 'Mariesii'**.
Viburnum odoratissimum Ker-Gawl. SWEET VIBURNUM.
Viburnum odoratissimum var. **awabuki** (K. Koch) Zabel. Syn.: *V. awabuki*.
Viburnum opulus L. EUROPEAN CRANBERRY BUSH. GUELDER ROSE.
Viburnum opulus 'Nanum'.
Viburnum opulus 'Roseum'. Syn.: *V. opulus* 'Sterile'. EUROPEAN SNOWBALL. SNOWBALL.
Viburnum opulus 'Sterile'. See **V. opulus 'Roseum'**.
Viburnum plicatum Thunb. Syn.: *V. tomentosum* var. *sterile*. JAPANESE SNOWBALL.
Viburnum plicatum 'Grandiflorum'?
Viburnum plicatum 'Lanarth'.

Viburnum plicatum 'Mariesii'. See **V. plicatum** forma **tomentosum** 'Mariesii'.
Viburnum plicatum 'Rowallane'.
Viburnum plicatum forma **tomentosum** (Thunb.) Rehd. Syn.: *V. tomentosum*. *V. plicatum* var. *tomentosum*. DOUBLE-FILE VIBURNUM.
Viburnum plicatum var. *tomentosum* Miq. See **V. plicatum** forma **tomentosum**.
Viburnum plicatum forma **tomentosum** 'Mariesii'. Syn.: *V. plicatum* 'Mariesii'. *V. tomentosum* var. *mariesii*.
Viburnum propinquum Hemsl.
Viburnum prunifolium L. BLACK HAW. NANNY-BERRY. STAGBUSH. SWEET HAW.
Viburnum × rhytidocarpum Lemoine. [**V. buddleifolium** × **V. rhytidophyllum**].
Viburnum × rhytidophylloides Suringar. [**V. lantana** × **V. rhytidophyllum**].
Viburnum × rhytidophylloides 'Willowwood'.
Viburnum rhytidophyllum Hemsl. LEATHERLEAF VIBURNUM.
Viburnum rigidum Vent.
Viburnum robustum. Listed name for **V. tinus 'Robustum'**.
Viburnum rufidulum Raf. SOUTHERN BLACK HAW. BLUE HAW. RUSTY NANNY-BERRY.
Viburnum setigerum Hance. Syn.: *V. theiferum*. TEA VIBURNUM.
Viburnum suspensum Lindl. SANDANKWA VIBURNUM.
Viburnum theiferum Rehd. See **V. setigerum**.
Viburnum tinus L. LAURUSTINUS.
Viburnum tinus 'Compactum'.
Viburnum tinus 'Lucidum'. Syn.: *V. lucidum*. SHINING LAURUSTINUS.
Viburnum tinus 'Nanum'.
Viburnum tinus 'Pomona'?
Viburnum tinus 'Pomona Dwarf'?
Viburnum tinus 'Robustum'. ROBUST VIBURNUM. ROUND-LEAF VIBURNUM.
Viburnum tinus 'Spring Blossom'.
Viburnum tinus 'Spring Bouquet'.
Viburnum tinus 'Variegatum'. VARIEGATED LAURUSTINUS.
Viburnum tomentosum Thunb. See **V. plicatum** forma **tomentosum**.
Viburnum tomentosum var. *mariesii* Veitch. See **V. plicatum** forma **tomentosum** 'Mariesii'.
Viburnum tomentosum var. *sterile* Dipp. See **V. plicatum**.
Viburnum trilobum Marsh. CRANBERRY BUSH. CRANBERRY TREE. HIGHBUSH CRANBERRY.
Viburnum utile Hemsl.
Viburnum wrightii Miq. WRIGHT VIBURNUM. LEATHERLEAF.
Vigna caracalla (L.) Verdc. Syn.: *Phaseolus caracalla*. SNAIL FLOWER. CORKSCREW FLOWER. SNAIL BEAN. CARACOL. SNAIL VINE.
Viguiera deltoidea Gray var. **parishii** (Greene) Vasey & Rose.
Vinca major L. MYRTLE. PERIWINKLE.
Vinca minor L. MYRTLE. PERIWINKLE.
VIOLET, PHILIPPINE: **Barleria cristata**.
Virgilia capensis (L.) Poir. Syn.: *V. oroboides*.
Virgilia divaricata Adamson.
Virgilia oroboides (Bergius) Salter. See **V. capensis**.
VIRGINIA CREEPER: **Parthenocissus quinquefolia**.
Vitex agnus-castus L. CHINESE CHASTE TREE. BLUE CHASTE TREE. CHASTE TREE.
Vitex agnus-castus 'Alba'.
Vitex agnus-castus 'Latifolia'.
Vitex lucens T. Kirk. NEW ZEALAND CHASTE TREE.
Vitex macrophylla. Listed name for **V. agnus-castus 'Latifolia'**.
Vitex negundo L.

Vitis antarctica (Vent.) Benth. See **Cissus antarctica**.
Vitis baudiniana F. J. Muell. See **Cissus antarctica**.
Vitis californica Benth. CALIFORNIA WILD GRAPE.
Vitis capensis Burm. f. See **Rhoicissus capensis**.
Vitis coignetiae Pulliat. CRIMSON GLORY VINE.
Vitis girdiana Munson.
Vitis henryana Hemsl. See **Parthenocissus henryana**.
Vitis heterophylla Thunb. See **Ampelopsis brevipedunculata** var. **maximowiczii**.
Vitis hypoglauca (Gray) F. J. Muell. See **Cissus hypoglauca**.
Vitis mandaianum. Listed name for **Cissus rhombifolia 'Mandaiana'**.
Vitis rhombifolia (Vahl) Baker. See **Cissus rhombifolia**.
Vitis vinifera L. WINE GRAPE. EUROPEAN GRAPE.
Vitis vinifera 'Painters'?
Vitis vinifera 'Purpurea'. TINTED GRAPE.
Vitis voinierana Baltet. See *Tetrastigma voinieranum*.

W

WALKING-STICK, DEVIL'S: **Aralia spinosa**.
  HARRY LAUDER'S: **Corylus avellana 'Contorta'**.
WALNUT, ARIZONA: **Juglans major**.
  BLACK: **Juglans nigra**.
  CALIFORNIA BLACK: **Juglans hindsii**.
  ENGLISH: **Juglans regia**.
  PARADOX: **Juglans 'Paradox'**.
  PERSIAN: **Juglans regia**.
  SOUTHERN CALIFORNIA BLACK: **Juglans californica**.
Washingtonia filamentosa (Fenzi) Kuntze. See **W. filifera**.
Washingtonia filifera (L. Linden) H. Wendl. Syn.: *W. filamentosa*. *Brahea filamentosa*. CALIFORNIA FAN PALM.
Washingtonia gracilis Parish. See **W. robusta**.
Washingtonia robusta H. Wendl. Syn.: *W. gracilis*. *W. sonorae*. SONORAN FAN PALM. MEXICAN FAN PALM.
Washingtonia sonorae S. Wats. See **W. robusta**.
WATTLE, BLACK: **Acacia mearnsii**.
  BLUE: **Acacia cyanophylla**.
  BLUE-LEAF: **Acacia cyanophylla**.
  BROOM: **Acacia calamifolia**.
  CEDAR: **Acacia terminalis**.
  COASTAL: **Acacia cyclops**.
  COOTAMUNDRA: **Acacia baileyana**.
  FROSTY: **Acacia pruinosa**.
  GOLDEN: **Acacia pycnantha**.
  GOLDEN-RAIN: **Acacia prominens**.
  GOSSAMER: **Acacia floribunda**.
  GRACEFUL: **Acacia decora**.
  GREEN: **Acacia decurrens**.
  QUEENSLAND GOLDEN: **Acacia podalyrifolia**.
  SILVER: **Acacia baileyana**.
  SILVER: **Acacia dealbata**.
  SYDNEY: **Acacia floribunda**.
  SYDNEY GOLDEN: **Acacia longifolia**.
WAXBERRY: **Myrica cerifera**.
WAXBERRY: **Symphoricarpos albus**.
WAXFLOWER, GERALDTON: **Chamelaucium uncinatum**.
WAX PLANT: **Hoya carnosa**.
  MINIATURE: **Hoya bella**.
  VARIEGATED: **Hoya carnosa 'Variegata'**.
WAX TREE: **Rhus succedanea**.
WAYFARING TREE: **Viburnum lantana**.

Weigela 'Boskoop Glory'.
Weigela 'Bristol Ruby'. See **W. florida 'Bristol Ruby'**.
Weigela 'Bristol Snowflake'. See **W. florida 'Bristol Snowflake'**.
Weigela 'Cardinal'. See **W. 'Newport Red'**.
Weigela coraeensis Thunb.
Weigela 'Eva Rathke'. See **W. florida 'Eva Rathke'**.
Weigela 'Eva Supreme'.
Weigela 'Fairy'?
Weigela 'Floreal'. See **W. praecox 'Floreal'**.
Weigela floribunda (Siebold & Zucc.) K. Koch.
Weigela florida (Bunge) A. DC. Syn.: *W. rosea*.
Weigela florida 'Alba'.
Weigela florida 'Bristol Ruby'.
Weigela florida 'Bristol Snowflake'.
Weigela florida 'Eva Rathke'.
Weigela florida 'Java Red'.
Weigela florida 'Rubra'?
Weigela florida 'Variegata'.
Weigela 'Ideal'.
Weigela 'Le Printemps'. See **W. praecox 'Le Printemps'**.
Weigela middendorffiana (Trautv. & C. A. Mey.) K. Koch.
Weigela 'Newport Red'.
Weigela praecox (Lemoine) L. H. Bailey.
Weigela praecox 'Floreal'.
Weigela praecox 'Le Printemps'.
Weigela 'Rhode Island Red'. See **W. 'Newport Red'**.
Weigela rosea Lindl. See **W. florida**.
Weigela rosea 'Variegata'. See **W. florida 'Variegata'**.
Weigela 'Vanicek'. See **W. 'Newport Red'**.
WESTRINGIA, ROSEMARY BUSH: **Westringia rosmariniformis**.
Westringia rosmariniformis Sm. ROSEMARY BUSH WESTRINGIA.
WHIN: **Ulex europaeus**.
Whipplea modesta Torr. YERBA-DE-SELVA.
WHITETHORN, CHAPARRAL: **Ceanothus leucodermis**.
  COAST: **Ceanothus incanus**.
  MOUNTAIN: **Ceanothus cordulatus**.
Widdringtonia schwarzii M. T. Mast. AFRICAN CEDAR.
Wigandia caracasana HBK.
Wigandia kunthii Choisy.
WILGA: **Geijera parviflora**.
WILLOW, ALASKA BLUE: **Salix purpurea**.
  ARCTIC: **Salix arctica**.
  ARROYO: **Salix lasiolepis**.
  AUSTRALIAN: **Geijera parviflora**.
  BAY: **Salix pentandra**.
  BAY-LEAF: **Salix pentandra**.
  BRITTLE: **Salix fragilis**.
  CORKSCREW: **Salix matsudana 'Tortuosa'**.
  CRACK: **Salix fragilis**.
  DESERT: **Chilopsis linearis**.
  DESERT: **Pittosporum phillyraeoides**.
  DRAGON-CLAW: **Salix matsudana 'Tortuosa'**.
  FLORIST'S: **Salix caprea**.
  FLOWERING: **Chilopsis linearis**.
  FRENCH PUSSY: **Salix caprea**.
  GLOBE: **Salix matsudana 'Umbraculifera'**.
  GLOBE NAVAJO: **Salix matsudana 'Navajo'**.
  GOAT: **Salix caprea**.
  GOLDEN: **Salix alba** var. **vitellina**.
  GOLDEN WEEPING: **Salix alba** var. **tristis**.
  HANKOW: **Salix matsudana**.
  LAUREL: **Salix pentandra**.
  LAUREL-LEAF: **Salix pentandra**.
  NIOBE: **Salix × blanda**.
  PEKIN: **Salix matsudana**.
  PINK PUSSY: **Salix caprea**.
  PURPLE: **Salix purpurea**.
  PUSSY: **Salix discolor**.
  RINGLEAF: **Salix babylonica 'Crispa'**.
  ROSE-GOLD PUSSY: **Salix gracilistyla**.
  SCOULER: **Salix scouleriana**.
  THURLOW WEEPING: **Salix × elegantissima**.
  TWISTED HANKOW: **Salix matsudana 'Tortuosa'**.
  WEEPING: **Salix babylonica**.
  WEEPING: **Salix × blanda**.
  WEEPING: **Salix × elegantissima**.
  WHITE: **Salix alba**.
  YELLOW: **Salix alba** var. **vitellina**.
WINEBERRY, NEW ZEALAND: **Aristotelia serrata**.
WINGNUT, CHINESE: **Pterocarya stenoptera**.
WINTERBERRY, JAPANESE: **Ilex serrata**.
WINTER CREEPER, BIG-LEAF: **Euonymus fortunei** var. **vegeta**.
  COMMON: **Euonymus fortunei** var. **radicans**.
  SILVER QUEEN: **Euonymus fortunei 'Silver Queen'**.
WINTERGREEN: **Gaultheria procumbens**.
  ALPINE: **Gaultheria humifusa**.
  OREGON: **Gaultheria ovatifolia**.
WINTER'S BARK: **Drimys winteri**.
WINTERSWEET: **Acokanthera oblongifolia**.
WINTERSWEET: **Chimonanthus praecox**.
WINTERSWEET: **Chimonanthus praecox 'Grandiflorus'**.
WIRE VINE: **Muehlenbeckia complexa**.
  CREEPING: **Muehlenbeckia axillaris**.
Wistaria. See **Wisteria**.
WISTERIA, CHINESE: **Wisteria sinensis**.
  JAPANESE: **Wisteria floribunda**.
  PINK: **Wisteria floribunda 'Rosea'**.
  RHODESIAN: **Bolusanthus speciosus**.
  SILKY: **Wisteria venusta**.
  SILKY: **Wisteria venusta 'Violacea'**.
WISTERIA TREE: **Bolusanthus speciosus**.
  SCARLET: **Sesbania punicea**.
Wisteria 'Carnea'. See **W. floribunda 'Carnea'**.
Wisteria chinensis DC. See **W. sinensis**.
Wisteria floribunda (Willd.) DC. JAPANESE WISTERIA.
Wisteria floribunda 'Alba'.
Wisteria floribunda 'Carnea'.
Wisteria floribunda 'Geisha'.
Wisteria floribunda 'Issai'. See **W. floribunda 'Praecox'**.
Wisteria floribunda 'Longissima'.
Wisteria floribunda 'Longissima Alba'.
Wisteria floribunda 'Macrobotrys'. Syn.: *W. multijuga*.
Wisteria floribunda 'Praecox'.
Wisteria floribunda 'Rosea'. PINK WISTERIA.
Wisteria floribunda 'Royal Purple'.
Wisteria floribunda 'Violacea Plena'.
Wisteria longissima. Listed name for **W. floribunda 'Longissima'**.
Wisteria multijuga Van Houtte. See **W. floribunda 'Macrobotrys'**.
Wisteria 'Royal Purple'. See **W. floribunda 'Royal Purple'**.
Wisteria sinensis (Sims) Sweet. Syn.: *W. chinensis*.
  CHINESE WISTERIA.
Wisteria sinensis 'Alba'.
Wisteria sinensis 'Caroline'.?
Wisteria sinensis 'Purpurea'.
Wisteria venusta Rehd. & Wils. SILKY WISTERIA.
Wisteria venusta 'Alba'. See **W. venusta**.
Wisteria venusta 'Violacea'. SILKY WISTERIA.
WITCH HAZEL: See HAZEL.
WONGA-WONGA VINE: **Pandorea pandorana**.

WOODBINE: **Lonicera periclymenum**.
   DUTCH: **Lonicera periclymenum** var. **belgica**.
WOOLLYBUTT, CAMDEN: **Eucalyptus macarthurii**.
WORMWOOD, BEACH: **Artemisia stellerana**.
   COMMON: **Artemisia absinthium**.
   FRINGED: **Artemisia frigida**.
   ROMAN: **Artemisia pontica**.

Xanthoceras sorbifolium Bunge. HYACINTH SHRUB. YELLOW HORN.
Xanthorrhoea arborea R. Br. BLACKBOY. GRASS TREE.
Xanthorrhoea preissii Endl. BLACKBOY.
Xanthorrhoea quadrangulata F. J. Muell. BLACKBOY. GRASS TREE.
Xanthosma sagittifolium Schott. BLUE ELEPHANT-EAR.
Xanthosma violaceum Schott. PURPLE ELEPHANT-EAR. BLUE TARO.
Xerophyllum tenax (Pursh) Nutt. ELK GRASS. BEAR GRASS.
   SQUAW GRASS. FIRE LILY.
XYLOSMA, SHINY: **Xylosma congestum**.
Xylosma congestum (Lour.) Merrill. Syn.: *Myroxylon senticosum*. *Xylosma racemosum*. *X. senticosum*.
   SHINY XYLOSMA.
Xylosma congestum **'Compactum'**.
Xylosma racemosum (Miq.) Siebold & Zucc. See **X. congestum**.
Xylosma senticosum Hance. See **X. congestum**.
Xylosma senticosum 'Compactum'. See **X. congestum 'Compactum'**.

YADAKE: **Sasa japonica**.
YAE-SHIDARE-HIGAN: **Prunus subhirtella 'Pendula Plena Rosea'**.
YANGTAO: **Actinidia chinensis**.
YATE, BUSHY: **Eucalyptus lehmannii**.
   FLAT-TOPPED: **Eucalyptus occidentalis**.
   TREE: **Eucalyptus cornuta**.
   WARTED: **Eucalyptus megacornuta**.
YAUPON: **Ilex vomitoria**.
   DWARF: **Ilex vomitoria 'Nana'**.
YELLOWBELLS: **Tecoma stans**.
YELLOW HORN: **Xanthoceras sorbifolium**.
YELLOW-JACKET: **Eucalyptus bloxsomei**.
YELLOWWOOD: **Rhodosphaera rhodanthema**.
   AMERICAN: **Cladrastis lutea**.
YERBA BUENA: **Satureja douglasii**.
YERBA-DE-SELVA: **Whipplea modesta**.
YERBA SANTA: **Eriodictyon crassifolium**.
YESTERDAY-AND-TODAY: **Brunfelsia pauciflora 'Eximia'**.
YESTERDAY-TODAY-AND-TOMORROW: **Brunfelsia australis**.
YESTERDAY-TODAY-AND-TOMORROW: **Brunfelsia pauciflora 'Eximia'**.
YESTERDAY-TODAY-AND-TOMORROW: **Brunfelsia pauciflora 'Floribunda'**.
YEW, BROOM: **Taxus baccata 'Erecta'**.
   BROWN'S: **Taxus × media 'Brownii'**.
   CANADA: **Taxus canadensis**.
   CHINESE PLUM: **Cephalotaxus fortunei**.
   COLUMNAR JAPANESE: **Taxus cuspidata 'Columnaris'**.
   COMPACT JAPANESE: **Taxus cuspidata 'Nana'**.
   CUSHION JAPANESE: **Taxus cuspidata 'Densa'**.
   DOVASTON: **Taxus baccata 'Dovastoniana'**.
   DWARF JAPANESE: **Taxus cuspidata** var. **nana**.
   ENGLISH: **Taxus baccata**.
   GOLDEN: **Taxus baccata 'Aurea'**.
   GOLDEN IRISH: **Taxus baccata 'Fastigiata Aurea'**.
   HALLORAN: **Taxus × media 'Halloriana'**.
   HARRINGTON PLUM: **Cephalotaxus harringtonia**.
   HATFIELD: **Taxus × media 'Hatfieldii'**.
   HICKS: **Taxus × media 'Hicksii'**.
   HILL'S PYRAMIDAL: **Taxus × media 'Hillii'**.
   IRISH: **Taxus baccata 'Stricta'**.
   JAPANESE: **Podocarpus macrophyllus**.
   JAPANESE: **Taxus cuspidata**.
   JAPANESE PLUM: **Cephalotaxus harringtonia** var. **drupacea**.
   KOREAN: **Cephalotaxus harringtonia**.
   KOREAN PLUM: **Cephalotaxus harringtonia**.
   PACIFIC: **Taxus brevifolia**.
   SILVER GREEN: **Taxus baccata 'Silver Green'**.
   SPREADING ENGLISH: **Taxus baccata 'Repandens'**.
   SPREADING JAPANESE: **Taxus cuspidata 'Expansa'**.
   UPRIGHT ENGLISH: **Taxus baccata 'Erecta'**.
   UPRIGHT JAPANESE: **Taxus cuspidata**.
   WESTERN: **Taxus brevifolia**.
YUCCA, BANANA: **Yucca baccata**.
   BLUE: **Yucca baccata**.
   BULB-STEM: **Yucca elephantipes**.
   GIANT: **Yucca elephantipes**.
   MOHAVE: **Yucca schidigera**.
   RED: **Hesperaloe parviflora**.
   SMALL SOAPWEED: **Yucca glauca**.
   SOAPTREE: **Yucca elata**.
   SPANISH-BAYONET: **Yucca baccata**.
   TEXAS: **Yucca faxoniana**.
Yucca aloifolia L. DAGGER PLANT. SPANISH-BAYONET.
Yucca aloifolia **'Variegata'**. VARIEGATED SPANISH-BAYONET.
Yucca australis (Engelm.) Trel.
Yucca baccata Torr. BANANA YUCCA. BLUE YUCCA. DATIL. SPANISH-BAYONET YUCCA.
Yucca brevifolia Engelm. JOSHUA TREE.
Yucca brevifolia var. *jaegerana* McKelvey. JAEGER JOSHUA TREE.
Yucca elata Engelm. SOAPTREE YUCCA. SOAPWEED.
Yucca elephantipes Regel. Syn.: *Y. gigantea*. BULB-STEM YUCCA. GIANT YUCCA.
Yucca faxoniana (Trel.) Sarg. TEXAS YUCCA.
Yucca filamentosa L. ADAM'S NEEDLE.
Yucca filifera. Listed name for **Y. australis**.
Yucca flaccida Haw.
Yucca gigantea Lem. See **Y. elephantipes**.
Yucca glauca Nutt. ex J. Fraser. SMALL SOAPWEED YUCCA. SOAPWEED.
Yucca gloriosa L. SPANISH-DAGGER.
Yucca gloriosa 'Alicia'?
Yucca gloriosa 'Variegata'?
Yucca harrimaniae Trel.
Yucca macrocarpa?
Yucca mohavensis Sarg. See **Y. schidigera**.
Yucca pendula Groenl. See **Y. recurvifolia**.
Yucca recurvifolia Salisb. Syn.: *Y. pendula*.
Yucca schidigera Roezl ex Ortgies. Syn.: *Y. mohavensis*. MOHAVE YUCCA.
Yucca schottii Engelm.
Yucca torreyi Shafer.
Yucca whipplei Torr. OUR LORD'S CANDLE.
YULAN: **Magnolia heptapeta**.

Yushania aztecorum McClure & E. W. Sm. MEXICAN WEEPING BAMBOO. OTATE. Plants called *Arthrostylidium longifolium* belong here.

Zamia debilis Ait. Syn.: Z. *media*.
Zamia fischeri Miq.
Zamia floridana A. DC. See **Z. pumila**.
Zamia furfuracea L. f. See **Z. pumila**.
Zamia media Jacq. See **Z. debilis**.
Zamia pumila L. Syn.: Z. *floridana*. Z. *furfuracea*. Z. *umbrosa*. COMPTIE. COONTIE. FLORIDA ARROWROOT. SAGO CYCAD. SEMINOLE BREAD.
Zamia umbrosa Small. See **Z. pumila**.
Zanthoxylum americanum L. PRICKLY ASH. TOOTHACHE TREE.
Zanthoxylum piperitum DC. JAPAN PEPPER.
Zauschneria californica K. Presl. CALIFORNIA FUCHSIA.
Zauschneria cana Greene.
ZEBRA PLANT: **Aphelandra squarrosa**.
ZELKOVA, JAPANESE: **Zelkova serrata**.
   SAWLEAF: **Zelkova serrata**.
Zelkova serrata Makino. JAPANESE ZELKOVA. SAWLEAF ZELKOVA. JAPANESE KEAKI.
Zenobia pulverulenta (Bartr. ex Willd.) Pollard.
Ziziphus: Sometimes, but not originally, spelled *Zizyphus*.
Ziziphus jujuba Mill. CHINESE DATE. CHINESE JUJUBE. JUJUBE.
Zizyphus. See **Ziziphus**.

APPENDIX

Authorities for the Binomials

To save space, only the last names of authors, or abbreviations of their names, are used in connection with the binomial (generic and specific) names. The following list gives the last name or abbreviation, full name, dates, and country of work, where known.

ABERCONWAY. Henry Duncan McLaren, Lord Aberconway. 1879-1953. England.
ABRAMS. LeRoy Abrams. 1874-1956. U.S.A.
ADAMS. Joseph Edison Adams. 1903-   . U.S.A.
ADAMSON. Robert Stephen Adamson. 1885-1965. England, South Africa.
AHRENDT. Leslie Walter Allan Ahrendt. 1903- England.
AIRY SHAW. Herbert Kenneth Airy Shaw. 1902- England.
AIT. William Aiton. 1731-1793. England.
AIT. f. William Townsend Aiton (son). 1766-1849. England.
AITCH. James Edward Tierney Aitchison. 1836-1898. England.
ALEF. Friedrich Georg Christoph Alefeld. 1820-1872. Germany.
ALLAN. Harry Howard Barton Allan (Henry Howard Allan). 1882-1957. New Zealand.
ALSTON. Arthur Hugh Garfit Alston. 1902-1958. England.
AMES, L. M. Lawrence M. Ames. 1900-1966. U.S.A.
ANDERSON, E. Edgar Anderson. 1897-1969. U.S.A.
ANDERSON, T. Thomas Anderson. 1832-1870. Scotland, India.
ANDERSON-HENRY. Isaac Anderson-Henry. 1800-1884. Scotland.
ANDERSS. Nils Johan Andersson. 1821-1880. Sweden.
ANDR. Henry C. Andrews. 1752-1830. England.
ANDR., C. Cecil Rollo Payton Andrews. 1870-1951. Australia.
ANDRÉ. Edouard François André. 1840-1911. France.
ARCANG. Giovanni Arcangeli. 1840-1921. Italy.
ARMSTR., J. B. John B. Armstrong. 1850-1926. New Zealand.
ARN. George Arnott [sometimes Arnold] Walker Arnott. 1799-1868. Scotland.
ARNOLD. Johann Franz Xaver Arnold. Fl. 1785. Austria.
ARRUDA DA CAMARA. Manuel Arruda da Camara. 1752-1810. Brazil.
ASCHERS. Paul Friedrich August Ascherson. 1834-1913. Germany.
ASCHERS. & GRAEBN. Paul Friedrich August Ascherson. Karl Otto Robert Peter Paul Graebner.
ASHBY. Edwin Ashby. 1861-1941. Australia.
ASHE. William Willard Ashe. 1872-1932. U.S.A.
ASHWIN. Margot Bernice Ashwin (later Forde). 1935- New Zealand.
AUBLET. Jean Baptiste Christophe Fusée Aublet. 1720-1778. France.
AUD. John James Laforest Audubon. 1780-1851. U.S.A.
AUDIBERT. Urbain Audibert. 1789-1846. France.
AUGUSTIN. Augustin. Fl. 1854.

BABINGTON. Charles Cardale Babington. 1808-1895. England.
BACIG. Rimo Charles Bacigalupi. 1901-   . U.S.A.
BACKER. Cornelis Andries Backer. 1874-1963. Netherlands.
BACKH. James Backhouse, 1825-1900. England.
BAILEY, D. K. Dana K. Bailey. Contemporary. U.S.A.
BAILEY, F. M. Frederick Manson Bailey. 1827-1895. Australia.
BAILEY, J. F. John Frederick Bailey. 1866-1938. Australia.
BAILEY, L. H. Liberty Hyde Bailey. 1858-1954. U.S.A.
BAILL. Henri Ernest Baillon. 1827-1895. France.
BAKER. John Gilbert Baker. 1834-1920. England.
BAKER, E. G. Edmund Gilbert Baker. 1864-1949. England.
BAKER, M. S. Milo Samuel Baker. 1868-1961. U.S.A.
BAKER, R. T. Richard Thomas Baker. 1854-1941. Australia.
BAL. Benedict Balansa. 1825-1891. France.
BALD. Antonio Baldacci. 1867-1950. Italy.
BALF. John Hutton Balfour. 1808-1884. Scotland.
BALF. f. Isaac Bayley Balfour. 1853-1922. Scotland.
BALF. f. & FORREST. Isaac Bayley Balfour. George Forrest.
BALF. f. & W. W. SM. Isaac Bayley Balfour. William Wright Smith.
BALF. f. & F. K. WARD. Isaac Bayley Balfour. Francis Kingdon Ward.
BALL. Carleton Roy Ball. 1873-1958. U.S.A.
BALTET. Charles Baltet. 1830-1908. France.
BANKS. Joseph Banks. 1743-1820. England.
BANKS & SOLAND. Joseph Banks. Daniel Carl Solander.
BARBIER. Albert Barbier. 1845-1931. France.
BARB.-RODR. João Barbosa Rodrigues. 1842-1909. Brazil.
BARNEBY. Rupert Charles Barneby. 1911- . England, U.S.A.
BARNÉOUD. François Marius Barnéoud. 1821- ? . France.
BARRATT. Joseph Barratt. 1796-1882. England, U.S.A.
BARTL. Friedrich Gottlieb Bartling. 1798-1875. Germany.
BARTL. & H. L. WENDL. Friedrich Gottlieb Bartling. Heinrich Ludolf Wendland.
BARTLETT. Harley Harris Bartlett. 1886-1960. U.S.A.
BARTR. William Bartram. 1729-1823. U.S.A.
BATAL. Aleksandr Fedorovich Batalin. 1847-1896 [1898?]. Russia.
BATSCH. August Johan Georg Karl Batsch. 1761-1802. Germany.
BAUMG. Johann Christian Gottlob Baumgarten. 1765-1843. Germany.
BAUSCH. Jan Bausch. 1917-   . Netherlands.
BAXTER. William Baxter. Fl. 1820's-1830's. England, Australia.
BEAN. William Jackson Bean. 1863-1947. England.
BEAUVOIS. Ambroise Marie François Joseph Palisot de Beauvois. 1755-1820. France.
BECC. Odoardo Beccari. 1843-1920. Italy.
BEDDOME. Richard Henry Beddome. 1830-1911. England, India.
BEISSNER. Ludwig Beissner. 1853-1927. Germany.
BELLAIR. Georges Adolphe Bellair. 1860- ? . France.
BELLAIR & ST.-LÉGER. Georges Adolphe Bellair. Léon Saint-Léger.
BENN., J. John Joseph Bennett. 1801-1876. England.
BENSON, L. Lyman Benson. 1909-   . U.S.A.
BENTH. George Bentham. 1800-1884. England.

BENTH. & HOOK. f. George Bentham. Joseph Dalton Hooker.
BERCHT. Friedrich von Berchtold. 1781-1876. Czechoslovakia.
BERCHT. & J. PRESL. Friedrich von Berchtold. Jan Swatopluk Presl.
BERCKMANS. Prosper Jules Alphonse Berckmans. 1830-1910. Belgium, U.S.A.
BERG. Otto Karl Berg. 1815-1866. Germany.
BERGER. Alwin Berger. 1871-1931. Germany.
BERGIUS. Peter Jonas Bergius. 1730-1790. Sweden.
BERGMANS. Johannes Baptista Bergmans. 1892-   . Netherlands.
BERL. Jean Louis Berlandier. 1805-1851. France.
BERNH. Johann Jacob Bernhardi. 1774-1850. Germany.
BERTH. Sabin Berthelot. 1794-1880. France, Canary Islands.
BERTHAULT. Pierre Berthault. Fl. 1918-1930. France.
BERTOLONI, G. Guiseppe Bertoloni. 1804-1878. Italy.
BERTRAND, C. Charles Eugene Bertrand. 1851-1917. France.
BESSER. Willibald Swibert Joseph Gottlieb von Besser. 1784-1842. Poland.
BETCHE. Ernst Betche. 1851-1913. Australia.
BIEB. Friedrich August Marschall von Bieberstein. 1768-1826. Germany, Russia.
BINNEND. Simon Binnendijk. 1821-1883. Netherlands.
BIRDSEY. Monroe Roberts Birdsey. 1922-   . U.S.A.
BISSET. James Bisset. 1843-1911. Scotland, England.
BISSET & S. MOORE. James Bisset. Spencer Le Marchant Moore.
BITTER. Friedrich August Georg Bitter. 1873-1927. Germany.
BIVONA. Antonio Bivona-Bernardi. 1778-1834. Italy.
BLACK, J. M. John McConnell Black. 1855-1951. Australia.
BLACKBURN. Benjamin Coleman Blackburn. 1908-   . U.S.A.
BLAKE, S. F. Sydney Fay Blake. 1892-1959. U.S.A.
BLAKE, S. T. Stanley Thatcher Blake. 1910-1973. Australia.
BLAKELY. William Faris Blakely. 1875-1940. Australia.
BLAKELY & STEEDMAN. William Faris Blakely. Henry Steedman.
BLANCO. Manuel Blanco. 1780-1845. Philippines.
BLUME. Karl Ludwig Blume. 1796-1862. Netherlands.
BODIN. Nils Gustaf Bodin. Fl. 1798. Sweden.
BOERLAGE. Jacob Gijsbert Boerlage. 1849-1900. Netherlands, Java.
BOIS. Désiré Georges Jean Marie Bois. 1856-1946. France.
BOIS & BERTHAULT. Désiré Georges Jean Marie Bois. Pierre Berthault.
BOISS. Edmond Pierre Boissier. 1810-1885. Switzerland.
BOISS. & BAL. Edmond Pierre Boissier. Benedict Balansa.
BOISS. & GAILLARDOT. Edmond Pierre Boissier. Charles Gaillardot.
BOJER. Wenzel [Wencelas] Bojer. 1797[1800?]-1856. Czechoslovakia.
BOLUS. Harry Bolus. 1834-1911. South Africa.
BOLUS & WOLLEY-DOD. Harry Bolus. Anthony Hurt Wolley-Dod.
BOLUS, L. Harriet Margaret Louisa Kensit Bolus. 1877-1970. South Africa.
BOMHARD. Miriam Lucile Bomhard. 1898-1952. U.S.A.
BONG. August Heinrich Gustav Bongard. 1786-1839. Germany, Russia.
BONPL. Aimé Jacques Alexandre Bonpland. 1773-1858. France, South America.
BOOM. Boudewijn Karel Boom. 1903-   . Netherlands.
BORKH. Moritz Balthasar Borkhausen. 1760-1806. Germany.

BORY. Jean Baptiste Geneviève Marcellin Bory de Saint-Vincent. 1778-1846. France.
BOSSE. Julius Friedrich Wilhelm Bosse. 1788-1864. Germany.
BOUCHARLAT. Boucharlat. 1807-1893. France.
BOUCHÉ. Carl David Bouché. 1809-1881. Germany.
BOUTELJE. Julius B. Boutelje. Fl. 1954.
BR., J. E John Ednie Brown. 1848-1899. Australia.
BR., N. E. Nicholas Edward Brown. 1849-1934. England.
BR., R. Robert Brown. 1773-1858. Scotland, Australia, England.
BR., R. W. Roland Wilbur Brown. 1893-1961. U.S.A.
BR., S. Stewardson Brown. 1867-1921. U.S.A.
BRANDEGEE. Townsend Stith Brandegee. 1843-1925. U.S.A.
BRANDEGEE, K. Katherine Brandegee (Mary Katherine Layne Curran). 1844-1920. U.S.A.
BRANDIS. Dietrich Brandis. 1824-1907. Germany, England, India.
BRAUN, A. Alexander Carl Heinrich Braun. 1805-1877. Germany.
BREMEKAMP. Cornelis Elisa Bertus Bremekamp. 1888-   . Netherlands.
BRIOT. Charles Briot. 1804-1888. France.
BRIQ. John Isaac Briquet. 1870-1931. Switzerland.
BRITT. Nathaniel Lord Britton. 1859-1934. U.S.A.
BRITT. & MILLSPAUGH. Nathaniel Lord Britton. Charles Frederick Millspaugh.
BRITT. & ROSE. Nathaniel Lord Britton. Joseph Nelson Rose.
BRITT. & SHAFER. Nathaniel Lord Britton. John Adolf Shafer.
BRITTEN. James Britten. 1846-1924. England.
BRONGN. Adolphe Theodore Brongniart. 1801-1876. France.
BROT. Felix da Silva Avellar Brotero. 1744-1828. Portugal.
BROUSS. Pierre Marie Auguste Broussonet. 1761-1807. France.
BSP. Nathaniel Lord Britton. Emerson Ellick Sterns. Justus Ferdinand Poggenburg.
BUCHANAN. John Buchanan. 1819-1898. New Zealand.
BUCHH. John Theodore Buchholz. 1888-1951. U.S.A.
BUCH.-HAM. Francis Buchanan-Hamilton (Francis Hamilton). 1762-1829. England, India.
BUC'HOZ. Pierre Joseph Buc'hoz. 1731-1807. France.
BUCKLEY. Samuel Botsford Buckley. 1809-1884. U.S.A.
BUEK. Heinrich Wilhelm Buek. 1796-1879. Germany.
BULL. William Bull. 1828-1902. England.
BUNGE. Alexander von Bunge. 1803-1890. Russia.
BUNTING. George Sydney Bunting. 1927-   . U.S.A.
BUR. Edouard Bureau. 1830-1918. France.
BUR. & K. SCHUM. Edouard Bureau. Karl Moritz Schumann.
BURBIDGE. Frederick William Thomas Burbidge. 1847-1905. Ireland, England.
BURCHARD. Oscar Burchard. Fl. 1909-1931. Germany, Canary Islands.
BURCHELL. William John Burchell. 1781-1863. England, South Africa.
BURGSD. Friedrich August Ludwig von Burgsdorf. 1747-1802. Germany.
BURKW. & SKIPW. Burkwood. Skipwith. Contemporary. England.
BURM. Johannes Burman. 1706-1779. Netherlands.
BURM. f. Nikolaus Laurens Burman (son). 1734-1793. Netherlands.
BURMEISTER, H. Hermann Carl Conrad Burmeister. 1807-1891. Argentina.

BURRET. Karl Ewald Maximilian Burret. 1883-1964. Germany.
BURTT, B. L. Brian Lawrence Burtt. 1913-   . England, Scotland.
BURTT, B. L. & R. M. SM. Brian Lawrence Burtt. Rosemary Margaret Smith.
BURV. Fréderic Burvenich. 1857-1917. Belgium.
BYHOUWER. Jans Tijs (Ali) Pieter Byhouwer (Bijhouwer). 1898-1938. Netherlands, Malaya.

CABRERA. Angel L. Cabrera. 1908-   . Argentina.
CALDERON. Cleofé E. Calderon. Contemporary. Argentina, U.S.A.
CAMB. Jacques Cambessèdes. 1799-1863. France.
CAMBAGE. Richard Hind Cambage. 1859-1928. Australia.
CAMUS, E. G. Edmond Gustave Camus. 1852-1915. France.
CARO. José A. Caro. Contemporary. Argentina.
CARRIÈRE. Elie Abel Carrière. 1816-1896. Argentina.
CARRUTHERS. William Carruthers. 1830-1922. England.
CARSE. Harry Carse. 1857-1930. New Zealand.
CASS. Alexandre Henri Gabriel Cassini. 1781-1832. France.
CAV. Antonio José Cavanilles. 1745-1804. Spain.
CHABAUD. J. B. Chabaud. 1833-1915. France.
CHAIX. Dominique Chaix. 1730-1799. France.
CHALL. Richard Westman Challinor. 1871-1951. Australia.
CHALL., CHEEL, & PENF. Richard Westman Challinor. Edwin Cheel. Arthur Raymond Penfold.
CHAM. Ludlof Adelbert von Chamisso (Louis Charles Adelaide Chamisseau de Boncourt). 1781-1838. Germany.
CHAM. & SCHLECHTEND. Ludlof Adelbert von Chamisso. Diederich Franz Leonhard von Schlechtendal.
CHAMBERLAIN. Charles Joseph Chamberlain. 1863-1943. U.S.A.
CHAPM. Alvan Wentworth Chapman. 1809-1899. U.S.A.
CHEEL. Edwin Cheel. 1872-1951. Australia.
CHEESEM. Thomas Frederick Cheeseman. 1846-1923. New Zealand.
CHEESMAN. Ernest Entwisle Cheesman. 1888-   . England.
CHENG. Wan-Chun Cheng. 1903-   . China.
CHITT. Frederick James Chittenden. 1873-1950. England.
CHODAT. Robert Hippolyte Chodat. 1865-1934. Switzerland.
CHODAT & VISCHER. Robert Hippolyte Chodat. Wilhelm Vischer.
CHOISY. Jacques Denis Choisy. 1799-1859. Switzerland.
CHRIST. Konrad Hermann Heinrich Christ. 1833-1933. Switzerland.
CHRISTENSEN. Carl Frederick Albert Christensen. 1872-1942. Denmark.
CHRISTM. Gottlieb Friedrich Christmann. 1752-1836. Germany.
CLARKE, C. B. Charles Baron Clarke. 1832-1906. England, India.
CLARKE, W. B. Walter Bosworth Clarke. 1879-1953. U.S.A.
CLAUSEN, R. T. Robert Theodore Clausen. 1911-   . U.S.A.
CLEMENCEAU. C. Clemenceau. Fl. 1866-1872. France.
CLEMENTI. Giuseppi C. Clementi. 1812-1873. Italy.
CLEMENTS. Frederic Edward Clements. 1875-1945. U.S.A.
COCHET. Pierre Charles Marie Cochet. 1866-1936. France.
COCKAYNE. Leonard C. Cockayne. 1855-1934. New Zealand.
COCKAYNE & ALLAN. Leonard C. Cockayne. Harry Howard Barton Allan.
COCKAYNE & LAING. Leonard C. Cockayne. Robert M. Laing.

CODD. Leslie Edward Wastell Codd. 1908-   . South Africa.
COE, E. F. Ernest F. Coe. 1867-1951. U.S.A.
COGN. Célestin Alfred Cogniaux. 1841-1916. France.
COGN. & MARCHAL. Célestin Alfred Cogniaux. Elie Marchal.
COLENSO. William Colenso. 1811-1899. New Zealand.
COLL. Henry Collet. 1836-1901. England.
COLL. & HEMSL. Henry Collet. William Botting Hemsley.
COLLA. Luigi Aloysius Colla. 1766-1848. Italy.
COLLADON. Louis Theodore Frederic Colladon. 1792-1862. Switzerland.
COMBER. Harold Frederick Comber. 1897-1969. England, South America, Tasmania, U.S.A.
COMPTON. Robert Harold Compton. 1886- ? . South Africa.
COOK., O. F. Orator Fuller Cook. 1867-1949. U.S.A.
CORNER. Edred John Henry Corner. 1906-   . England.
CORREA. José Francisco Correa da Serra. 1751-1823. Portugal.
CORREVON. Louis Henri Correvon. 1854-1939. Switzerland.
COSTANTIN. Julien Noël Costantin. 1857-1936. France.
COSTANTIN & BOIS. Julien Noël Costantin. Désiré Georges Jean Marie Bois.
COSTE. Hippolyte Jacques Coste. 1858-1924. France.
COSTE & SOULIÉ. Hippolyte Jacques Coste. Jean André Soulié.
COULT. John Merle Coulter. 1851-1928. U.S.A.
COURT. Albert Bertram Court. 1927-   . Australia.
COURTIN. Albert Courtin. Fl. 1850-1858. Germany.
COV. Frederick Vernon Coville. 1867-1937. U.S.A.
COV. & BRITT. Frederick Vernon Coville. Nathaniel Lord Britton.
COWAN. John Macqueen Cowan. 1891-1960. Scotland.
COWAN & DAVIDIAN. John Macqueen Cowan. Hagop Haroutune Davidian.
CRAIB. William Grant Craib. 1882-1933. England.
CRANTZ. Heinrich Johann Nepomuk von Crantz. 1722-1797. Austria.
CRÉPIN. François Crépin. 1830-1903. Belgium.
CUNN., A. Allan Cunningham. 1791-1838. England, Australia.
CUNN., A. & FRASER. Allan Cunningham. Charles Fraser.
CUNN., R. Richard Cunningham. 1793-1835. England, Australia.
CURRAN. Mary Katherine Layne Curran. See K. Brandegee.
CURTIS. William Curtis. 1746-1799. England.

DALLIM. William Dallimore. 1871-1959. England.
DALLIM. & A. B. JACKSON. William Dallimore. Albert Bruce Jackson.
DAMMER. Carl Lebrecht Udo Dammer. 1860-1920. Germany.
DANDY. James Edgar Dandy. 1903-1976. England.
DANSEREAU. Pierre Dansereau. 1911-   . Canada.
DARBY. John Darby. 1804-1877. U.S.A.
DAVIDIAN. Hagop Haroutune Davidian. 1907-   . Scotland.
DAVIDS., A. Anstruther Davidson. 1860-1932. U.S.A.
DAVIS. Peter Hadland Davis. 1918-   . Scotland.
DAVY. Joseph Burtt Davy. 1870-1940. England, U.S.A., South Africa.
DC. Augustin Pyramus de Candolle. 1778-1841. Switzerland.
DC., A. Alphonse Louis Pierre Pyramus de Candolle. 1806-1893. Switzerland.

DC., C. Anne Casimir Pyramus de Candolle. 1836-1918. Switzerland.
DEANE. Henry Deane. 1847-1924. England, Australia.
DEANE & MAIDEN. Henry Deane. Joseph Henry Maiden.
DECNE. Joseph Decaisne. 1807-1882. Belgium, France.
DECNE. & PLANCH. Joseph Decaisne. Jules Émile Planchon.
DECORSE. J. Decorse. Fl. 1912.
DECORSE & POISSON. J. Decorse. Poisson.
DEFLERS. Albert Deflers. 1841-1921. France.
DEGEN. Árpád von Degen. 1866-1934. Hungary.
DEGEN & BALD. Árpád von Degen. Antonio Baldacci.
DEGENER. Otto Degener. 1899- . U.S.A.
DEHNHARDT. Friedrich Dehnhardt. 1787-1870. Italy.
DE LANNOY. De Lannoy. Fl. 1863. France.
DE NOTARIS. Guiseppe de Notaris. 1805-1877. Italy.
DESF. Réné Louiche Desfontaines. 1750-1833. France.
DESMARAIS. Yves Desmarais. 1918- . Canada.
DESMOUL. Charles Robert Alexandre Desmoulins. 1797-1875. France.
DESR. Louis Auguste Joseph Desrousseaux. 1753-1838. France.
DESV. Auguste Nicaise Desvaux. 1784-1856. France.
DE WILD. Émile Auguste Joseph de Wildeman. 1866-1947. Belgium, Africa.
DE WILD. & T. DURAND. Émile Auguste Joseph de Wildeman. Theophile Alexis Durand.
DE WIT. Hendrik Cornelius Dirk de Wit. 1909- Netherlands, Malaysia.
DICKSON. Edward Dalzell Dickson. ? -1900. England, Turkey.
DIECK. Georg Dieck. 1847-1925. Germany.
DIELS. Friedrich Ludwig Emil Diels. 1874-1945. Germany.
DIELS & GILG. Friedrich Ludwig Emil Diels. Ernst Gilg.
DIETRICH, A. Albert Gottfried Dietrich. 1795-1856. Germany.
DIETRICH, D. David Nathanael Friedrich Dietrich. 1799-1888. Germany.
DIPP. Ludwig Dippel. 1827-1914. Germany.
DODE. Louis Albert Dode. 1875-1945. France.
DÖLL. Johann Christoph Döll. 1808-1885. Germany.
DOMBEY. Joseph Dombey. 1742-1796. France.
DOMBRAIN. Henry Honywood D'Ombrain. 1818-1905. England.
DOMIN. Karel Domin. 1882-1953. Czechoslovakia.
DON, D. David Don. 1799-1841. Scotland, England.
DON, G. George Don. 1798-1856. England.
DONN. James Donn. 1758-1813. England.
DOUGL. David Douglas. 1799-1834. England.
DOUGL. & LAMB. David Douglas. Aylmer Bourke Lambert.
DRAKE. Emmanuel Drake del Castillo. 1855-1904. France.
DRESS. William John Dress. 1918- . U.S.A.
DRUCE. George Claridge Druce. 1850-1932. England.
DRUDE. Carl Georg Oskar Drude. 1852-1933. Germany.
DRUDE & GRISEB. Carl Georg Oskar Drude. August Heinrich Rudolph Grisebach.
DRUMMOND, J. James Drummond. 1784-1863. Scotland, Australia.
DRYANDER. Jonas Dryander. 1748-1810. Sweden, England.
DUBY. Jean Étienne Duby. 1798-1885. Switzerland.
DUDL. William Russel Dudley. 1849-1911. U.S.A.
DUG. Armando Dugand. 1906-1971. Colombia.
DUHAMEL. Henri Louis Duhamel de Monceau. 1700-1781. France.
DULFER. Hans Dulfer. Contemporary. Austria.
DUM.-COURS. Georges Louis Marie Dumont de Courset. 1746-1824. France.
DÜMMER. Richard Arnold Dümmer. 1887-1922. Africa, England.
DUMORT. Barthélemy Charles Joseph Dumortier. 1797-1878. Belgium.
DUNAL. Michel Felix Dunal. 1789-1856. France.
DUNKLE. Meryl Byron Dunkle. 1888- . U.S.A.
DUNN. Stephen Troyte Dunn. 1868-1938. England.
DURAND, E. Elias Magloire Durand. 1794-1873. U.S.A.
DURAND, T. Theophile Alexis Durand. 1855-1912. Belgium, Africa.
DURANDE. Jean François Durande. 1732-1794. France.
DURAZZ. Antonio Durazzini. Fl. 1772. Italy.
DURIEU. Michel Charles Durieu de Maisonneuve. 1796-1878. France.
DU ROI. Johann Philipp Du Roi. 1741-1785. Germany.
DUTHIE. John Firminger Duthie. 1845-1922. England.
DYER. William Turner Thiselton-Dyer. 1843-1928. England.

EADE. George William Eade. 1905- . U.S.A.
EASTW. Alice Eastwood. 1859-1953. U.S.A.
EBERM. Carl Heinrich Ebermaier. 1802-1870. Germany.
ECKLON. Christian Friedrich Ecklon. 1795-1868. Denmark, South Africa.
EHRENB., C. A. Carl (Karl) August Ehrenberg. 1801-1849. Germany.
EHRH., J. F. Jakob Friedrich Ehrhart. 1742-1795. Germany.
EICHL. August Wilhelm Eichler. 1839-1887. Germany.
ELLIOTT. Stephen Elliott. 1771-1830. U.S.A.
ELLIS. John Ellis. 1711[1705?, 1710?]-1776. England.
ELMER. Adolf Daniel Edward Elmer. 1870-1942. U.S.A.
ENDL. Stephen Ladislaus Endlicher. 1804-1849. Austria.
ENGELM. George Engelmann. 1809-1884. U.S.A.
ENGLER. Heinrich Gustav Adolph Engler. 1844-1930. Germany.
ENGLER & GRAEBN. Heinrich Gustav Adolph Engler. Karl Otto Robert Peter Paul Graebner.
ENGLER & WARB. Heinrich Gustav Adolph Engler. Otto Warburg.
EPLING. Carl Clawson Epling. 1894-1968. U.S.A.
ESCALANTE. Escalante. Contemporary. Argentina.
ESCHSCH. Johann Friedrich Eschscholtz. 1793-1831. Russia, Germany.
EWAN. Joseph Andorfer Ewan. 1909- . U.S.A.
EXELL. Arthur Wallis Exell. 1901- . England.

FEDDE. Friedrich Karl Georg Fedde. 1873-1942. Germany.
FENZI. Emanuele Orazio Fenzi (Later known as Francesco Franceschi). 1843-1924. Italy, U.S.A., Libya.
FERN. Merritt Lyndon Fernald. 1873-1950. U.S.A.
FINET. Achille Eugene Finet. 1862-1913. France.
FINET & GAGNEP. Achille Eugene Finet. François Gagnepain.
FISCH. Friederich Ernst Ludwig von Fischer. 1782-1854. Russia.
FITZG., W. William Vincent Fitzgerald. ? -1929. Australia.
FLORIN. Carl Rudolf Florin. 1894-1965. Sweden.
FLORIN & BOUTELJE. Carl Rudolf Florin. Julius B. Boutelje.
FORBES. James Forbes. 1773-1861. England.
FORD. Neridah Clifton Ford. Fl. 1950. Australia.
FORD & VICKERY. Neridah Clifton Ford. Joyce Winifred Vickery.

FORREST. George Forrest. 1873-1932. Scotland.
FORREST & DIELS. George Forrest. Friedrich Ludwig Emil Diels.
FORSSK. Pehr (Peter) Forsskål. 1732-1768. Sweden, Egypt, Arabia.
FORST., G. Johann Georg Adam Forster (son). 1754-1794. Germany.
FORST., J. R. Johann Reinhold Forster. 1729-1798. Germany.
FORST., J. R. & G. FORST. Johann Reinhold Forster. Johann Georg Adam Forster.
FORTUNE. Robert Fortune. 1812-1880. England.
FOURCADE. Henri Georges (Henry George) Fourcade. ? -1948. South Africa.
FOURNIER. Eugene Pierre Nicolas Fournier. 1834-1884. France.
FRAHM. G. F. Frahm. Fl. 1898. Germany.
FRANCESCHI. See Emanuele Orazio Fenzi.
FRANCH. Adrien René Franchet. 1834-1900. France.
FRANCH. & SAV. Adrien René Franchet. Paul Amadée Ludovic Savatier.
FRANCO. João Manuel Antonio Paes do Franco. 1921- Portugal.
FRASER, C. Charles Fraser. ?1788-1831. Australia.
FRASER, J. John Fraser. 1750-1811. England.
FRÉM. John Charles Frémont. 1813-1890. U.S.A.
FRESENIUS. Johann Baptist Georg Wolfgang Fresenius. 1808-1866. Germany.
FREYER. Heinrich Freyer. 1802-1866. Austria.
FRYX. Paul Arnold Fryxell. 1927- . U.S.A.

GABLE. Joseph Benson Gable. 1886-1972. U.S.A.
GAERTNER. Joseph Gaertner. 1732-1791. Germany.
GAERTNER, C. F. Carl Friederich von Gaertner. 1754-1825. Germany.
GAGNEP. François Gagnepain. 1866-1952. France.
GAILLARDOT. Charles Gaillardot. 1814-1883. France.
GALEOTTI. Henri Guillaume Galeotti. 1814-1858. Belgium.
GALPIN. Ernest Edward Galpin. 1858-1941. South Africa.
GAMBLE. James Sykes Gamble. 1847-1925. England.
GAND. Michel Gandoger. 1850-1925. France.
GARDNER, C. A. Charles Austin Gardner. 1896-1970. Australia.
GARDNER, G. George Gardner. 1812-1849. England.
GAUD.-BEAUP. Charles Gaudichaud-Beaupré. 1789-1854. France.
GAUDIN. Jean François Gottlieb Philippe Gaudin. 1766-1833. Switzerland.
GAY. Claude Gay. 1800-1873. France, Chile.
GENTRY, A. Alwyn Howard Gentry. 1945- . U.S.A.
GEORGE. Alex S. George. Contemporary. Australia.
GERRARD. William Tyrer Gerrard. ? -1866. South Africa.
GIBSON, D. Dorothy Nash Gibson. 1921- . U.S.A.
GILG. Ernst Gilg. 1867-1933. Germany.
GILLIES. John Gillies. 1792-1834. Scotland, Chile, Argentina.
GILLIES & HOOK. John Gillies. William Jackson Hooker.
GIORDANO. Guiseppe Camillo Giordano. 1841-1901. Italy.
GLAZEBR. Thomas Kirkland Glazebrook. 1780-1855. England.
GLEDITSCH. Johann Gottlieb Gleditsch. 1714-1826. Germany.
GLOGAU. Arthur Glogau. Fl. 1928.
GMELIN, J. F. Johann Friedrich Gmelin. 1748-1804. Germany.

GMELIN, J. G. Johann Georg Gmelin. 1709-1755. Germany, Siberia.
GOEPP. Heinrich Robert Goeppert (Göppert). 1800-1884. Germany.
GORD. George Gordon. 1806-1879. England.
GRAEBN. Karl Otto Robert Peter Paul Graebner. 1871-1933. Germany.
GRAHAM, R. Robert Graham. 1786-1845. Scotland.
GRANT, A. L. Adele Lewis Grant. 1881-1967. U.S.A.
GRAY. Asa Gray. 1810-1888. U.S.A.
GREEN, M. L. Mary Leticia Green (Mrs. T. A. Sprague). 1886-1978. England.
GREEN, P. S. Peter S. Green. 1920- . England.
GREENE. Edward Lee Greene. 1843-1915. U.S.A.
GREENM. Jesse More Greenman. 1867-1951. U.S.A.
GREV. Robert Kaye Greville. 1794-1866. Scotland.
GREV. & BALF. Robert Kaye Greville. John Hutton Balfour.
GRIFFITH, W. William Griffith. 1810-1845. England, India.
GRIGNAN. G. T. Grignan. Fl. 1903-1914. France.
GRISEB. August Heinrich Rudolph Grisebach. 1814-1879. Germany.
GRISEB & H. WENDL. August Heinrich Rudolph Grisebach. Hermann Wendland.
GROENL. Johannes Groenland. 1824-1891. France.
GROOTEND. Hermann Johannes Grootendorst. 1911- Netherlands.
GUILFOYLE. William Robert Guilfoyle. 1840-1912. England, Australia.
GUILLAUM. André Guillaumin. 1885-1952. France.
GUSSONE. Giovanni Gussone. 1787-1866. Italy.
GUTHRIE. Francis Guthrie. 1831-1899. England, South Africa.
GUTHRIE & BOLUS. Francis Guthrie. Harry Bolus.

HAENKE. Thaddaeus Haenke. 1761-1816. Czechoslovakia.
HALÁCSY. Eugen von Halácsy. 1842-1913. U.S.A.
HALL. Harvey Monroe Hall. 1874-1932. U.S.A.
HALL & CLEMENTS. Harvey Monroe Hall. Frederic Edward Clements.
HAMET. Raymond Hamet. 1890- . France.
HANCE. Henry Fletcher Hance. 1827-1886. England, China.
HAND.-MAZZ. Heinrich von Handel-Mazzetti. 1882-1940. Austria.
HANSEN. Niels Ebbesen Hansen. 1866-1945. U.S.A.
HARA. Hiroshi Hara. 1908- . Japan.
HARIOT. Paul Auguste Hariot. 1854-1917. France.
HARMS. Hermann August Theodor Harms. 1870-1942. Germany.
HARPER, F. Francis Harper. 1886- . U.S.A.
HARROW. Robert Lewis Harrow. 1867-1954. England.
HARTWEG. Karl Theodor Hartweg. 1812-1871. Germany, Mexico, U.S.A.
HARV. William Henry Harvey. 1811-1866. Ireland.
HARV. & SOND. William Henry Harvey. Otto Wilhelm Sonder.
HASSK. Justus Karl Hasskarl. 1811-1894. Germany.
HAW. Adrian Hardy Haworth. 1768-1833. England.
HAYATA. Bunzo Hayata. 1874-1934. Japan.
HAYNE. Friedrich Gottlob Hayne. 1763-1832. Germany.
HBK. Friedrich Wilhelm Heinrich Alexander von Humboldt. Aimé Jacques Alexandre Bonpland. Carl Sigismund Kunth.
HELLER. Amos Arthur Heller. 1867-1944. U.S.A.
HEMSL. William Botting Hemsley. 1843-1924. England.

HEMSL. & WILS. William Botting Hemsley. Ernest Henry Wilson.
HENDERS., L. F. Louis Fourniquet Henderson. 1853-1942. U.S.A.
HENDERS., M. D. Mayda Doris Henderson. 1928- Africa.
HENKEL. Johann Baptist Henkel. 1815-1871. Germany.
HENKEL & W. HOCHST. Johann Baptist Henkel. Wilhelm Christian Hochstetter.
HENRY. Augustine Henry. 1857-1930. Ireland, China.
HENRY, L. Louis Henry. 1853-1903. France.
HENSLOW. John Stevens Henslow. 1796-1861. England.
HERDER. Ferdinand Godofried (Gotofred) Theobald Maximilian von Herder. 1828-1896. Germany, Russia.
HÉRINCQ. François Hérincq. 1820-1891. France.
HERRMANN. Johann Herrmann. 1738-1800. France.
HESSE. Hermann Albrecht Hesse. 1852-1937. Germany.
HEYNE, B. Benjamin Heyne. ? -1819. England, India.
HEYNE, K. Karel Heyne. 1877-1947. Netherlands.
HEYWOOD. Vernon Hilton Heywood. 1927-   . England.
HIBB. James Shirley Hibberd. 1825-1890. England.
HIERN. William Philip Hiern. 1839-1925. England.
HIERON. Georg Hans Emo Wolfgang Hieronymus. 1846-1921. Germany, Argentina.
HILDEBR. Friedrich Hermann Gustav Hildebrand. 1835-1915. Germany.
HILL, D. D. Hill. 1847-1929. U.S.A.
HILL, J. John Hill. 1716-1775. England.
HILL, W. Walter Hill. 1820-1904. Australia.
HITCHC., C. L. Charles Leo Hitchcock. 1902-   . U.S.A.
HJELMQUIST. Hakon Hjelmquist. 1905-   . Sweden.
HOCHR. Bénédict Pierre Georges Hochreutiner. 1873-1959. Switzerland.
HOCHST. Christian Ferdinand Hochstetter. 1787-1860. Germany.
HOCHST., W. Wilhelm Christian Hochstetter. 1825-1881. Germany.
HOESS. Franz Hoess. 1756-1840. Austria.
HOFFM., J. Johann Joseph Hoffmann. 1805-1878. Germany.
HOFFM., J. & H. SCHULT. Johann Joseph Hoffmann. H. Schultes.
HOFFM., K. Kaethe (Käthe) Hoffmann. Fl. 1910-1931. Germany.
HOFFMANNSEGG. Johann Centurius von Hoffmannsegg. 1766-1849. Germany.
HOLTT. Richard Eric Holttum. 1895-   . England.
HOLTT. & STANDL. Richard Eric Holttum. Paul Carpenter Standley.
HONDA. Masaji Honda. 1897-   . Japan.
HOOK. William Jackson Hooker. 1785-1865. England.
HOOK. & ARN. William Jackson Hooker. George Arnott Walker Arnott.
HOOK. & HARV. William Jackson Hooker. William Henry Harvey.
HOOK. f. Joseph Dalton Hooker (son). 1817-1911. England.
HOOK f. & HARV. Joseph Dalton Hooker. William Henry Harvey.
HOOK. f. & THOMS. Joseph Dalton Hooker. Thomas Thomson.
HOOPES. Josiah Hoopes. 1832-1904. U.S.A.
HOOVER. Robert Francis Hoover. 1913-1970. U.S.A.
HOOVER & ROOF. Robert Francis Hoover. James Roof.
HOPE. John Hope. 1725-1786. Scotland.
HORNEM. Jens Wilken Hornemann. 1770-1841. Denmark.
HORNIBR. Murray Hornibrook. 1874-1949. England.

HORNSTEDT. Claës [Claudius] Frederic Hornstedt. 1758-1809. Sweden.
HORT. Horticulture. Hortorum. Literally, of the garden. Placed after names current among horticulturists; frequently indicates garden or unknown origin.
HOUTT. Maarten Houttuyn. 1720-1798. Netherlands.
HOUZEAU DE LEHAIE. Jean Houzeau de Lehaie. 1820-1888. Belgium.
HOWELL, J. T. John Thomas Howell. 1903-   . U.S.A.
HOWELL, T. J. Thomas Jefferson Howell. 1842-1912. U.S.A.
HOYT. Roland Stewart Hoyt. ? -1968. U.S.A.
HU. Shiu-Ying Hu. 1910-   . China, U.S.A.
HU & CHENG. Shiu-Ying Hu. Wan-Chun Cheng.
HUBBARD, F. T. Frederick Tracy Hubbard. 1875-1962. U.S.A.
HUDS. William Hudson. 1730-1793. England.
HUEGEL. Carl Alexander Anselm von Huegel (Hügel). 1794-1870. Austria.
HULL. John Hull. 1761-1813. England.
HUMB. Friedrich Wilhelm Heinrich Alexander von Humboldt. 1769-1859. Germany.
HUMB. & BONPL. Friedrich Wilhelm Heinrich Alexander von Humboldt. Aimé Jacques Alexandre Bonpland.
HUTCHINSON. John Hutchinson. 1884-1972. England.
HYLANDER. Nils Hylander. 1904-1970. Sweden.

INGHAM. Norman D. Ingham. Fl. 1908. U.S.A.
INGRAM, J. John William Ingram. 1924-   . U.S.A.
ISELY. Duane Isely. 1918-   . U.S.A.
ITO. Tokutaro Ito. 1868-1941. Japan.

JACK. William Jack. 1795-1822. Scotland, Malaya.
JACKMAN. George Jackman. 1837-1887. England.
JACKSON, A. B. Albert Bruce Jackson. 1876-1947. England.
JACKSON, A. B. & DALLIM. Albert Bruce Jackson. William Dallimore.
JACKSON, B. D. Benjamin Daydon Jackson. 1846-1927. England.
JACOBSEN. Hermann Johannes Heinrich Jacobsen. 1898-   . Germany.
JACQ. Nicolaus Joseph Jacquin. 1727-1817. Austria.
JACQ. f. Joseph Franz von Jacquin. 1766-1839. Austria.
JAEG. Hermann Jaeger. 1815-1890. Germany.
JALAS. Arvo Jaakko Juhani Jalas. 1920-   . Finland.
JAMES. Edwin James. 1797-1861. U.S.A.
JANCHEN. Erwin Janchen. 1882-   . Austria.
JEPS. Willis Linn Jepson. 1867-1946. U.S.A.
JEPS. & WIESL. Willis Linn Jepson. Albert Everett Wieslander.
JESSOP. John Peter Jessop. 1939-   . South Africa, Australia.
JOHNSON, L. Lawrence Alexander Sidney Johnson. 1925-   . Australia.
JOHNST., I. M. Ivan Murray Johnston. 1898-1960. U.S.A.
JOHNST., M. C. Marshall Corning Johnston. 1930- U.S.A.
JOHNSTONE. George Henry Johnstone. 1881-1960. England.
JONES, G. N. George Neville Jones. 1903-1970. U.S.A.
JONKER, A. Anni Margriet Emma Jonker-Verhoef. 1920-   . Netherlands.
JONKER, F. Frederik Pieter Jonker. 1912-   . Netherlands.

JONKER, A. & F. Anni Margriet Emma Jonker-Verhoef. Frederik Pieter Jonker.
JUNGHUHN. Franz Wilhelm Junghuhn. 1809-1864. Germany.
JUSS. Antoine Laurent de Jussieu. 1748-1836. France.

KACHE. Paul Kache. 1882-1945. Germany.
KALMB. George Anthony Kalmbacher. 1897- . U.S.A.
KARST. Gustav Karl Wilhelm Hermann Karsten. 1817-1908. Austria.
KARST. & TRIANA. Gustav Karl Wilhelm Hermann Karsten. José Jerónimo Triana.
KARWINSKI. Wilhelm Friedrich Karwinski von Karwin. 1780-1855. Germany.
KASAPLIGIL. Baki Kasapligil. 1918- . Turkey, U.S.A.
KAULFUSS. Georg Friedrich Kaulfuss. 1786-1830. Germany.
KAUSEL. Eberhard Maximilano Leopoldo Otto Kausel. 1910-1972. Chile.
KEARNEY. Thomas Henry Kearney. 1874-1956. U.S.A.
KECK. David Daniels Keck. 1903- . U.S.A.
KELL. Albert Kellogg. 1813-1887. U.S.A.
KENSIT. See L. Bolus, née Kensit.
KER-GAWL. John Bellenden Ker (John Ker Bellenden. John Ker Gawler before 1804). 1764-1842. England.
KESSELRING. Jakob Kesselring. 1835-1909.
KIAERSK. Hjalmar Frederik Christian Kiaerskou. 1835-1900. Denmark.
KING. George King. 1840-1909. England, India.
KIRCHN. Georg Kirchner. 1837-1885. Germany.
KIRK, T. Thomas Kirk. 1828-1898. England, New Zealand.
KITAIBEL. Paul Kitaibel. 1757-1817. Hungary.
KITAMURA. Siro Kitamura. 1906- . Japan.
KLATT. Friedrich Wilhelm Klatt. 1825-1897. Germany.
KLOTZ. Gerhard Klotz. 1928- . Germany.
KLOTZSCH. Johann Friedrich Klotzsch. 1805-1860. Germany.
KNIGHT, H. Henry Knight. 1834-1896. England.
KNIGHT, J. Joseph Knight. 1781-1855. England.
KOBUSKI. Clarence Emmeren Kobuski. 1900-1963. U.S.A.
KOCH, K. Karl (Carl) Heinrich Emil Koch. 1809-1879. Germany.
KOCH, K. & AUGUSTIN. Karl (Carl) Heinrich Emil Koch. Augustin.
KOCH, K. & H. SELLO. Karl (Carl) Heinrich Emil Koch. Herman Ludwig Sello.
KOCH, W. D. J. Wilhelm Daniel Joseph Koch. 1771-1849. Germany.
KOEHNE. Bernhard Adalbert Emil Koehne. 1848-1918. Germany.
KOEHNE & LINGELSH. Bernhard Adalbert Emil Koehne. Alexander von Lingelsheim.
KOENIG. Carl Dietrich Eberhard Koenig. 1774-1851. Germany, England.
KOENIG, J. Johann Gerhard Koenig. 1728-1785. Latvia, India.
KOERN. Friedrich August Koernicke [Körnicke]. 1828-1908. Russia.
KOERTE. Franz Koerte [Körte]. 1782-1845. Germany.
KOHANKIE. Henry Kohankie. 1887- . U.S.A.
KOIDZ., G. Gen'ichi Koidzumi. 1883-1953. Japan.
KOM. Vladimir Leontyevitch Komarov. 1869-1946. Russia.
KOMATSU. Shunzo Komatsu. 1879-1932. Japan.
KORSH. Sergei Ivanowitsch Korshinsky. 1861-1900. Russia.
KOSTERMANS. André Joseph Guillaume Henri Kostermans. 1907- . Indonesia.

KOTSCHY. Carl Georg Theodor Kotschy. 1813-1866. Austria.
KRÄNZLIN. Friedrich Wilhelm Ludwig Kränzlin. 1847-1934. Germany.
KRAPOVICKAS. Antonio Krapovickas. 1921- . Argentina.
KRASSER. Fridolin Krasser. 1863-1922. Czechoslovakia.
KRAUSS, C. F. Christian Ferdinand Friedrich von Krauss. 1812-1890. Germany.
KRUKOFF. Boris Alexander Krukoff. 1898- . U.S.A.
KRUKOFF & BARNEBY. Boris Alexander Krukoff. Rupert Charles Barneby.
KRUKOFF & MOLDENKE. Boris Alexander Krukoff. Harold Norman Moldenke.
KUNTH. Carl Sigismund Kunth. 1788-1850. Germany.
KRÜSSM. Gerd Krüssmann. 1910- . Germany.
KUNTH & BOUCHÉ. Carl Sigismund Kunth. Carl David Bouché.
KUNTZE. Carl Ernst Otto Kuntze. 1843-1907. Germany.
KURZ. Wilhelm Sulpiz Kurz. 1834-1878. Germany, East Indies.

L. Carolus Linnaeus (Carl von Linné). 1707-1778. Sweden.
L. f. Carl von Linné (son). 1741-1783. Sweden.
LABILL. Jacques Julien Houtton de Labillardière. 1755-1834. France.
LAG. Mariano Lagasca y Segura. 1776-1839. Spain.
LAGERH. Nils Gustav Lagerheim. 1860-1926. Sweden.
LAING. Robert M. Laing. Contemporary. New Zealand.
LAM. Jean Baptiste Pierre Antoine de Monet de Lamarck. 1744-1829. France.
LAMB. Aylmer Bourke Lambert. 1761-1842. England.
LAMBERTYE. Leonce de Lambertye. 1810-1877. France.
LANDON. John W. Landon. Contemporary. U.S.A.
LANGE, J. Johan Martin Christian Lange. 1818-1898. Denmark.
LARRÉATEGUI. Joseph Dionisio Larréategui. Fl. 1795-1805. Mexico.
LAUCHE. Wilhelm Lauche. 1827-1882. Germany.
LAUTH. Thomas L. Lauth. 1758-1826. France.
LAVALL. Pierre Alphonse Martin Lavallée. 1836-1884. France.
LAWS., C. Charles Lawson. 1794-1873. Scotland.
LAWS., P. Peter Lawson. Fl. 1770-1821. Scotland.
LAWS., P. & C. LAWS. Peter Lawson. Charles Lawson.
LAXM. Erich Laxmann. 1737-1796. Russia.
LEANDRI. Jacques Leandri. 1903- . France.
LEDEB. Karl Friedrich von Ledebour. 1785-1851. Germany.
LEGRAND, D. Carlos Maria Diego Enrique Legrand. 1901- . Uruguay.
LEGRAND, D. & KAUSEL. Carlos Maria Diego Enrique Legrand. Eberhard Maximilano Leopoldo Otto Kausel.
LEHM. Johann Georg Christian Lehmann. 1792-1860. Germany.
LE JOLIS. Auguste François Le Jolis. 1823-1904. France.
LEM. Charles Antoine Lemaire. 1801-1871. Belgium.
LEMMON. John Gill Lemmon. 1832-1908. U.S.A.
LEMOINE. Pierre Louis Victor Lemoine or Emil Lemoine (son).
LEMOINE, E. Emil Lemoine. 1862-1943. France.
LEMOINE, V. Pierre Louis Victor Lemoine. 1823-1911. France.
LENNÉ. Peter Joseph Lenné. 1789-1866.
LESCUYER. O. Lescuyer. Fl. 1855-1872. France.
LESS. Christian Friedrich Lessing. 1810-1862. Germany.

LÉV. Auguste Abel Hector Léveillé. 1863-1918. France.
LÉV. & VANIOT. Auguste Abel Hector Léveillé. Eugene Vaniot.
LEX. Juan Martinez de Lexarza. 1785-1824. Mexico.
L'HÉR. Charles Louis L'Hériter de Brutelle. 1746-1800. France.
LI, H. L. Hui-Lin Li. 1911-   . U.S.A.
LIEBM. Friedrik Michael Liebmann. 1813-1856. Denmark, Mexico.
LINDAU. Gustav Lindau. 1866-1923. Germany.
LINDB, S. Sextus Otto Lindberg. 1835-1889. Sweden.
LINDEN. Jean Jules Linden. 1817-1898. Belgium.
LINDEN & ANDRÉ. Jean Jules Linden. Edouard François André.
LINDEN, L. Lucien Linden. Fl. 1881-1896.
LINDL. John Lindley. 1799-1865. England.
LINDL. & PAXT. John Lindley. Joseph Paxton.
LINGELSH. Alexander von Lingelsheim. 1874-1937. Germany.
LINK. Johann Heinrich Friedrich Link. 1767-1851. Germany.
LINK & OTTO. Johann Heinrich Friedrich Link. Christoph Friedrich Otto.
LITTLE. Elbert Luther Little, Jr. 1907-   . U.S.A.
LLAVE. Canónigo Pablo de La Llave. 1773-1833. Mexico.
LLAVE & LEX. Canónigo Pablo de La Llave. Juan Martinez de Lexarza.
LLOYD. Robert Michael Lloyd. 1938-   . U.S.A.
LOCKW. Tommie Earl Lockwood. 1941-1975. U.S.A.
LODD. Conrad Loddiges. 1732-1826. George Loddiges (son). 1784-1846. England.
LOES. Ludwig Eduard Theodor Loesener. 1865-1941. Germany.
LOISEL. Jean Louis Auguste Loiseleur-Deslongchamps. 1774-1849. France.
LORENTZ. Paul Günther Lorentz. 1835-1881. Switzerland, Uruguay, Argentina.
LOUD. John Claudius Loudon. 1783-1843. England.
LOUR. João de [Juan] Loureiro. 1710[?1717]-1791[?1796]. Portugal, Indochina.
LÖVE, Á. Áskell Löve. 1916-   . U.S.A.
LÖVE, Á. & D. Áskell Löve. Doris Benta Maria Löve.
LÖVE, D. Doris Benta Maria Löve. 1918-   . U.S.A.
LUEHM. John George Luehmann. 1843-1904. Australia.

MACB. James Francis Macbride. 1891-1976. U.S.A.
MACFADY. James Macfadyen. 1798-1850. Scotland, Jamaica.
MACKAY. James Townsend Mackay. 1775-1862. Ireland.
MAIDEN. Joseph Henry Maiden. 1859-1925. Australia.
MAIDEN & BETCHE. Joseph Henry Maiden. Ernst Betche.
MAIDEN & BLAKELY. Joseph Henry Maiden. William Faris Blakely.
MAIDEN & CAMBAGE. Joseph Henry Maiden. Richard Hind Cambage.
MAIRE. René Charles Joseph Ernest Maire. 1878-1949. Algeria.
MAKINO. Tomitaro Makino. 1861-1957. Japan.
MAKINO & SHIBATA. Tomitaro Makino. Keita Shibata.
MAKINO & UCHIDA. Tomitaro Makino. Shigetaro Uchida.
MANDA. W. A. Manda. Fl. 1889-1897. U.S.A.
MANETTI. Guiseppe Manetti. 1831-1858. Italy.
MANTEN. Jacob [Jack] Manten. 1898-1958. Canada.
MARCHAL. Elie Marchal. 1839-1923. Belgium.
MARCHANT. William James Marchant. 1886-1952. England.
MARLIAC. Joseph [Bory] Latour-Marliac. 1830- ? .

MARLOTH. Hermann Wilhelm Rudolf Marloth. 1855-1931. Germany, South Africa.
MARNOCK. Robert Marnock. 1800-1889. England.
MARQUAND. Cecil Victor Boley Marquand. 1897-1943. England.
MARSH. Humphrey Marshall. 1722-1801. U.S.A.
MARSILI. Giovanni Marsili. 1727-1795. Italy.
MART. Karl (Carl) Friedrich Philipp von Martius. 1794-1868. Germany.
MARTENS, M. Martin Martens. 1797-1863. Belgium.
MARTENS, M. & GALEOTTI. Martin Martens. Henri Guillaume Galeotti.
MARTINEZ. Maximino Martinez. 1888-1964. Mexico.
MASF. Ramón Masferrer y Arquimbau. 1850-1884. Spain, Philippines.
MASLIN. Bruce R. Maslin. Contemporary. Australia.
MASON. Herbert Louis Mason. 1896-   . U.S.A.
MAST., M. T. Maxwell Tylden Masters. 1833-1907. England.
MATHIEU. Louis Mathieu. 1793-1867. France.
MATSUM. Jinzo Matsumura. 1856-1928. Japan.
MAXIM. Carl (Karl) Johann Maximowicz. 1827-1891. Russia.
MAXIM. & RUPR. Carl Johann Maximowicz. Franz Joseph Ruprecht.
MAYR. Heinrich Mayr. 1856-1911. Germany.
MAZEL. Mazel. Fl. 1880.
McCLELLAND. John McClelland. 1800[?1805]-1883. England.
McCLINT. Elizabeth May McClintock. 1912-   . U.S.A.
McCLURE. Floyd Alonso McClure. 1897-1970. U.S.A.
McKELVEY. Susan Adams (née Delano) McKelvey. 1883-1965. U.S.A.
McMINN. Howard Earnest McMinn. 1892-1963. California.
McMURTRIE. Henry McMurtrie. 1793-1865. U.S.A.
MEDIC. Friedrich Casimir Medicus. 1736-1808. Germany.
MEDVEDEV. Iakov Sergeevich Medvedev [Medwedew]. 1847-1923. Russia.
MEISSN. Carl Friedrich Meissner (Meisner). 1800-1874. Switzerland.
MELANDER. Leonard William Melander. 1893-   . U.S.A.
MELANDER & EADE. Leonard William Melander. George William Eade.
MELVILLE. Ronald Melville. 1903-   . England.
MERRIAM. Clinton Hart Merriam. 1855-1942. U.S.A.
MERRILL. Elmer Drew Merrill. 1876-1956. U.S.A.
MERRILL & L. M. PERRY. Elmer Drew Merrill. Lily May Perry.
METTENIUS. Georg Heinrich Mettenius. 1823-1866. Germany.
MEUNISSIER. Auguste Alexandre Meunissier. 1876-1947. France.
MEY., C. A. Carl Anton von Meyer. 1795-1855. Russia.
MEY., E. H. Ernst Heinrich Friedrich Meyer. 1791-1858. Germany.
MEY., F. G. Frederick Gustav Meyer. 1917-   . U.S.A.
MICHELI. Marc Micheli. 1844-1902. Switzerland.
MICHX. André Michaux. 1746-1802. France, U.S.A.
MICHX. f. François André Michaux (son). 1770-1855. France, U.S.A.
MIERS. John Miers. 1789-1879. England.
MIKAN. Johann Christian Mikan. 1769-1844. Czechoslovakia
MILL. Phillip Miller. 1691-1771. England.
MILL., J. John Miller (Johann Sebastian Miller). 1715-?1790. England.
MILLSPAUGH. Charles Frederick Millspaugh. 1854-1923. U.S.A.

MIQ. Friedrich Anton Wilhelm Miquel. 1811-1871. Netherlands.
MIRB. Charles François Brisseau de Mirbel. 1776-1854. France.
MITF. Algernon Bertram Freeman-Mitford, Lord Rosedale. 1837-1916. England.
MOÇ. José Mariano Moçiño Suares Losada. 1757-182?. Mexico.
MOÇ. & SESSÉ. José Mariano Moçiño Suares Losada. Martin de Sessé y Lacasta.
MOENCH. Konrad (Conrad) Moench. 1744-1805. Germany.
MOLDENKE. Harold Norman Moldenke. 1909-   . U.S.A.
MOLINA. Juan Ignacio Molina. 1737-1829. Spain.
MONTIN. Lars Jonasson Montin. 1723-1785. Sweden.
MOON. Alexander Moon. ? -1825. England.
MOORE, C. Charles Moore. 1820-1905. Australia.
MOORE, C. & F. J. MUELL. Charles Moore. Ferdinand Jacob Heinrich von Mueller.
MOORE, H. E. Harold Emery Moore, Jr. 1917-   . U.S.A.
MOORE, L. B. Lucy B. Moore. 1906-   . New Zealand.
MOORE, S. Spencer Le Marchant Moore. 1851-1931. England.
MOORE, T. Thomas Moore. 1821-1887. England.
MOORE, T. & JACKMAN. Thomas Moore. George Jackman.
MOQ. Christian Horace Bénédict Alfred Moquin-Tandon. 1804-1863. France.
MORDANT DE LAUNAY. Jean Claude Michel Mordant de Launay. 1750-1816. France.
MORELET. Pierre Marie Arthur Morelet. 1809-1892. France, Algeria.
MORICAND. Moïse Étienne (Stefano) Moricand. 1799-1854. Switzerland.
MORIS. Guiseppe Giacinto Moris. 1796-1869. Italy.
MORIS & DE NOTARIS. Guiseppe Giacinto Moris. Guiseppe de Notaris.
MORITZI. Alexander Moritzi. 1806-1850. Switzerland.
MORONG. Thomas Morong. 1827-1894. U.S.A.
MORR., C. Charles François Antoine Morren. 1807-1858. Belgium.
MORR., C. & DECNE. Charles François Antoine Morren. Joseph Decaisne.
MORR., E. Charles Jacques Edouard Morren. 1833-1886. Belgium.
MOTTET. Séraphin Joseph Mottet. 1861-1930. France.
MOUILLEFERT. Pierre Mouillefert. 1846-1903. France.
MUDIE. Robert Mudie. 1777-1842. England.
MUELL., F. J. Ferdinand Jacob Heinrich von Mueller. 1825-1896. Australia.
MUELL., F. J. & J. E. BR. Ferdinand Jacob Heinrich von Mueller. John Ednie Brown.
MUELL., F. J. & DRUDE. Ferdinand Jacob Heinrich von Mueller. Carl Georg Oskar Drude.
MUELL., P. J. Philipp Jakob Mueller. 1832-1889. France.
MUELL.-ARG. Jean Mueller (Argovensis, i.e., of Aragau). 1828-1896. Switzerland.
MUENCHH. Otto von Muenchhausen. 1716-1774. Germany.
MUHLENB. Gotthilf Henry Ernest Muhlenberg (formerly Gotthilf Heinrich Ernst Muehlenberg). 1756-1817. U.S.A.
MUNRO. William Munro. 1818-1880. England.
MUNSON. Thomas Volney Munson. 1843-1913. U.S.A.
MUNZ. Philip Alexander Munz. 1892-1974. U.S.A.
MUNZ & I. M. JOHNST. Philip Alexander Munz. Ivan Murray Johnston.
MURAV. O. A. Muravjova. Fl. 1936. Russia.
MURR., A. Andrew Murray. 1812-1878. Scotland.
MURR., E. Albert Edward Murray. 1935-   . U.S.A.
MURR., J. Johan Andreas Murray. 1740-1791. Sweden.

NADEAUD. Jean Nadeaud [Nadeau]. 1834-1898. France.
NAKAI. Takenoshi Nakai. 1882-1952. Japan.
NAUD. Charles Naudin. 1815-1899. France.
NÉE. Luis Née. Fl. 1791. Spain.
NEES. Christian Gottfried Daniel Nees von Esenbeck. 1776-1858. Germany.
NEES & EBERM. Christian Gottfried Daniel Nees von Esenbeck. Carl Heinrich Ebermaier.
NEILREICH. August Neilreich. 1803-1871. Austria.
NELS., A. Aven Nelson. 1859-1952. U.S.A.
NELS., J. John Nelson. Fl. 1866. England.
NEWB. John Strong Newberry. 1822-1892. U.S.A.
NICHOLS. George Nicholson. 1847-1908. England.
NIEDENZU. Franz Josef Niedenzu. 1857-1937. Germany.
NITSCHE. Walter Nitsche. 1883-   . Germany.
NOIS. Louis Claude Noisette. 1772-1849. France.
NORLINDH. Nils Tycho Norlindh. 1906-   . Sweden.
NUTT. Thomas Nuttall. 1786-1859. England, U.S.A.
NYMAN. Carl Fredrik Nyman. 1820-1893. Sweden.

O'BRIEN. James O'Brien. 1842-1930. England.
OEDER. Ceorg Christian Oeder. 1728-1791. Denmark.
OERST. Anders Sandøe Oersted. 1816-1872. Denmark, Costa Rica, Colombia.
OHWI. Jisaburo Ohwi, 1905-   . Japan.
OKEN. Lorenz Oken (né Okenfuss). 1779-1851. Switzerland.
OLIVER, D. Daniel Oliver. 1830-1917. England.
OLIVER, G. A. Guillaume Antoine Oliver. 1756-1814. France.
OLIVER, W. Walter Reginald Brook Oliver. 1883-1957. New Zealand.
OPIZ. Philipp Maximilian Opiz. 1787-1858. Czechoslovakia.
ORT. Casimiro Gómez Ortega. 1740-1818. Spain.
ORTGIES. Karl Eduard Ortgies. 1829-1916. Switzerland.
OSBECK. Pehr Osbeck. 1723-1805. Sweden.
OSBORN. Arthur Osborn. 1878-1964. England.
OSTENFELD. Carl Emil Hansen Ostenfeld. 1873-1931. Denmark.
OTTLEY. Alice Maria Ottley. 1882-   . U.S.A.
OTTO. Christoph Friedrich Otto. 1783-1856. Germany.
OTTO & A. DIETRICH. Christoph Friedrich Otto. Albert Gottfried Dietrich.

PALIBIN. Ivan Vladimirovich Palibin. 1872-1949. Russia.
PALL. Peter Simon Pallas. 1741-1811. Russia.
PAMP. Renato Pampanini. 1875-1949. Italy.
PANČIĆ. Josef Pančić. 1814-1888. Yugoslavia.
PARDÉ. Leon Gabriel Charles Pardé. 1865-1943. France.
PARISH. Samuel Bonsall Parish. 1839-1923. U.S.A.
PARL. Filippo Parlatore. 1816-1877. Italy.
PARRY. Charles Christopher Parry. 1823-1890. U.S.A.
PAU. Carlos Pau y Español. 1857-1937. Spain.
PAV. José Antonio Pavón. 1754-1844. Spain.
PAX. Ferdinand Albin Pax. 1858-1942. Germany.
PAX & K. HOFFM. Ferdinand Albin Pax. Kaethe (Käthe) Hoffmann.
PAXT. Joseph Paxton. 1802-1865. England.
PEARCE. Sydney Albert Pearce. 1906-   . England.
PEARSON, H. Henry Harold Welch Pearson. 1870-1916. England.

PEARSON, R. Robert Hooper Pearson. 1866-1918. England.
PENF. Arthur Raymond Penfold. 1890-   . Australia.
PENN. Francis Whittier Pennell. 1886-1952. U.S.A.
PERROTTET. Georges Samuel Perrottet. 1793-1870. Switzerland.
PERRY, L. M. Lily May Perry. 1895-   . U.S.A.
PERS. Christiaan Hendrik Persoon. 1755-1837. South Africa, Germany, France.
PFITZER. Ernst Hugo Heinrich Pfitzer. 1846-1906. Germany.
PHILLIPS. Edwin Percy Phillips. 1884-1967. South Africa.
PHILLIPS & HUTCHINSON. Edwin Percy Phillips. John Hutchinson.
PICHON. Marcel Pichon. 1921-1954. France.
PIERRE. Jean Baptiste Louis Pierre. 1833-1905. France.
PIKE. A. V. Pike. Fl. 1946. England.
PILG. Robert Knud Friedrich Pilger. 1876-1953. Germany.
PILLANS. Neville Stuart Pillans. 1884-1964. South Africa.
PIPER. Charles Vancouver Piper. 1867-1926. U.S.A.
PITCHER. J. R. Pitcher. Fl. 1889-1897. U.S.A.
PITCHER & MANDA. J. R. Pitcher. W. A. Manda.
PLANCH. Jules Émile Planchon. 1833-1900. France.
POEPP. Eduard Friedrich Poeppig. 1798-1868. Austria.
POEPP. & ENDL. Eduard Friedrich Poeppig. Stephen Ladislaus Endlicher.
POGGENBURG. Justus Ferdinand Poggenburg. 1840-1893. U.S.A.
POHL. Johann Baptist Emanuel Pohl. 1782-1834. Austria.
POIR. Jean Louis Marie Poiret. 1755-1834. France.
POISSON. Poisson. Fl. 1912.
POIT. Pierre Antoine Poiteau. 1766-1854. France.
POLAKOWSKY. Hellmuth Polakowsky. 1847-1917. Costa Rica.
POLE EVANS. Illtyd Buller Pole Evans. 1879-1968. South Africa.
POLLARD. Charles Louis Pollard. 1872-1945. U.S.A.
POURR. Pierre André Pourret de Figeac. 1754-1818. France.
PRAIN. David Prain. 1857-1944. England.
PREISS. Ludwig Preiss. 1811-1883. Australia.
PRESL, J. Jan Swatopluk Presl. 1791-1849. Czechoslovakia.
PRESL, J. & K. PRESL. Jan Swatopluk Presl. Karel Boriweg Presl.
PRESL, K. Karel Boriweg Presl. 1794-1852. Czechoslovakia.
PRESTON. Isabella Preston. 1881-1965. Canada.
PRITZEL, E. Ernst Georg Pritzel. 1875-1946. Germany.
PUISSANT. Pierre A. Puissant. 1831-1911. Belgium, U.S.A.
PULLIAT. Victor Pulliat. 1827-1866. France.
PURKYNĚ. Emanuel Purkyně. 1831-1882. Austria.
PURPUS, C. Carl Albert Purpus. 1851-1941. Germany, Mexico.
PURSH. Fredrick Traugott Pursh (Friedrich Pursch). 1774-1820. Germany, U.S.A., England.

RADDI. Giuseppe Raddi. 1770-1829. Italy.
RADLK. Ludwig Adolph Timotheus Radlkofer. 1829-1927. Germany.
RAEUSCH. Ernest Adolph Raeuschel. Fl. 1772-1797. Germany.
RAF. Constantine Samuel Rafinesque-Schmaltz. 1783-1840. U.S.A.
RAFFILL. Charles Percival Raffill. 1876-1951. England.
RAMOND. Louis François Elisabeth Ramond de Carbonieres. 1753-1827. France.
RAOUL. Étienne Fiacre Louis Raoul. 1815-1852. Australia.
RCHB. Heinrich Gottlieb Ludwig Reichenbach. 1793-1879. Germany.

RCHB. f. Heinrich Gustav Reichenbach (son). 1823-1889. Germany.
RCHB. f. & ZOLLINGER. Heinrich Gustav Reichenbach. Heinrich Zollinger.
READER. Felix Maximillian Reader. 1850-1911. Australia.
RECHINGER. Karl Heinz Rechinger. 1906-   . Austria.
REDOUTÉ. Pierre Joseph Redouté. 1761-1840. France.
REGEL. Eduard August von Regel. 1815-1892. Germany, Russia.
REGEL & HERDER. Eduard August von Regel. Ferdinand Godofried Theobald Maximilian von Herder.
REGEL & KOERN. Eduard August von Regel. Friedrick August Koernicke [Körnicke].
REHD. Alfred Rehder. 1863-1949. Germany, U.S.A.
REHD. & WILS. Alfred Rehder. Ernest Henry Wilson.
REICHARD. Johann Jacob Reichard. 1743-1782. Germany.
REICHE. Karl Friedrich (Carlos Federico) Reiche. 1860-1929. Chile.
RÉMY. Jules (Ezechiel Julius) Rémy. 1826-1893. France.
RENDLE. Alfred Barton Rendle. 1865-1938. England.
RETZIUS. Anders Johan Retzius. 1742-1821. Sweden.
REYNOLDS. Gilbert Westacott Reynolds. 1895-1967. South Africa.
RICH., A. Achille Richard (son). 1794-1852. France.
RICH., L. Louis Claude Marie Richard. 1754-1821. France.
RICHARDSON. John Richardson. 1787-1865. England.
RIGHTER. Francis Irving Righter. 1897-1970. U.S.A.
RISSO. Joseph Antoine Risso. 1777-1845. France.
RISSO & POIT. Joseph Antoine Risso. Pierre Antoine Poiteau.
RIVERS. Thomas Rivers. 1798-1877. England.
RIVIÈRE, A. Marie Auguste Rivière. 1821-1877. France.
RIVIÈRE, A. & C. Marie Auguste Rivière. Charles Marie Rivière.
RIVIÈRE, C. Charles Marie Rivière. 1845- ? . France.
ROBINSON, B. L. Benjamin Lincoln Robinson. 1864-1935. U.S.A.
ROBS., N. Norman Keith Bonner Robson. 1928-   . England.
ROCK. Joseph Francis Charles Rock. 1884-1962. U.S.A.
RODWAY. Leonard Rodway. 1853-1936. Australia.
ROEMER. Johann Jakob Roemer. 1763-1819. Switzerland.
ROEMER & SCHULT. Johann Jakob Roemer. Joseph August Schultes.
ROEMER, M. J. Max J. Roemer. Fl. 1835-1847. Germany.
ROEZL. Benito Roezl. 1824-1855. Czechoslovakia, South America, Mexico.
ROLFE. Robert Allen Rolfe. 1855-1921. England.
RONN. Karl Ronniger. 1871-1954. Austria.
ROOF. James Roof. 1910-   . U.S.A.
ROSCOE. William Roscoe. 1753-1831. England.
ROSE. Joseph Nelson Rose. 1862-1928. U.S.A.
ROSE & I. M. JOHNST. Joseph Nelson Rose. Ivan Murray Johnston.
ROSS, R. Robert Ross. 1912-   . England.
ROSTER. Roster. Fl. 1914. Italy.
ROTH. Albrecht Wilhelm Roth. 1757-1834. Germany.
ROTTL. Johann Peter Rottler. 1749-1836. Denmark.
ROULEAU. Joseph Albert Ernest Rouleau. 1916-   . Canada.
ROURKE. John Rourke. Contemporary. South Africa.
ROYLE. John Forbes Royle. 1798-1858. India, England.
ROXB. William Roxburgh. 1759-1815. Scotland, India.
RUDOLPHI, K. Karl Asmund Rudolphi. 1771-1832. Germany.
RUIZ. Hipolito Ruiz Lopez. 1764-1815. Spain.

RUIZ & PAV. Hipolito Ruiz Lopez. José Antonio Pavón.
RUPPIUS. Heinrich Bernhard Ruppius. 1688-1719. Germany.
RUPR. Franz Joseph Ruprecht. 1814-1870. Czechoslovakia, Russia.
RUSBY. Henry Hurd Rusby. 1855-1940. U.S.A.
RUSSELL. Paul George Russell. 1889-1963. U.S.A.
RYDB. Per Axel Rydberg. 1860-1931. U.S.A.

SABINE. Joseph Sabine. 1770-1837. England.
SAFFORD. William Edwin Safford. 1859-1926. U.S.A.
SALISB. Richard Anthony Salisbury. 1761-1829. England.
SALM-DYCK. Joseph Maria Franz Anton Hubert Ignaz Salm-Reifferscheid-Dyck. 1773-1861. Germany.
SALTER. Terrence MacLeane Salter. 1883-1969. South Africa.
SALZM. Philipp Salzmann. 1781-1853. France.
SANDER. Henry Frederick Conrad Sander. 1847-1920. Germany, England.
SANDWITH. Noel Yvri Sandwith. 1901-1965. England.
SARG. Charles Sprague Sargent. 1841-1927. U.S.A.
SAV. Paul Amadée Ludovic Savatier. 1830-1891. France.
SCHAUER. Johan Konrad Schauer. 1813-1848. Germany.
SCHEFFER. Rudolph Herman Christiaan Carel Scheffer. 1844-1880. Netherlands, Indonesia.
SCHEIDW. Michel Joseph François Scheidweiler. 1799-1861. Germany.
SCHELLE. Ernest Schelle. 1864-1929. Germany.
SCHIEDE. Christian Julius Wilhelm Schiede. 1798-1836. Mexico.
SCHINDLER. Anton Karl Schindler. 1879-1964. Germany.
SCHINZ. Hans Schinz. 1858-1941. Switzerland.
SCHLECHTEND. Diederich Franz Leonhard von Schlechtendal. 1794-1866. Germany.
SCHLECHTEND. & CHAM. Diederich Franz Leonhard von Schlechtendal. Ludolf Adelbert von Chamisso.
SCHLECHTER. Friedrich Richard Rudolf Schlechter. 1872-1925. Germany.
SCHMIDT, FR. Friedrich Schmidt. 1832-1908. Russia.
SCHMIDT, J. A. Johann Anton Schmidt. 1823-1905. Germany.
SCHNEEV. George Voorhlem Schneevoogt. 1775-1850. Netherlands.
SCHNEID. Camillo Karl Schneider. 1876-1951. Germany.
SCHOTT. Heinrich Wilhelm Schott. 1794-1865. Austria.
SCHOTT & ENDL. Heinrich Wilhelm Schott. Stephen Ladislaus Endlicher.
SCHOTT & KOTSCHY. Heinrich Wilhelm Schott. Carl Georg Theodor Kotschy.
SCHOUSBOE. Peder Kofod Anker Schousboe. 1766-1832. Denmark, Morocco.
SCHRAD. Heinrich Adolph Schrader. 1767-1836. Germany.
SCHRAD. & WENDL. Heinrich Adolph Schrader. Johann Christoph Wendland.
SCHRANK. Franz von Paula von Schrank. 1747-1835. Germany.
SCHRANK & MART. Franz von Paula von Schrank. Karl Friedrich Philipp von Martius.
SCHREBER. Johann Christian Daniel von Schreber. 1739-1810. Germany.
SCHREIB. Beryl Olive Schreiber (Beryl Schreiber Jespersen). 1911-1968. U.S.A.
SCHUBERT. Bernice Giduz Schubert. 1913-   . U.S.A.
SCHULT. Joseph August Schultes. 1773-1831. Austria.
SCHULT. & SCHULT. f. Joseph August Schultes. Julius Hermann Schultes.
SCHULT. f. Julius Hermann Schultes (son). 1804-1840. Austria.
SCHULT., H. H. Schultes. Fl. 1852.
SCHULTZ-BIP. Karl Heinrich Schultz "Bipontinus". 1805-1867. Germany.
SCHUM., K. Karl Moritz Schumann. 1851-1904. Germany.
SCHUSTER, J. Julius Schuster. 1886-   . Germany.
SCHWARZ, O. Otto Schwarz. 1900-   . Germany.
SCHWEIGG. August Friedrich Schweigger. 1783-1821. Germany.
SCHWEIGG. & KOERTE. August Friedrich Schweigger. Franz Koerte [Körte].
SCHWEINFURTH. Georg August Schweinfurth. 1836-1925. Africa.
SCHWERIN. Fritz Kurt Alexander von Schwerin. 1856-1934. Germany.
SCOP. Giovanni Antonio Scopoli. 1723-1788. Austria.
SEALY. Joseph Robert Sealy. 1907-   . England.
SEEM. Berthold Carl Seemann. 1825-1871. Germany.
SEEM. & H. WENDL. Berthold Carl Seemann. Hermann Wendland.
SELLO, H. Herman Ludwig Sello. 1800-1876.
SELLOW, F. Friedrich Sello [later Sellow]. 1789-1831. Germany.
SENDTN. Otto Sendtner. 1814-1859. Germany.
SER. Nicolas Charles Seringe. 1776-1858. France.
SESSÉ. Martin de Sessé y Lacasta. ? -1809. Mexico.
SESSÉ & MOÇ. Martin de Sessé y Lacasta. José Mariano Moçiño Suares Losada.
SEUBERT. Moritz August Seubert. 1818-1878. Germany.
SHAFER. John Adolf Shafer. 1863-1918. U.S.A.
SHARP. Ward McClintic Sharp. 1904-   . U.S.A.
SHAW. George Russell Shaw. 1848-1937. U.S.A.
SHIBATA. Keita Shibata. 1887-1949. Japan.
SHINN. Lloyd Herbert Shinners. 1918-1971. U.S.A.
SHORT. Thomas Kier Short. Fl. 1838-1861. England.
SIEBER. Franz Wilhelm Sieber. 1789-1844. Czechoslovakia.
SIEBOLD. Philipp Franz von Siebold. 1796-1866. Netherlands.
SIEBOLD & ZUCC. Philipp Franz von Siebold. Joseph Gerhard Zuccarini.
SIM. Thomas Robertson Sim. 1856-1938. South Africa.
SIMMONDS, A. Arthur Simmonds. 1892-1968. England.
SIMMONDS, N. W. Norman W. Simmonds. 1922-   . England.
SIMONKAI. Lajos tól Simonkai [né Simkowicz]. 1851-1910. Hungary.
SIMON-LOUIS. Leon Louis Simon-Louis. 1834-1913. France.
SIMPS. George Simpson. 1880-1952. New Zealand.
SIMPS. & J. S. THOMS. George Simpson. John Scott Thomson.
SIMS. John Sims. 1749-1831. England.
SKAN. Sidney Alfred Skan. 1870-1939. England.
SKEELS. Homer Collar Skeels. 1873-1934. U.S.A.
SKOTTSBERG. Carl Johan Frederik Skottsberg. 1880-1963. Sweden, Pacific Islands.
SLEUMER. Hermann Otto Sleumer. 1906-   . Germany, Argentina, Netherlands.
SM. James Edward Smith. 1759-1828. England.
SM., A. C. Albert Charles Smith. 1906-   . U.S.A.
SM., C. Christian Smith. 1785-1816. Norway, Africa.
SM., E. W. Elmer W. Smith. 1920-   . U.S.A.
SM., H. G. Henry George Smith. 1852-1924. Australia.
SM., J. D. John Donnell Smith. 1829-1928. U.S.A.
SM., J. D. & ROSE. John Donnell Smith. Joseph Nelson Rose.

SM., L. B. Lyman Bradford Smith. 1904-   . U.S.A.
SM., L. B. & SCHUBERT. Lyman Bradford Smith. Bernice Giduz Schubert.
SM., R. M. Rosemary Margaret Smith. 1933-   . England.
SM., W. W. William Wright Smith. 1875-1956. Scotland.
SMALL. John Kunkel Small. 1869-1938. U.S.A.
SODERSTROM. Thomas R. Soderstrom. Contemporary. U.S.A.
SODERSTROM & CALDERON. Thomas R. Soderstrom. Cleofé E. Calderon.
SOLAND. Daniel Carl Solander. 1733-1782. England.
SOLMS-LAUB. Hermann Maximilian Carl Ludwig Friedrich Solms-Laubach. 1842-1915. Germany.
SOND. Otto Wilhelm Sonder. 1812-1881. Germany.
SONN. Pierre Sonnerat. 1748-1814. France.
SOUL.-BOD. Étienne Soulange-Bodin. 1774-1846. France.
SOULIÉ. Jean André Soulié. 1868-1930. France.
SOUSTER. J. E. Souster. Contemporary. England.
SPACH. Edouard Spach. 1801-1879. France.
SPAE. Dieudonné Spae. 1819-1858[?1879]. Belgium.
SPAETH, F. L. Franz Ludwig Spaeth [Späth]. 1839-1913. Germany.
SPAETH, H. L. Hellmut Ludwig Spaeth [Späth]. 1885-1945. Germany.
SPEGAZZINI. Carlos Spegazzini. 1858-1926. Argentina.
SPONGBERG. Stephen Spongberg. Contemporary. U.S.A.
SPRAGUE. Thomas Archibald Sprague. 1877-1958. England.
SPRAGUE & SUMMERHAYES. Thomas Archibald Sprague. Victor Samuel Summerhayes.
SPRENG., A. Anton Sprengel. 1803-1851. Germany.
SPRENG., K. Kurt [Curt] Polycarp Joachim Sprengel. 1766-1833. Germany.
STANDL. Paul Carpenter Standley. 1884-1963. U.S.A.
STAPF. Otto Stapf. 1857-1933. England.
STAPF & DAVY. Otto Stapf. Joseph Burtt Davy.
STEARN. William Thomas Stearn. 1911-   . England.
STEEDMAN, H. Henry Steedman. 1866-1953. Australia.
STEENIS. Cornelis Gijsbert Gerrit Jan van Steenis. 1901-   . Netherlands.
STERNS. Emerson Ellick Sterns. 1846-1926. U.S.A.
STEUD. Ernst Gottlieb von Steudel. 1783-1856. Germany.
STEVEN. Christian von Steven. 1781-1863. Russia.
STEWART, J. L. John Lindsay Stewart. 1832-1873. Scotland, India.
ST.-HIL. Auguste de Saint-Hilaire. 1779-1853. France.
ST.-HIL. & NAUD. Auguste de Saint-Hilaire. Charles Naudin.
ST.-LÉGER. Léon Saint-Léger. Fl. 1899. France.
STOCKW. William Palmer Stockwell. 1898-1950. U.S.A.
STOCKW. & RIGHTER. William Palmer Stockwell. Francis Irving Righter.
STOKES. Jonathan Stokes. 1755-1831. England.
STOKES, S. Susan Gabriella Stokes. 1868-1954. U.S.A.
STUNTZ. Stephen Conrad Stuntz. 1875-1918. U.S.A.
SUCKOW. Georg Adolph Suckow. 1751-1813. Germany.
SUDW. George Bishop Sudworth. 1864-1927. U.S.A.
SUKACHEV. Vladimir Nikolaevich Sukachev [Sukatschew, Sukaczev]. 1880-1967. Russia.
SUMMERHAYES. Victor Samuel Summerhayes. 1897-1974. England.
SURINGAR. Jan Valckenier Suringar. 1864-1932. Netherlands.
SVENTENIUS. Eric R. Sventenius. 1910-1973. Canary Islands.
SWARTZ. Olof Peter Swartz. 1760-1818. Sweden.
SWEET. Robert Sweet. 1783-1835. England.

SWINGLE. Walter Tennyson Swingle. 1871-1952. U.S.A.
SYMON. David Eric Symon. 1920-   . Australia.
SZYSZ. Ignaz Szyszylowicz. 1857-1910. Poland.

TAGG. Harry Frank Tagg. ?1873-1933. Scotland.
TAKEDA. Hisayoshi Takeda. 1883-   . Japan.
TANAKA. Tyozaburo Tanaka. 1885-   . Japan.
TANFANI. Enrico Tanfani. 1848-1892. Italy.
TAUSCH. Ignaz Friedrich Tausch. 1793-1848. Czechoslovakia.
TAYLOR, N. Norman Taylor. 1883-   . U.S.A.
TAYLOR, T. Thomas Mayne Cunninghame Taylor. 1904-   . Canada.
TAYLOR, T. & VRUGTMAN. Thomas Mayne Cunninghame Taylor. Freek Vrugtman.
TEMPLE. F. L. Temple. Fl. 1885-1889. U.S.A.
TEN. Michele Tenore. 1780-1861. Italy.
TERÁN. Manuel de Mier y Terán. 1789-1832. Mexico.
TERÁN & BERL. Manuel de Mier y Terán. Jean Louis Berlandier.
TERRACCIANO, A. Achille Terracciano. 1862-1917. Italy.
TEYSM. Johannes Elias Teysmann [Teijsmann]. 1809-1882. Netherlands.
TEYSM. & BINNEND. Johannes Elias Teysmann. Simon Binnendijk.
THOMS. Thomas Thomson. 1817-1878. Scotland, India.
THOMS, J. S. John Scott Thomson. 1882-1943. New Zealand.
THORNB. John James Thornber. 1872-1962. U.S.A.
THORY. Claude Antoine Thory. 1759-1827. France.
THOUIN. André Thouin. 1747-1824. France.
THUNB. Carl Peter Thunberg. 1743-1828. Sweden, Japan, South Africa.
THURBER. George Thurber. 1821-1890. U.S.A.
THURST. Edgar Thurston. 1855-1935. England.
THWAITES. George Henry Kendrick Thwaites. 1812-1882. England.
TINEO. Vincenzo Tineo. 1791-1856. Italy.
TOD. Agostino Todaro. 1818-1892. Italy.
TOPF. Alfred Topf. Fl. 1851-1857. Germany.
TORR. John Torrey. 1796-1873. U.S.A.
TORR. & FRÉM. John Torrey. John Charles Frémont.
TORR. & GRAY. John Torrey. Asa Gray.
TRABUT. Louis Charles Trabut. 1853-1929. France, Algeria.
TRATT. Leopold Trattinick. 1764-1849. Austria.
TRAUTV. Ernst Rudolph von Trautvetter. 1809-1889. Russia.
TRAUTV. & C. A. MEY. Ernst Rudolph von Trautvetter. Carl Anton von Meyer.
TREL. William Trelease. 1857-1945. U.S.A.
TRIANA. José Jerónimo Triana. 1828[?1834]-1890. Colombia.
TRYON, R. Rolla Milton Tryon, Jr. 1916-   . U.S.A.
TUCKER. John Maurice Tucker. 1916-   . U.S.A.
TURCZ. Nikolai Stepanovich Turchaninov [Turczaninow]. 1796-1864. Russia.
TURRA. Antonio Turra. 1730-1796. Italy.
TURRILL. William Bertram Turrill. 1898-1961. England.
TUSSAC. François Richard de Tussac. 1751-1837. France.
TUTIN. Thomas Gaskell Tutin. 1908-   . England.

UCHIDA. Shigetaro Uchida. 1885-   . Japan.
ULBRICH. Oskar Eberhard Ulbrich. 1879-1952. Germany.
URBAN. Ignatz Urban. 1848-1931. Germany.
URSCH. Eugene Ursch. 1882-1962. France.
URSCH & LEANDRI. Eugene Ursch. Jacques Leandri.

VAHL. Martin Vahl. 1749-1804. Denmark.
VAN GEERT. Auguste Van Geert. 1818-1886. Belgium.
VAN HOUTTE. Louis van Houtte. 1810-1876. Belgium.
VANIOT. Eugene Vaniot. ? -1913. France.
VASEY. George Vasey. 1822-1893. U.S.A.
VASEY & ROSE. George Vasey. Joseph Nelson Rose.
VATKE. Georg Carl Wilhelm Vatke. 1849-1889. Germany.
VAUVEL. Léopold Eugène Vauvel. 1848-1915. France.
VEITCH. John Gould Veitch. 1839-1870. England.
VELL. José Marianno da Conceiçao Velloso [Vellozo]. 1742-1811. Portugal, Brazil.
VENT. Étienne Pierre Ventenat. 1757-1808. France.
VERDC. Bernard Verdcourt. 1925-   . England.
VERDOORN. Inez Clare Verdoorn. 1896-   . South Africa.
VERL., B. Pierre Bernard Lazare Verlot. 1836-1897. France.
VERL., J. Jean Baptiste Verlot. 1825-1891. France.
VIALA. Pierre Viala. 1859-1936. France.
VICKERY. Joyce Winifred Vickery. 1908-   . Australia.
VIDAL, S. Sebastian Vidal y Soler. 1842-1889. Spain, Philippines.
VIG., R. René Viguier. 1880-1931. France.
VIG., R. & GUILLAUM. René Viguier. André Guillaumin.
VILL. Dominique Villars. 1745-1814. France.
VILM., M. L. Maurice Levêque de Vilmorin. 1849-1918. France.
VILM., P. L. Philippe Levêque de Vilmorin. 1872-1917. France.
VILM. Maurice Levêque de Vilmorin or Philippe Levêque de Vilmorin.
VIS. Roberto de Visiani. 1800-1878. Italy.
VISCHER. Wilhelm Vischer. 1890-
VIVIANI. Domenico Viviani. 1772-1840. Italy.
VOGEL. Julius Rudolph Theodor Vogel. 1812-1841. Germany.
VOS. Cornelis de Vos. 1806-1895. Netherlands.
VOSS. Andreas Voss. 1857-1924. Germany.
VRUGTMAN. Freek Vrugtman. 1927-   . Canada.

WAHLENB. Göran Wahlenberg. 1780-1851. Sweden.
WALDSTEIN. Franz de Paula Adam von Waldstein-Wartemberg. 1759-1823. Hungary.
WALDSTEIN & KITAIBEL. Franz de Paula Adam von Waldstein-Wartemberg. Paul Kitaibel.
WALLICH. Nathaniel Wallich. 1786-1854. Denmark, India.
WALP. Wilhelm Gerhard Walpers. 1816-1853. Germany.
WALTER. Thomas Walter. 1740-1789. U.S.A.
WALTHER. Edward Eric Walther. 1892-1959. U.S.A.
WANGENH. Friedrich Adam Julius von Wangenheim. 1747-1800. Germany.
WARB. Otto Warburg. 1859-1938. Germany.
WARB., E. F. Edmund Fredric Warburg. 1908-1966. England.
WARD, F. K. Francis Kingdon Ward. 1885-1958. England.
WARDER. John Aston Warder. 1812-1883. U.S.A.
WARMING. Johannes Eugenius Bülow Warming. 1841-1924. Denmark.
WASSHAUSEN. Dieter Carl Wasshausen. 1938-   . U.S.A.
WASSHAUSEN & L. B. SM. Dieter Carl Wasshausen. Lyman Bradford Smith.
WATS., H. Hewett Cottrell Watson. 1804-1881. England.
WATS., P. W. Peter William Watson. 1761-1830. England.
WATS., S. Sereno Watson. 1826-1892. U.S.A.
WATS., W. William Watson. 1858-1925. England.
WEATHERBY. Charles Alfred Weatherby. 1875-1949. U.S.A.

WEBB. Philip Barker Webb. 1793-1854. England, Canary Islands.
WEBB & BERTH. Philip Barker Webb. Sabin Berthelot.
WEBB, D. A. David Allardice Webb. 1912-   . England.
WEBER, C. Jean-Germaine Claude Weber. 1922- France, Switzerland, U.S.A.
WEBSTER. Angus Duncan Webster. Fl. 1893-1920. England.
WEICK. Alphonse Weick. Fl. 1863. France.
WELW. Friedrich Martin Josef Welwitsch. 1806-1872. Austria.
WENDL. Johann Christoph Wendland. 1755-1828. Germany.
WENDL., H. Hermann Wendland [grandson of J. C. Wendland]. 1823[?1825]-1903. Germany.
WENDL., H. & DRUDE. Hermann Wendland. Carl Georg Oskar Drude.
WENDL., H. L. Heinrich Ludolph Wendland (son of J. C. Wendland). 1791-1869. Germany.
WENZ. Theodor Wenzig. 1824-1892. Germany.
WESM. Alfred Wesmael. 1821[?1832]-1905. Belgium.
WESTON. Richard Weston. 1733-1806. England.
WEYER. W. van der Weyer. Fl. 1920.
WHITE, C. T. Cyril Tenison White. 1890-1950. Australia.
WIEGAND. Karl McKay Wiegand. 1873-1942. U.S.A.
WIESL. Albert Everett Wieslander. 1890-   . U.S.A.
WIESL. & SCHREIB. Albert Everett Wieslander. Beryl Olive Schreiber.
WIGHT. Robert Wight. 1796-1872. England, India.
WIGHT & ARN. Robert Wight. George Arnott Walker Arnott.
WIGHT, W. F. William Franklin Wight. 1874-1954. U.S.A.
WILLD. Karl Ludwig Willdenow. 1765-1812. Germany.
WILLIAMS, I. Ion J. M. Williams. Contemporary. South Africa.
WILLIS. James Hamlyn Willis. 1910-   . Australia.
WILLK. Heinrich Moritz Willkomm. 1821-1895. Germany.
WILLK. & J. LANGE. Heinrich Moritz Willkomm. Johan Martin Christian Lange.
WILS. Ernest Henry Wilson. 1876-1930. England, U.S.A.
WINKLER. Hubert Winkler. 1875-1941. Germany.
WITTM. Max Carl Ludwig Wittmack. 1839-1929. Germany.
WOLF, C. B. Carl Brandt Wolf. 1905-1974. U.S.A.
WOLLEY-DOD. Anthony Hurt Wolley-Dod. 1861-1948. England.
WOOD. Alphonso Wood. 1810-1881. U.S.A.
WOODSON. Robert Everard Woodson, Jr. 1904-1963. U.S.A.
WOOLLS. William Woolls. 1814-1893. Australia.
WOOT. Elmer Otis Wooton. 1865-1945. U.S.A.
WRIGHT, C. H. Charles Henry Wright. 1864-1941. England.
WURDACK. John Wurdack. Contemporary. U.S.A.

YOUNG, R. A. Robert Armstrong Young. 1876- ? . U.S.A.

ZABEL. Hermann Zabel. 1832-1912. Germany.
ZENARI. Silvia Zenari. 1895-1956. Italy.
ZENKER. Jonathan Carl Zenker. 1799-1837. Germany.
ZOLLINGER. Heinrich Zollinger. 1818-1859. Switzerland, Java.
ZOLLINGER & MORITZI. Heinrich Zollinger. Alexander Moritzi.
ZUCC. Joseph Gerhard Zuccarini. 1797-1848. Germany.

REFERENCES

In checking nomenclature, we used the following standard references, as well as local floras and manuals from around the world, revisions and monographs, and checklists and registration lists of cultivars.

BAILEY, L. H.
    1928-1942. *The Standard Cyclopedia of Horticulture.* 3 vols. New York: The Macmillan Company.
    1949. *Manual of Cultivated Plants.* Rev. ed. New York: The Macmillan Company.

BAILEY, LIBERTY HYDE, HORTORIUM and STAFF
    1976. *Hortus Third.* New York: The Macmillan Company.

BEAN, W. J.
    1970-1976. *Trees and Shrubs Hardy in the British Isles.* 8th ed. D. L. Clarke, chief editor. 3 vols. A-Rh. London: John Murray.

CHITTENDEN, FRED J., ed.
    1951-1969. *The Royal Horticultural Society Dictionary of Gardening.* 4 vols. and supplement. Oxford: Clarendon Press.

DALLIMORE, WILLIAM, and A. BRUCE JACKSON
    1967. *A Handbook of Coniferae and Ginkgoaceae.* Revised by S. G. Harrison. New York: St. Martin's Press.

HILLIER, H. G., R. GARDNER, and R. LANCASTER
    1971. *Hilliers' Manual of Trees and Shrubs.* Winchester, England: Hillier & Sons.

HOSHIZAKI, BARBARA JOE
    1975. *Fern Growers Manual.* New York: Alfred A. Knopf.

JACOBSEN, HERMANN
    1970. *Das Sukkulenten Lexikon.* Jena, Germany: VEB Gustav Fischer Verlag.

KRÜSSMAN, GERD
    1960-1962. *Handbuch der Laubgehölze.* 2 vols. Berlin: Paul Parey.
    1972. *Handbuch der Nadelgehölze.* Berlin and Hamburg: Paul Parey.

LITTLE, ELBERT L., JR.
    1953. *Check List of Native and Naturalized Trees of the United States.* Agriculture Handbook No. 41. Washington, D.C.: U. S. Forest Service.

OUDEN, PIETER DEN, and B. K. BOOM
    1975. *Manual of Cultivated Conifers.* The Hague: M. Nijhoff.

REHDER, ALFRED
    1940. *Manual of Cultivated Trees and Shrubs.* 2nd ed. New York: The Macmillan Co.
    1949. *Bibliography of Cultivated Trees and Shrubs.* Jamaica Plain, Mass.: The Arnold Arboretum.

WELCH, H. J.
    1966. *Dwarf Conifers. A Complete Guide.* London: Faber and Faber.

WYMAN, DONALD
    1971. *Wyman's Gardening Encyclopedia.* New York: The Macmillan Company.